Digital Filtering in One and Two Dimensions

Design and Applications

Digital Filtering in One and Two Dimensions

Design and Applications

Robert King

Imperial College
London, England

Majid Ahmadi

University of Windsor
Windsor, Ontario, Canada

Raouf Gorgui-Naguib

University of Newcastle upon Tyne
Newcastle upon Tyne, England

Alan Kwabwe

Imperial College
and Bankers Trust Company
London, England

and

Mahmood Azimi-Sadjadi

Colorado State University
Fort Collins, Colorado

Springer Science+Business Media, LLC

Library of Congress Cataloging in Publication Data

Digital filtering in one and two dimensions: design and applications / Robert King . . . [et al.].

p. cm.

Includes bibliographies and index.

1. Digital filters (Mathematics) 2. Signal processing – Digital techniques. 3. Transformations (Mathematics) I. King, Robert Ashford.

QA297.D54 1989 88-30268

621.38′043 – dc19 CIP

DOI 10.1007/978-1-4899-0918-3

Originally published by Plenum Press, New York in 1989

MyCopy version of the original edition 1989

To

Gillian
Jocelyne
May
Yetunde
Sheida

Preface

This book has been conceived to extend the generally published work on one- and two-dimensional digital filters in order to include some of the more recently developed ideas. It is intended to supplement and build on the classical books which cover the fundamental concepts of the topic. As a consequence of this, the basic theory is stated in a compact manner and is not developed thoroughly, as this would result in considerable duplication of existing books. The main theme of the book has been to provide a comprehensive background to the methods available for the realization of both recursive and nonrecursive digital filters, and to give an insight into some of the more recent implementation procedures.

The book is planned to cover one- and two-dimensional systems in parallel, showing the techniques which are applicable in both areas, and also the limitations and constraints necessary when a one-dimensional technique is extended to systems of higher dimensionality. The theme of the book commences with several chapters on the design of filter transfer functions to meet given specifications. This is followed by a discussion of methods of implementing these in a practical system and the limitations imposed as a result of noise and finite word length. Finally, a discussion of some applications is included.

The first chapter includes a brief review of the theory which is necessary in the remainder of the book; it covers some of the fundamental topics of convolution, Z transformation, and Fourier transforms without any attempt at mathematical rigor or completeness. These topics are already well covered in a number of existing texts.

This is followed by three chapters which provide a detailed study of the techniques available for the design of both nonrecursive and recursive filter transfer functions. In the design of nonrecursive digital filters, the use of windows in both one and two dimensions is considered. This is followed by a discussion of the frequency sampling technique and the use of iterative methods for the design of nonrecursive filters. Finally, the use of frequency transformations in two dimensions is studied.

A chapter follows which briefly reviews the conditions for stability of one-dimensional systems and the techniques available for the stabilization of inherently unstable filters. The extension to two dimensions is taken to considerable length in order to show the greatly increased complexity of the problem and the various attempts which have been made to solve the two-dimensional stabilization problem for a range of filter functions. A description of several different methods of generating stable polynomials is offered next.

This is followed by a chapter on recursive filters in both one and two dimensions, in which the objective is to design a filter to fit a given specification with no danger of incurring problems of instability. Most of the techniques discussed here are based in the frequency domain; however, a short outline of some of the more profitable space-domain two-dimensional filter design methods is included.

The next chapter gives an overall view of the types of errors which occur in digital filters as a result of the discrete nature of the process; the finite word length representing the values of data and filter coefficients; and the errors which occur as a result of overflow in the multiplication process. The limit cycle oscillations which may be generated as a result of quantization in the system are also considered.

The following three chapters cover some methods of implementing the designed transfer functions in terms of the basic multipliers, adders, and delay elements. The first technique considered is that whereby an analogue filter is designed as a ladder network and then transformed by means of a set of linear equations into a form analogous to wave equations. These may now be digitized, giving a direct implementation as a digital system. This retains the advantage of the good sensitivity of analogue structures.

The second approach is via the state-space representation of filter functions. In this form, it is relatively simple to obtain a general transformation matrix which may be chosen to realize a large body of implementations of the transfer functions; these different realizations may have the advantage of low noise, consistent with limited dynamic range, minimum number of elements, or other properties.

A third approach to the implementation of systems processing large arrays of data is that of breaking the data into subunits and processing these blocks individually, aggregating them at the output to give the complete resultant image. Several methods of performing this are considered.

The next chapter digresses to consider the implementation of a filter function by transforming the data into some alternative domain, and a number of Fourier-like methods, mainly in number systems defined in a finite field, are considered. Some of the newer systems, such as polynomial transforms and *p*-adic transforms, are studied.

There follows a chapter on image modeling; this is concerned with the introduction of two-dimensional causal, semicausal, and noncausal autoregressive moving-average models and the related issues of parameter estimation, spectral factorization, stability, and convergence. The applications are in the restoration of satellite and radar images.

The final two chapters are devoted to a consideration of the general philosophy of the application of digital filters to practical problems. The first of these two chapters deals with the general methodology of applying digital filters to real problems, and the types of filter functions needed in many cases. The second chapter takes four case studies and shows how filtering may be used to improve or enhance those properties of an image which are of interest in a given application.

We should like to record our thanks and gratitude to a number of our colleagues, both past and present, who have either knowingly or unwittingly contributed to this book. In particular, we should mention Drs. G. A. Jullien, V. Ramachandran, and S. Udpa, who read and commented on parts of the manuscript and suggested modifications and improvements. The assistance of Ahmet Kayran, Nahy Nassar, Amar Ali, Desmond McLernon, Nasser Nasrabadi, Behrouz Nowrouzian, Karim Henein, and H. Lee is also gratefully acknowledged.

Finally, our gratitude goes to Christine Lane and Shirley Ovellette, who produced an excellent typescript from our less than perfect manuscripts.

Contents

Explanatory List of Abbreviations

AR	autoregressive
ARMA	autoregressive moving average
BHP	broad sense Hurwitz polynomial
BIBO	bounded-input, bounded-output
CNTT	complex number theoretic transform
CRT	Chinese Remainder Theorem
CT	computerized tomography
DFT	discrete Fourier transform
DR	dynamic range
FFT	fast Fourier transform
FIR	finite impulse response
FNT	Fermat number transform
HP	Hurwitz polynomial
HVS	human visual system
IDFT	inverse discrete Fourier transform
IIR	infinite impulse response
LHP	left-half plane
LP	low pass
MA	moving average
MIMO	multi-input, multi-output
MNT	Mersenne number transform
MVR	minimum variance representation
NHP	narrow sense Hurwitz polynomial
NSHP	nonsymmetrical half-plane
NTT	number theoretic transform
PAT	p-adic transform
PLSI	planar least-squares inverse
PPS	phase-plane symmetry
PSF	point spread function

PT	polynomial transform
RLC	reverse limit cycle
SDF	spatial density function
SHP	strict Hurwitz polynomial
SISO	single-input, single-output
	summable-input, summable-output (Chapter 3)
VSHP	very strict Hurwitz polynomial

1

Discrete Time/Space Linear Filtering

1.1. INTRODUCTION

Digital filters in one, two, and many dimensions form a large class of systems used nowadays for the processing of both analogue and digital signals. We shall be considering in this book those digital systems classified as linear and time-invariant. Their properties, design, implementation, and applications will be discussed.

In one dimension the most usual application of digital filters is in the time domain, where they may be applied to the modification of speech or music signals either to enhance the intelligibility or reduce the noise, or, in the case of music, to generate entirely new sounds. In addition, any signal, such as a television picture, which is transmitted serially in time, may be processed by a one-dimensional (1-D) digital filter.

In two dimensions, the manipulation of data is commonly referred to as image processing. This is because such data are usually displayed as spatial images for human evaluation or appreciation, even though the two dimensions may not necessarily both be spatial. In signal processing, a dimension can mean any physical domain in which a signal is defined. Time, space, and frequency are examples of such domains, and any combination of these is allowed as a coordinate system.

There are three general branches of image processing. The most important of these, digital filtering, is the main theme of this book. The remaining two aspects, image encoding and image analysis, will not be discussed at length.

Two-dimensional (2-D) digital filters are computational algorithms that transform 2-D input sequences of numbers into 2-D output sequences of numbers according to prespecified rules, hence yielding some desired modifications to the characteristics of the input sequences. Applications of 2-D digital filters cover a wide spectrum, the object being usually either enhancement of an image to make it more acceptable to the human eye, or removal of the effects of some degradation mechanisms, or separation of features for easier identification and measurement by human or machine.

Many important applications of 2-D digital filters have been in the field of space technology. Here, digitally processed satellite images are used in monitoring environmental effects, earth resources, and urban land use. In such applications, 2-D digital filters enhance or reduce boundaries, remove low-frequency shading effects, reduce noise, and correct for distortions inherent to the imaging systems employed.

There are also applications in medicine and biology. Two-dimensional X-ray films are digitally processed to reduce the contents of the low spatial frequencies and, by doing so, fracture lines and other features with large high-frequency components become easier to detect. This procedure is sometimes followed or preceded by constrast enhancement and noise reduction. Additional biomedical uses include removal of scan lines in radioisotope scanning, low-frequency background noise reduction in photomicrographs, and digital processing of acoustical holograms.

Seismic prospecting is one area in which data are acquired as 2-D sequences that are not images in the conventional sense. Seismic detectors are placed at intervals both along and across an area, and the digitized outputs of the detectors after an explosion form a 2-D data array. This array is then processed to minimize the effects of multiple reflections and wind-induced noise, and information about the subsurface structure of the locality is then readily obtained from the output of the filtering operation.

Geophysics is yet another science where 2-D digital signal processing is extensively employed. Similarly, magnetic and gravity measurements are processed digitally to reduce the effects of surface anomalies, thus facilitating the identification of large subsurface features.

The above is only a brief review of some of the many possible applications of 2-D filters. Such a discussion invariably omits a number of other applications, and does not attempt to speculate on the possible future additions to the list. However, it is evident from the above that any area in which 2-D data are encountered is also a possible field of application of 2-D digital filters.

1.2. DIGITAL SYSTEMS

A 1-D signal, considered as a function of time $x(t)$, may be sampled at a number of discrete periodic instants of time to form the discrete sequence $x(nT)$, where T is the interval between sampling instants. In order that all the information contained within the signal is retained in the sampled version, the sampling frequency must be chosen such that it is at least twice the highest frequency of the spectrum of the signal. The variable in the 1-D case is usually time, and the intersampling interval, T, is normally a time variable. This intersampling interval is frequently omitted in analysis and the signal considered as $x(n)$, where n is an integer.

In the 2-D situation, frequently a spatial image in two dimensions, the sampling intervals will be spatial, resulting in an array of pixels which may be defined by $x(m, n)$, where m and n are integers defined over the area of the image. In geophysical prospecting, images exist in which one dimension is spatial and the other temporal. However, the same representation as above may be used and the precise dimensionality of the axes is irrelevant in the analysis of the system.

Higher-dimensionality systems may be represented by discrete functions $x(m_1, m_2, m_3, \ldots, m_n)$ for an n-dimensional system. Since the analysis, design, and implementation of multidimensional systems differ little in principle from 2-D systems, they will not be considered in great detail, the results appropriate for systems of higher dimensionality being derived by extension from those for two dimensions.

1.2.1. Causality

A 1-D temporal system is causal if the output array is dependent on past values of input and not on future values. Thus a system is causal if, and only if, its impulse response $h(m)$ satisfies the condition

$$h(m) = 0, \qquad m < 0$$

In two and more dimensions the idea of causality may be extended to include all systems in which

$$h(m, n) = 0, \qquad m < 0 \quad \text{and} \quad n < 0$$

An alternative name for such systems is quadrantal or quarter-plane systems. When h is defined for a nonnegative region of support, it is known as a first-quadrant function denoted by

$${}^1h = \{{}^1h(m, n)\}, \qquad m \geq 0, \quad n \geq 0$$

Similarly second, third, and fourth quadrant functions may be defined:

$${}^2h = \{{}^2h(m, n)\}, \qquad m \leq 0, \quad n \geq 0$$

$${}^3h = \{{}^3h(m, n)\}, \qquad m \leq 0, \quad n \leq 0$$

$${}^4h = \{{}^4h(m, n)\}, \qquad m \geq 0, \quad n \leq 0$$

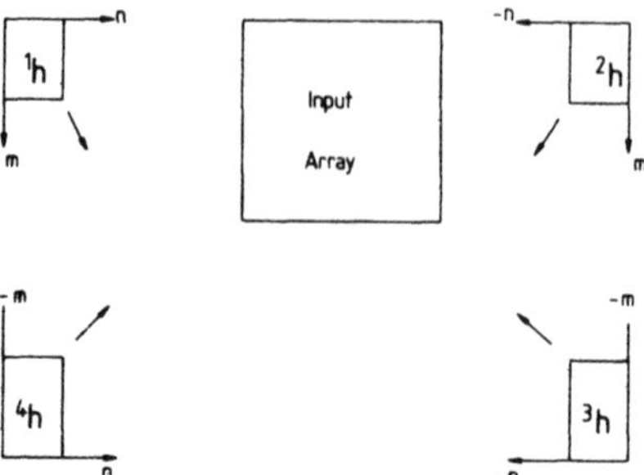

Figure 1.1. Recursion directions for the four-quadrantal filter functions.

Most two- and multidimensional systems require filter functions which are symmetrical in all directions and thus are noncausal. We shall see that it is not possible to design a four-quadrant filter which is of recursive form since, to compute the present value of the output, it would be necessary to use the output data at points both to the left and right of the observation points and, depending on the manner of computation, one of these data points will not have been previously determined. However, all network transfer functions, both causal and noncausal, may be formed from a summation of four-quadrantal filter functions. This is illustrated in Figure 1.1.

1.2.2. Recursion Directions of Finite Area Array[1]

A (causal) first-quadrant filter recurses in the $(+m, +n)$ direction. The causal recursion starts at the northwest corner of an input array.

A 2-D array that is nonzero for only a finite area in the spatial domain is referred to as a finite-area array.[2]

For a finite-area array $I(M \times N)$, the following array operations are used to obtain different directions of recursion:

1. 180° reflection in $(M + 1)/2$ axis (row reversal)

$$^{2}h(m, n) = {}^{1}h(M - m + 1, n)$$

2. Clockwise 180° rotation (row-column reversal)

$$^{3}h(m, n) = {}^{1}h(M - m + 1, N - n + 1)$$

3. 180° reflection in $(N + 1)/2$ axis (column reversal)

$$^{4}h(m, n) = {}^{1}h(m, N - n + 1)$$

for all m and n such that $m = 1, 2, \ldots, M$ and $n = 1, 2, \ldots, N$.

We note that when M or N is odd, the reflection takes place along one sample row (or column) and that this row (or column) is unchanged by the reflection, and when M or N is even, the axis of reflection lies midway between the centrally placed two rows (or columns).

One can thus obtain, from a first-quadrant filter, filters recursing in the $(-m, n)$, $(-m, -n)$, and $(m, -n)$ directions with ${}^2h(m, n)$, ${}^3h(m, n)$, and ${}^4h(m, n)$, respectively.

1.3. RECURSION EQUATIONS

In the simplest case of a digital nonrecursive system, the output sequence is defined as the weighted sum of the input sequence over a number of preceding samples. Thus, in one dimension, the output $y(n)$ written as a function of the input $u(n)$ is given by

$$y(n) = \sum_{k=0}^{N} a(k)u(n-k) \tag{1.1a}$$

where $a(k)$ is a weighting factor on the various inputs. In two dimensions this becomes

$$y(m, n) = \sum_{i=0}^{M} \sum_{j=0}^{N} a(i, j)u(m-i, n-j) \tag{1.1b}$$

In other systems, known as recursive systems, the output is not only a function of the input, but also of the previous values of the output. This may be written for one and two dimensions in the respective forms

$$y(n) = \sum_{k=0}^{N_1} a(k)u(n-k) - \sum_{k=1}^{N_2} b(k)y(n-k) \tag{1.2a}$$

and

$$y(m, n) = \sum_{i=0}^{M_1} \sum_{j=0}^{N_1} a(i, j)u(m-i, n-j) - \sum_{\substack{i=0 \\ i+j\neq 0}}^{M_2} \sum_{j=0}^{N_2} b(i, j)y(m-i, n-j) \tag{1.2b}$$

The objective of all digital filter design may be summarized in the solution of the above equations and the choice of the weighting coefficients, in order that the output sequence bears some specified relationship to the input.

1.3.1. Convolution Equations

The response of either of these systems, defined by equations (1.1) and (1.2), may be written as a convolution of the input with a function $h(n)$, the impulse response of the filter. In the recursive case, we may apply equation (1.2) successively, resulting in an infinite expansion for the output; in one dimension this is

$$\begin{aligned} y(n) &= \sum_{k=0}^{\infty} h(k)u(n-k) \\ &= h(n) * u(n) \end{aligned} \tag{1.3a}$$

and for a two-dimensional first-quadrant system

$$\begin{aligned} y(m, n) &= \sum_{i=0}^{\infty} \sum_{j=0}^{\infty} h(i, j)u(m-i, n-j) \\ &= h(m, n) * u(m, n) \end{aligned} \tag{1.3b}$$

where $h(n)$ and $h(m, n)$ are functions of the coefficients $a(k)$ and $b(k)$ or $a(i, j)$ and $b(i, j)$, respectively. Since the impulse response is infinite, an alternative name for a recursive filter is an infinite impulse response (IIR) filter.

The symbol $*$ is used to denote the one- or two-dimensional convolution operator.

For a nonrecursive system, the summations are of finite extent and are obtained directly from equations (1.1), namely,

$$y(n) = \sum_{k=0}^{N} h(k)u(n-k) \tag{1.4a}$$

and

$$y(m, n) = \sum_{i=0}^{M} \sum_{j=0}^{N} h(i, j)u(m-i, n-j) \tag{1.4b}$$

When compared with system (1.1), we see that the filter coefficients are identical to the convolution coefficients $h(k)$. For this reason these are alternatively known as finite impulse response (FIR) filters.

1.3.2. Impulse Response

The impulse function $\delta(n)$ of a 1-D system is defined as

$$\delta(n) = \begin{cases} 1, & n = 0 \\ 0, & \text{otherwise} \end{cases}$$

Similarly, for a 2-D system

$$\delta(m, n) = \begin{cases} 1, & m = 0, \quad n = 0 \\ 0, & \text{otherwise} \end{cases}$$

The impulse response of a filter may be obtained by putting $u(n)$ equal to $\delta(n)$ in equation (1.3a), and thus we see that the output sequence, $\{y(n)\}$, is identical to $\{h(n)\}$ and is the impulse response of a recursive system. Similarly, $\{h(m, n)\}$ is the 2-D impulse response.

For a nonrecursive filter, using equation (1.4) we see that the impulse response $\{h(n)\}$ is identical to the sequence of coefficients $\{a(k)\}$.

1.4. FOURIER TRANSFORM

An infinite causal (one-sided) discrete periodic signal $\bar{f}(n)$ may be represented by a linear combination of sinusoidal components whose periods are subperiods of the period N of the signal $\bar{f}(n)$. Thus

$$\bar{f}(n) = \frac{1}{N} \sum_{k=0}^{N-1} F(k) \exp(\mathrm{j}2\pi nk/N) \tag{1.5}$$

and the amplitudes $F(k)$ give the amplitude of the component of $\bar{f}(n)$ at the frequency $\omega = 2\pi k/N$ where

$$F(k) = \sum_{n=0}^{N-1} \bar{f}(n) \exp(-\mathrm{j}2\pi nk/N) \tag{1.6}$$

Now we may define the signal $f(n)$ as

$$f(n) = \begin{cases} \bar{f}(n), & n = 0, 1, \ldots, N-1 \\ 0, & n < 0 \quad \text{and} \quad n \geq N \end{cases} \tag{1.7}$$

We see that equations (1.5) and (1.6) may now be written respectively as

$$f(n) = \frac{1}{N} \sum_{k=0}^{N-1} F(k) \exp(\mathrm{j}2\pi nk/N) \qquad \text{for } n = 0, 1, \ldots, N-1 \tag{1.8}$$

and

$$F(k) = \sum_{n=0}^{N-1} f(n) \exp(-j2\pi nk/N) \qquad \text{for } k = 0, 1, \ldots, N-1 \tag{1.9}$$

Now the Fourier transform of $f(n)$ is given by

$$\bar{F}(\omega) = \sum_{n=0}^{N-1} f(n) \exp(-jn\omega) \tag{1.10}$$

which is a continuous function in ω. Comparison of equation (1.10) with (1.9) shows that $F(k)$ is a sampled version of the Fourier transform of $f(n)$ evaluated at the discrete frequency points $\omega = 2\pi k/N$.

Thus equation (1.9) gives the discrete Fourier transform (DFT) coefficients of the time-bounded sequence $f(n)$, and equation (1.8) gives the inverse relationship (IDFT).

Similar relations for the forward and inverse DFT exist in two dimensions, namely,

$$F(p, q) = \sum_{m=0}^{M-1} \sum_{n=0}^{N-1} f(m, n) \exp(-j2\pi mp/M - j2\pi nq/N) \tag{1.11}$$

and

$$f(m, n) = \frac{1}{MN} \sum_{p=0}^{M-1} \sum_{q=0}^{N-1} F(p, q) \exp(j2\pi mp/M + j2\pi nq/N) \tag{1.12}$$

One very valuable property of the 2-D DFT is that it may be evaluated by performing two successive sets of 1-D DFTs.

Equation (1.11) may be written in the form

$$F(p, q) = \frac{1}{M} \sum_{m=0}^{M-1} \exp(-j2\pi mp/M) \frac{1}{N} \sum_{n=0}^{N-1} f(m, n) \exp(-j2\pi nq/N)$$

$$= \frac{1}{M} \sum_{m=0}^{M-1} f_1(m, q) \exp(-j2\pi mp/M) \tag{1.13}$$

where

$$f_1(m, q) = \frac{1}{N} \sum_{n=0}^{N-1} f(m, n) \exp(-j2\pi nq/N) \tag{1.14}$$

and is a set of M 1-D DFTs.

Similarly, equation (1.13) represents a set of N 1-D DFTs along the other axis.

1.4.1. Periodicity

Equation (1.6) shows that $F(k)$ is periodic in k with period N, since

$$\exp(-j2\pi nk/N) = \exp[-j2\pi n(N+k)/N]$$

As a consequence, all the data of the transformed sequence are contained in the range $k = 0$ to $N - 1$, and the DFT is seen to be a transformation from N data points in the time domain to N data points in the frequency domain.

Similarly, a 2-D DFT is periodic in both dimensions of period M and N; it represents an $M \times N$ point transformation of an $M \times N$ point spatial array.

In the majority of situations the array is square, and thus equations (1.11) and (1.12) may be simplified by setting $M = N$.

1.4.2. Symmetry

From equation (1.6) we obtain

$$\begin{aligned} F(-k) &= \sum_{n=0}^{N-1} \bar{f}(n) \exp(j2\pi nk/N) \\ &= F^*(k) \end{aligned} \tag{1.15a}$$

where F^* is the complex conjugate of F; this follows from the fact that $f(n)$ is a real sequence. Thus

$$|F(-k)| = |F(k)| \tag{1.15b}$$

and

$$\arg[F(-k)] = -\arg[F(k)] \tag{1.15c}$$

namely, the magnitude of the DFT is symmetric and the phase angle antisymmetric.

In two dimensions we may similarly observe that

$$|F(-p,-q)| = |F(p,q)| \tag{1.16a}$$

and

$$\arg[F(-p,-q)] = -\arg[F(p,q)] \tag{1.16b}$$

Thus, a 2-D DFT has central symmetry of its magnitude function and central antisymmetry of its phase.

1.4.3. Translation

It is almost trivial to show that, if the time axes of a 1-D sequence are shifted by a distance n_0, the Fourier transform is multiplied by an exponential function. Thus, if the sequence $f(n)$ has a Fourier transform $F(k)$, then the DFT of the sequence $f(n - n_0)$ is

$$F(k) \exp(-j2\pi k n_0/N) \tag{1.17a}$$

In two dimensions the DFT of the shifted array $f(m - m_0, n - n_0)$ is

$$F(p, q) \exp(-j2\pi p m_0/M - j2\pi q n_0/N) \tag{1.17b}$$

These may both be proved by direct application of equations (1.9) and (1.11).

One useful application of this last theorem is to move the spatial origin from the corner of an array to the center. Considering an array, $f(m, n)$, of size $M \times N$, we may move the origin to one corner, and thus the array is represented by $f(m - M/2, n - N/2)$ which has a DFT

$$(-1)^{p+q} F(p, q) \tag{1.18}$$

1.4.4. Rotation

It may be shown that rotation of a 2-D spatial array about the origin by an angle θ will result in a rotation of the Fourier transform by an identical angle.

We assume that the spatial coordinates (m, n) are transformed by the rotation to (m', n') where

$$m' = m \cos\theta + n \sin\theta \qquad \text{and} \qquad n' = -m \sin\theta + n \cos\theta \tag{1.19}$$

It should be noted that m' and n' must be rounded to the nearest integer value. Direct substitution in equation (1.11) shows that the new frequency domain coordinates (p', q') will be given by

$$p' = p \cos\theta + q \sin\theta \qquad \text{and} \qquad q' = -p \sin\theta + q \cos\theta \tag{1.20}$$

showing that the Fourier transform has been rotated by the same angle θ.

This may be extended to multidimensional systems, where rotation may take place in n-dimensional space.

1.4.5. Circularly Symmetric Arrays

In many 2-D spatial problems, an input signal at some point in space will cause an output which is identical at all points at a fixed distance from the given point. This implies that the impulse response of the system is equal at all points equidistant from the spatial origin. Such a system is termed circularly symmetric. By consideration of the rotation property described in Section 1.4.4, it may be seen that the Fourier transform is also circularly symmetrical.

1.4.6. Relation between Discrete and Continuous Fourier Transforms

The Fourier transform of a discrete sequence is, as stated by equation (1.10), a continuous function of frequency. The discrete Fourier transform is equal to the values of the continuous Fourier transform at the appropriate frequency sampling points. For band-limited signals the reverse is also true, and the continuous Fourier transform may be determined uniquely from the DFT.

1.5. *Z* TRANSFORMS[3]

The Z transform of a one-dimensional sequence $f(n)$ is defined as

$$F(z) = \sum_{n=0}^{\infty} f(n) z^{-n} \tag{1.21}$$

and this is defined over a region of convergence in the z plane. The inverse Z transform is defined as

$$f(n) = \frac{1}{2\pi j} \oint_C F(z) z^{n-1}\, dz \tag{1.22}$$

where C is an appropriate closed contour in the z domain.

We note that this is the single-sided Z transform where it is assumed that $f(n)$ is a causal sequence.

In two dimensions, the corresponding forward and inverse Z transforms for the more customary noncausal sequences are the two-sided transforms given respectively by

$$F(z_1, z_2) = \sum_{i=-\infty}^{\infty} \sum_{j=-\infty}^{\infty} f(i,j) z_1^{-i} z_2^{-j} \tag{1.23}$$

and

$$f(i,j) = \frac{1}{4\pi^2} \oint_{C_1} \oint_{C_2} F(z_1, z_2) z_1^{i-1} z_2^{j-1}\, dz_1\, dz_2 \tag{1.24}$$

where C_1 and C_2 are appropriate closed contours in the z_1 and z_2 domains.

Table 1.1. *Z* Transforms

$f(n)$	$F(z)$
$f(n-m)u_{-1}(n-m)$	$z^{-m}F(z)$
$f(n-1)$	$z^{-1}F(z)+f(-1)$
$f(n-m)$	$z^{-m}F(z)+z^{-m+1}f[-1]+\cdots+f[-m]$
$a^n f(n)$	$F(z/a)$
$u_0(n)$; impulse	1
$u_{-1}(n)=1$; step	$\dfrac{z}{z-1}$
$u_{-2}(n)=n$; ramp	$\dfrac{z}{(z-1)^2}$
a^n; exponential	$\dfrac{z}{z-a}$
$n(n-1)/2$	$\dfrac{z}{(z-1)^3}$
n^2	$\dfrac{z^2+z}{(z-1)^3}$
na^{n-1}	$\dfrac{z}{(z-a)^2}$
$\cos n\omega T$	$\dfrac{z^2-z\cos\omega T}{z^2-2z\cos\omega T+1}$
$\sin n\omega T$	$\dfrac{z\sin\omega T}{z^2-2z\cos\omega T+1}$
$\dfrac{n(n-1)\cdots(n-m+1)}{m!}a^{n-m}$	$\dfrac{z}{(z-a)^{m+1}}$

In two dimensions and higher, the region of convergence of the summation in expression (1.23) is difficult to establish.

Table 1.1 gives the *Z*-transform relationships for a few simple analytic functions.

1.5.1. *Z* Transform of Delayed Function

If $F(z)$ is the Z transform of $f(n)$, it may be seen that the Z transform $F'(z)$ of $f(n-1)$ is

$$F'(z) = z^{-1}F(z) + f(0) \tag{1.25}$$

and of $f(n-k)$ is

$$F^{(k)}(z) = z^{-k}F(z) + [z^{-k+1}f(0) + z^{-k+2}f(-1) + \cdots + f(-k+1)] \tag{1.26}$$

where $f(0), f(-1), \ldots, f(-k+1)$ are boundary conditions or, in the time domain, initial conditions.

Similar expressions exist in two dimensions involving the boundary conditions along both axes, namely, $f(0, j)$ and $f(i, 0)$ for all i and j, such that $i = 0, 1, \ldots, M-1$ and $j = 0, 1, \ldots, N-1$.

For zero boundary conditions, equation (1.26) reduces to

$$F^{(k)}(z) = z^{-k}F(z) \tag{1.27}$$

1.5.2. *Z* Transform of Convolution

From the definitions of convolution and Z transform, it may be shown that, if two sequences $u(n)$ and $h(n)$ are convolved to give $y(n)$ as given by equation (1.3), their Z transforms are multiplied. Thus

$$Y(z) = H(z)U(z) \tag{1.28}$$

where $Y(z)$, $H(z)$, and $U(z)$ are the Z transforms of $y(n)$, $h(n)$, and $u(n)$.

Similarly, in two dimensions, if

$$y(m, n) = h(m, n) * u(m, n)$$

their respective Z transforms are related by

$$Y(z_1, z_2) = H(z_1, z_2)U(z_1, z_2) \tag{1.29}$$

1.5.3. Inverse *Z* Transform

The inverse Z transform is formally defined by equation (1.22) or (1.24). However, the evaluation of this contour integral for high-order functions in one dimension is cumbersome and in two dimensions it is usually impossible. Other simpler techniques are available for finding the inverse Z transform of all rational 1-D functions and most rational 2-D functions, although the result will not necessarily be in closed form.

In the case of a 1-D function $F(z)$, it is possible to factorize the denominator and then to express the function $F(z)/z$ as the sum of a number of partial fractions. Each of these may be identified in the time domain with a single exponential function of time, giving a closed-form solution to the inverse Z transform. In two dimensions, factorization of the denominator is not possible in the general case. Thus

$$\frac{F(z)}{z} = \sum_i \frac{a_i}{z - p_i}$$

where p_i and a_i are either both real or occur in pairs of complex conjugates. From this we obtain

$$F(z) = \sum_i \frac{a_i z}{z - p_i}$$

and by reference to Table 1.1 we may express the time sequence as $f(n) = \sum_i a_i p_i^n$ assuming p_i to be real.

EXAMPLE 1.1.

$$F(z) = \frac{z+3}{(z+2)(z+4)}$$

$$\frac{F(z)}{z} = \frac{3}{8z} - \frac{1}{4(z+2)} - \frac{1}{8(z+4)}$$

Thus

$$F(z) = \frac{3}{8} - \frac{z}{4(z+2)} - \frac{z}{8(z+4)}$$

and

$$f(n) = (\tfrac{3}{8} - \tfrac{1}{4}2^n - \tfrac{1}{8}4^n)u_0(n)$$

An alternative method, which is applicable to both one and two dimensions and also for nonrational functions, is to expand the function as a power series in z (or z_1 and z_2) either by the binomial theorem or by a Taylor's series. Each term may then be individually transformed to give the elements of the required 1-D or 2-D time/space array, or analytically as the sum of a sequence of delayed impulse responses. In certain cases, a closed-form solution of the series may be obtained.

EXAMPLE 1.2.

$$\begin{aligned} F(z_1, z_2) &= \frac{2z_1z_2}{z_1 + 2z_2 + 2z_1z_2} \\ &= (1 + z_1^{-1} + z_2^{-1}/2)^{-1} \\ &= 1 - (z_1^{-1} + z_2^{-1}/2) + (z_1^{-1} + z_2^{-1}/2)^2 - \cdots \\ &= 1 - z_1^{-1} - z_2^{-1}/2 + z_1^{-2} + z_1^{-1}z_2^{-1} + z_2^{-2}/4 + \cdots \end{aligned}$$

Hence

$$\begin{aligned} f(m, n) = {} & u_0(m, n) - u_0(m-1, n) - u_0(m, n-1)/2 + u_0(m-2, n) \\ & + u_0(m-1, n-1) + u_0(m, n-2)/4 - \cdots \end{aligned}$$

1.5.4. Relation between Z Transform and Discrete Fourier Transform

Comparison between equations (1.21) and (1.9), or (1.23) and (1.11), shows that the discrete Fourier transform evaluated over the interval from 0 to $N-1$, in one dimension, is identical to the Z transform in which z is replaced by $\exp(j2\pi k/N)$ at the appropriate sampling instants represented by integer values of k. The DFT is thus identical to the Z transform evaluated at discrete points around a circle of unit radius in the z domain.

Similarly, the 2-D DFT may be obtained from the Z transform by evaluation at a grid of discrete points on the boundary of the bidisc in the (z_1, z_2) domain defined by $|z_1^{-1}| = 1$ and $|z_2^{-1}| = 1$.

1.6. SOLUTION OF RECURSION EQUATIONS

The recursion equation (1.1) or (1.2) may be solved either in the time (or space) domain, or in one of the transformed domains.

1.6.1. Z-Transform Solution—Transfer Function

Taking the Z transform of equation (1.1a) and assuming zero boundary conditions, we have

$$Y(z) = \sum_{k=0}^{N} z^{-k} a(k) U(z) \tag{1.30}$$

We may define a transfer function $H(z)$ such that

$$\begin{aligned} H(z) &= Y(z)/U(z) \\ &= \sum_{k=0}^{N} z^{-k} a(k) \end{aligned} \tag{1.31a}$$

and using equation (1.4a) we obtain

$$H(z) = \sum_{k=0}^{N} z^{-k} h(k) \tag{1.31b}$$

It is thus seen that the transfer function is the Z transform of the impulse response.

A solution may therefore be obtained by determining the Z transform of the impulse response that is equal to the transfer function and multiplying it by the Z transform of the input sequence. The inverse Z transform of this product will give the output sequence.

Similarly, by taking the Z transform of equation (1.2a) and rearranging we obtain

$$\sum_{k=0}^{N_2} z^{-k} b(k) Y(z) = \sum_{k=0}^{N_1} z^{-k} a(k) U(z) \tag{1.32}$$

and again a transfer function may be defined as

$$\begin{aligned} H(z) &= Y(z)/U(z) \\ &= \sum_{k=0}^{N_1} z^{-k} a(k) \Big/ \sum_{k=0}^{N_2} z^{-k} b(k) \end{aligned} \tag{1.33a}$$

and the solution of equation (1.2a) may proceed similar to the above approach. Furthermore, taking the Z transforms of equation (1.3a) we may write

$$Y(z) = \sum_{k=0}^{\infty} z^{-k} h(k) U(z)$$

and thus

$$H(z) = \sum_{k=0}^{\infty} z^{-k} h(k) \tag{1.33b}$$

again showing that, for a recursive system, the transfer function is the Z transform of the impulse response.

Similar relations may be obtained for 2-D systems in which the transfer function $H(z_1, z_2)$ is seen to be the 2-D Z transform of the impulse response $h(m, n)$.

Thus, by taking the Z transform of equation (1.1b), we obtain

$$\begin{aligned} H(z_1, z_2) &= Y(z_1, z_2)/U(z_1, z_2) \\ &= \sum_{i=0}^{M} \sum_{j=0}^{N} z_1^{-i} z_2^{-j} h(i, j) \end{aligned} \tag{1.34a}$$

and similarly for the recursive system of equations (1.2b)

$$H(z_1, z_2) = \frac{\sum_{i=0}^{M_1} \sum_{j=0}^{N_1} z_1^{-i} z_2^{-j} a(i, j)}{\sum_{i=0}^{M_2} \sum_{j=0}^{N_2} z_1^{-i} z_2^{-j} b(i, j)} \tag{1.34b}$$

1.6.2. Frequency Response

The system of equations (1.3) and (1.4) may be solved alternatively in the frequency domain. This is achieved by taking the discrete Fourier transform of the relevant equation, resulting in expressions very similar to those obtained above for the Z transform. In this case we may define a network function

$$H(e^{j\omega T}) = Y(e^{j\omega T})/U(e^{j\omega T}) \tag{1.35a}$$

where $H(e^{j\omega T})$, $Y(e^{j\omega T})$, and $U(e^{j\omega T})$ are the discrete Fourier transforms of the corresponding time functions; $H(e^{j\omega T})$ is the frequency response of the network, giving both magnitude and phase as a function of frequency.

Function $H(e^{j\omega T})$ is thus seen to be the DFT of the impulse response $h(n)$. This relationship holds for both recursive and nonrecursive systems. A corresponding situation exists in two dimensions. The frequency response of a 2-D system is obtained similarly:

$$H(e^{j\omega_1 T_1}, e^{j\omega_2 T_2}) = Y(e^{j\omega_1 T_1}, e^{j\omega_2 T_2})/U(e^{j\omega_1 T_1}, e^{j\omega_2 T_2}) \tag{1.35b}$$

Thus the solution of system (1.1) or (1.2) may be obtained directly by deriving the Fourier transform of the input and the system impulse response and taking the IDFT of their product.

It may further be noted by comparing equations (1.33a) and (1.35a) that the frequency response of a network, $H(e^{j\omega T})$, or in two dimensions $H(e^{j\omega_1 T_1}, e^{j\omega_2 T_2})$, may be obtained from the network function $H(z)$ or $H(z_1, z_2)$ by the substitution $z = \exp(j\omega T)$ or, in two dimensions, $z_1 = \exp(j\omega_1 T_1)$ and $z_2 = \exp(j\omega_2 T_2)$.

1.6.2.1. *Properties of Frequency Response*

In Section 1.4.2, it has been shown that the amplitude and phase functions of a 1-D system are symmetrical and antisymmetrical, respectively. For a 2-D system also, the corresponding functions have central symmetry and antisymmetry.

A zero-phase filter function is one in which the phase is zero at all points. In the case of a 2-D system, this imposes the additional constraint

$$|F(\omega_1, \omega_2)| = |F(\omega_1, -\omega_2)| = |F(-\omega_1, \omega_2)| = |F(-\omega_1, -\omega_2)| \tag{1.36a}$$

or, in terms of the Z transform, the network transfer function may be written as

$$H(z_1, z_2) = F(z_1, z_2)F(z_1^{-1}, z_2)F(z_1^{-1}, z_2^{-1})F(z_1, z_2^{-1}) \tag{1.36b}$$

1.7. FAST FOURIER TRANSFORM

One of the most powerful and popular techniques for evaluating the DFT of a sequence is the fast Fourier transform (FFT) algorithm. The underlying principle in the FFT algorithm is the factorization of N into a number, m, of comparatively small prime factors. The classical work on this was done by Cooley and Tukey[4] and was based on the earlier work of Good.[5] Since then this has been extended, particularly to mixed radix systems.[6-11] Although the Cooley-Tukey algorithm may be used on a 1-D array of any length N, considerable attention has been devoted to the case for which $N = 2^k$, where k is an integer, since it is readily applicable to programming on computers.

The 1-D DFT may be expressed in the form

$$F(k) = \sum_{n=0}^{N-1} f(n) W_N^{nk} \qquad \text{for } k = 0, 1, \ldots, N-1 \tag{1.37}$$

where

$$W_N = \exp(-j2\pi/N) \tag{1.38}$$

If we write the DFT sequence as a vector F defined by

$$F = [F(0)\ F(1) \cdots F(N-1)]^t \tag{1.39a}$$

where superscript t stands for the transpose, and the time sequence as a similar vector f,

$$f = [f(0)\ f(1) \cdots f(N-1)]^t \tag{1.39b}$$

then equation (1.37) assumes the form

$$F = Af \tag{1.40}$$

where A is an $N \times N$ matrix having (k, n)th element $a_{(k,n)}$ given by

$$a_{(k,n)} = W_N^{kn(\text{mod } N)} \tag{1.41}$$

where (mod N) indicates that calculation is to be performed modulo N.

As it stands, equation (1.40) will involve about $(N-1)^2$ complex multiplications and $N(N-1)$ complex additions. The factorization of N will allow the matrix A to be decomposed into m sparse matrices, and hence the number of multiplications may be reduced considerably.

1.7.1. Decimation-in-Time Algorithm

From (1.38) we see that W_N^{nk} is periodic of period N in both the n and k variables, since

$$W_N^{(n+mN)k} = W_N^{nk} W_N^{mkN} = W_N^{nk} \, \mathrm{e}^{-\mathrm{j}2\pi mk} = W_N^{nk} \tag{1.42}$$

and similarly

$$W_N^{n(k+lN)} = W_N^{nk} \tag{1.43}$$

The suffix N in W_N emphasizes this periodicity.

Let us break $f(n)$ into two sequences: $f_1(n)$ containing all even values of n, and $f_2(n)$ containing all the odd points. Thus

$$\left.\begin{aligned} f_1(n) &= f(2n) \\ f_2(n) &= f(2n+1) \end{aligned}\right\} \qquad n = 0, 1, \ldots, (N-2)/2 \tag{1.44}$$

Then the N-point DFT is

$$\begin{aligned} F(k) &= \sum_{\substack{n=0 \\ \text{(even)}}}^{N-1} f(n) W_N^{nk} + \sum_{\substack{n=0 \\ \text{(odd)}}}^{N-1} f(n) W_N^{nk} \qquad \forall k = 0, 1, \ldots, N-1 \\ &= \sum_{n=0}^{(N-2)/2} f(2n) W_N^{2nk} + \sum_{n=0}^{(N-2)/2} f(2n+1) W_N^{(2n+1)k} \\ &= \sum_{n=0}^{(N-2)/2} f_1(n) W_{N/2}^{nk} + W_N^k \sum_{n=0}^{(N-2)/2} f(n) W_{N/2}^{nk} \end{aligned} \tag{1.45}$$

since

$$W_N^2 = \left[\exp\left(\mathrm{j}\frac{2\pi}{N}\right)\right]^2 = \exp\left(\mathrm{j}\frac{2\pi}{N/2}\right) = W_{N/2} \tag{1.46}$$

Thus

$$F(k) = F_1(k) + W_N^k F_2(k) \qquad \forall k = 0, 1, \ldots, N-1 \tag{1.47}$$

We note that functions F_1 and F_2 are $(N/2)$-point DFTs and thus they are defined for $k = 0, 1, \ldots, (N-2)/2$. To obtain values for the full range of

k, we note that F_1 and F_2 are periodic of period $N/2$, and hence

$$\left.\begin{aligned} F_1(k - N/2) &= F_1(k) \\ F_2(k - N/2) &= F_2(k) \end{aligned}\right\} \quad \forall k = N/2, \ldots, N-1 \tag{1.48}$$

It is also noteworthy that $W_N^{k+N/2} = W_N^k \exp(j\pi) = -W_N^k$. Therefore

$$\begin{aligned} F(k) &= F_1(k) + W_N^k F_2(k), & 0 \leq k \leq N/2 - 1 \\ &= F_1(k - N/2) - W_N^{k-N/2} F_2(k - N/2), & N/2 \leq k \leq N - 1 \end{aligned} \tag{1.49}$$

This process for an eight-point DFT is shown in Figure 1.2a.

The outputs of the two four-point DFT blocks are combined according to equations (1.49); the butterfly symbol as shown in Figure 1.2b indicates that the upper right-hand output node gives the sum of the two input nodes at the left and the lower output node gives the difference; the triangle symbol indicates multiplication by a factor.

It may be noted that the output array occurs in its correct sequential order, while the input array is in a different order. We shall look at this matter shortly.

Now each of these $(N/2)$-point DFTs may again be broken into two $(N/4)$-point DFTs. Considering the top DFT block, we may write

$$\left.\begin{aligned} F_1(k) &= X_1(k) + W_{N/2}^k Y_1(k), & 0 \leq k \leq N/4 - 1 \\ &= X_1(k - N/4) - W_{N/2}^{k-N/4} Y_1(k - N/4), & N/4 \leq k \leq N/2 - 1 \end{aligned}\right\} \tag{1.50}$$

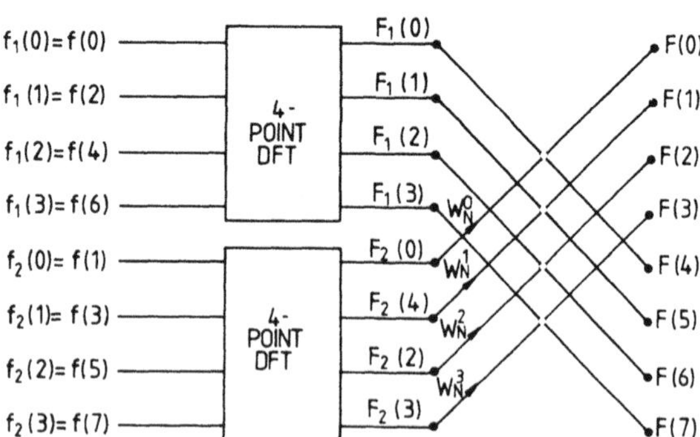

Figure 1.2a. Eight-point DFT broken into two four-point DFTs using the decimation-in-time algorithm.

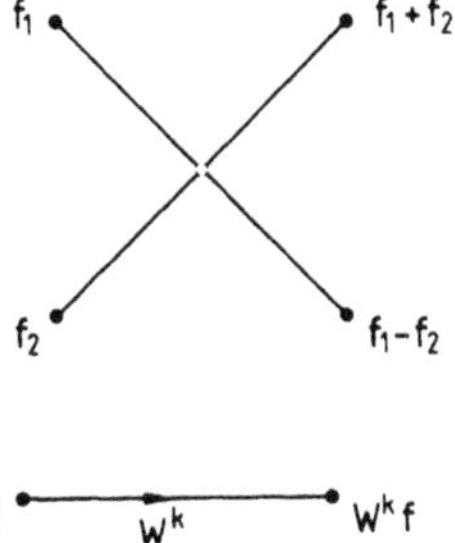

Figure 1.2b. Butterfly and multiplier symbols used in FFT diagrams.

Consideration of the case where $N = 8$ gives the realization of Figure 1.2c.

This process may be repeated until we obtain a two-point DFT, as in the example above. This two-point DFT may be evaluated directly from

$$X_1(k) = \sum_{n=0}^{1} x_1(n) W_2^{nk}$$

Thus

$$X_1(0) = x_1(0) + x_1(1) W_2^0 \qquad \text{and} \qquad X_1(1) = x_1(0) + x_1(1) W_2^1 \quad (1.51)$$

or

$$X_1(0) = x_1(0) + x_1(1)$$

$$X_1(1) = x_1(0) - x_1(1) \quad (1.52)$$

and hence this involves no multiplications, only additions. This is shown in Figure 1.2d. Putting these together gives the complete implementation

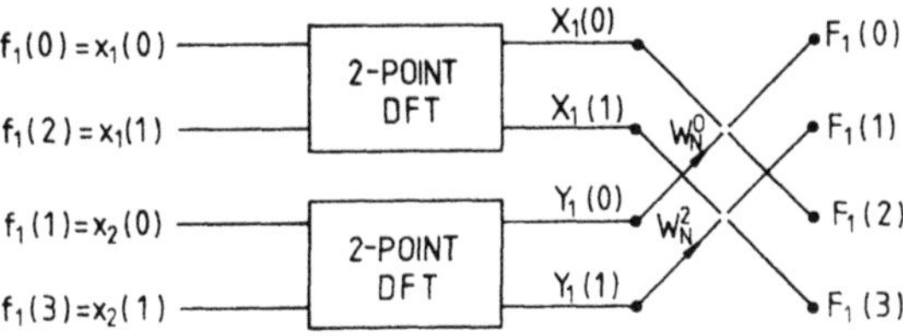

Figure 1.2c. Four-point DFT broken into two two-point DFTs using the decimation-in-time algorithm.

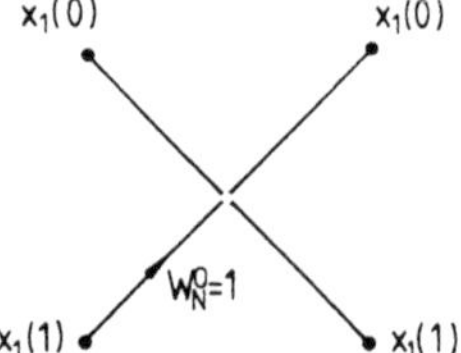

Figure 1.2d. Implementation of a two-point DFT.

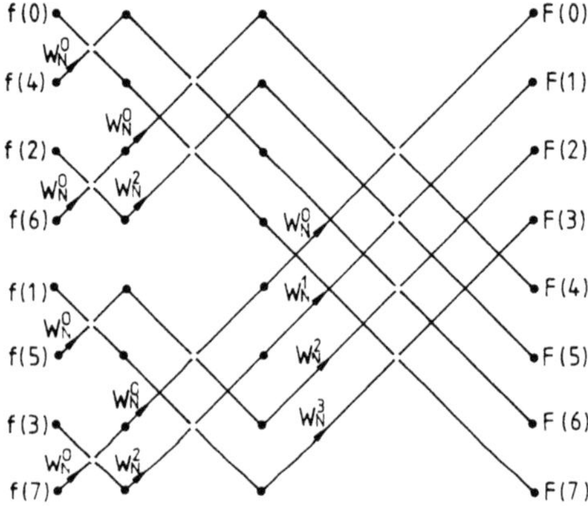

Figure 1.3. Complete flow diagram for an eight-point DFT using the decimation-in-time algorithm.

of Figure 1.3 for the decimation-in-time algorithm for an eight-point DFT.

1.7.2. Decimation-in-Frequency Algorithm

An alternative approach is to split the time wave form into two parts by considering the first and second halves to form the two subseries. We thus define

$$\left.\begin{aligned} f_1(n) &= f(n) \\ f_2(n) &= f(n + N/2) \end{aligned}\right\} \qquad n = 0, 1, \ldots, (N-2)/2 \tag{1.53}$$

Taking the DFT gives

$$\begin{aligned} F(k) &= \sum_{n=0}^{(N-2)/2} f(n)\,W_N^{nk} + \sum_{n=N/2}^{N-1} f(n)\,W_N^{nk} \\ &= \sum_{n=0}^{(N-2)/2} f_1(n)\,W_N^{nk} + \sum_{n=0}^{(N-2)/2} f_2(n)\,W_N^{(n+N/2)k} \\ &= \sum_{n=0}^{(N-2)/2} [f_1(n) + W_N^{Nk/2} f_2(n)]\,W_N^{nk} \\ &= \sum_{n=0}^{(N-2)/2} [f_1(n) + e^{-j\pi k} f_2(n)]\,W_N^{nk} \end{aligned} \tag{1.54}$$

If we separate this into odd and even parts, we obtain

$$\begin{aligned} F(2k) &= \sum [f_1(n) + f_2(n)]\,W_N^{2nk} \\ &= \sum [f_1(n) + f_2(n)]\,W_{N/2}^{nk} \\ &= \sum x(n)\,W_{N/2}^{nk}, \qquad n = 0, 1, \ldots, (N-2)/2 \end{aligned} \tag{1.55a}$$

and

$$\begin{aligned} F(2k+1) &= \sum [f_1(n) - f_2(n)]\,W_N^{n(2k+1)} \\ &= \sum [\{f_1(n) - f_2(n)\}\,W_N^n]\,W_{N/2}^{nk} \\ &= \sum y(n)\,W_{N/2}^{nk} \end{aligned} \tag{1.55b}$$

which are two $N/2$-point DFTs of the two sequences $x(n)$ and $y(n)$ where

$$x(n) = f_1(n) + f_2(n) \qquad \text{and} \qquad y(n) = [f_1(n) - f_2(n)]\,W_N^n \tag{1.56}$$

This is illustrated for $N = 8$ in Figure 1.4.

This procedure may be repeated up to a total of $N/2$ decimations until one obtains a two-point DFT. The complete decimation-in-frequency algorithm for $N = 8$ is shown in Figure 1.5.

1.7.3. Computation of Inverse Discrete Fourier Transform

It is easily shown that the inverse DFT may be evaluated using the same algorithm. The inverse DFT is defined by

$$x(n) = \frac{1}{N} \sum_{k=0}^{N-1} X(k)\,W^{-nk} \tag{1.57a}$$

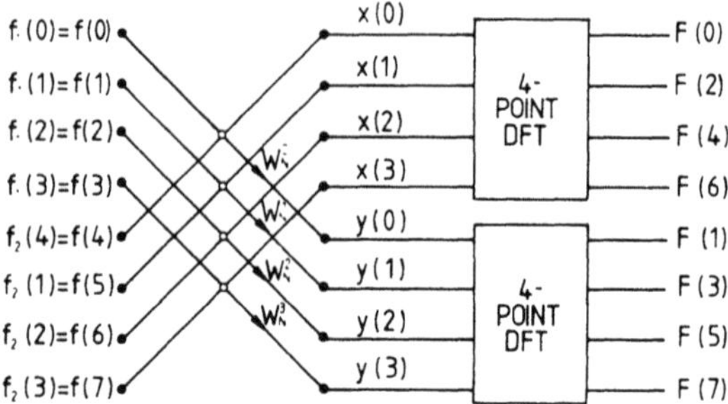

Figure 1.4. First stage of the decimation-in-frequency algorithm for an eight-point DFT.

On taking the complex conjugate of each side we have

$$x^*(n) = \frac{1}{N}\sum_{k=0}^{N-1} X^*(k)\,W^{nk}$$

leading to

$$x(n) = \frac{1}{N}\left[\sum_{k=0}^{N-1} X^*(k)\,W^{nk}\right]^* \tag{1.57b}$$

Thus the same algorithm will suffice for both forward and inverse DFTs.

1.7.4. Data Shuffling

As has been seen, the input data in the decimation-in-time algorithm needs to be reordered, and similarly the output data in the decimation-in-

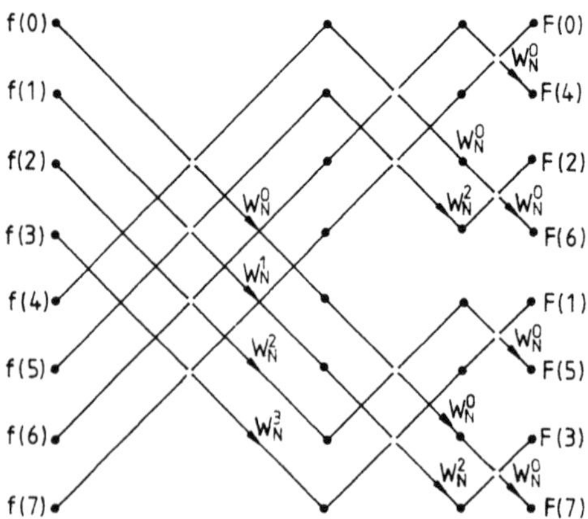

Figure 1.5. Complete flow diagram for an eight-point DFT using the decimation-in-frequency algorithm.

0 ⟶ 0	000 ⟶ 000
1 ⟶ 4	001 ⟶ 100
2 ⟶ 2	010 ⟶ 010
3 ⟶ 6	011 ⟶ 110
4 ⟶ 1	100 ⟶ 001
5 ⟶ 5	101 ⟶ 101
6 ⟶ 3	110 ⟶ 011
7 ⟶ 7	111 ⟶ 111

Figure 1.6. Example of data shuffling.

frequency algorithm. This data shuffling may be achieved by writing the index of each element in the shuffled data in binary form and noting that its position in the sequence is obtained in binary form by reversing the bit order. This is shown for an eight-bit sequence in Figure 1.6.

1.7.5. Number of Multiplications for One-Dimensional Fast-Fourier-Transform Algorithm

We assume that N, the size of the array, is factorized into m subarrays of prime factors, N_i, such that

$$N = \prod_{i=1}^{m} N_i \tag{1.58}$$

It may be shown that the total number of complex multiplications needed for the decomposition of the matrix A is approximately[9,11,12]

$$N_{\text{mult}} = N \sum_{i=1}^{m} N_i - (m+2)N + N/N_m + 1 \tag{1.59a}$$

and the total number of complex additions is

$$N_{\text{add}} = N \sum_{i=1}^{m} N_i - mN \tag{1.59b}$$

The latter expressions give a measure of the numerical complexity in evaluating a mixed radix FFT. These numbers of arithmetic processes are an order of magnitude less than those required for the direct evaluation.

When all the factors are identical and $N = p^m$, with p an odd prime, the total number of complex multiplications[9] may be reduced to

$$mN(p-1)^2/4p \tag{1.60}$$

Again, in the situation encountered with the most usual form of the FFT when $p = 2$ and $N = 2^m$, the number of complex multiplications may be reduced to approximately

$$N_{\text{mult}} = (N/2)\log_2 N \tag{1.61}$$

1.7.6. Two-Dimensional Fast Fourier Transform

It has been shown in equation (1.13) that a 2-D DFT may be computed by carrying out two 1-D DFTs successively. The same applies to the FFT algorithm; thus to obtain the FFT of a 2-D array, one first obtains a 1-D FFT along one dimension and then takes a second FFT of this resulting array along the second dimension.

1.7.7. Number of Multiplications for Two-Dimensional Fast Fourier Transform

The computation of the FFT of a 2-D array of size $D_1 \times D_2$ may be determined by an extension of the above. If both D_1 and D_2 are factorized into prime factors according to

$$D_1 = \prod_{i=1}^{m} d_{1i} \tag{1.62a}$$

and

$$D_2 = \prod_{i=1}^{n} d_{2i} \tag{1.62b}$$

then the number of multiplications in the FFTs carried out along the D_1 axis is

$$N_{\text{mult}(1)} = D_1 \sum_{i=1}^{m} d_{1i} - (m+2)D_1 + D_1/d_{1m} + 1$$

and along the D_2 axis

$$N_{\text{mult}(2)} = D_2 \sum_{i=1}^{n} d_{2i} - (n+2)D_2 + D_2/d_{2n} + 1$$

Hence the total number of complex multiplications for the 2-D array will be

$$N_{\text{FFT}} = N_{\text{mult}(1)}D_2 + N_{\text{mult}(2)}D_1 \tag{1.63}$$

In the particular case of a radix-2 transform, we have

$$N_{\text{FFT}} = (D_1 D_2/2) \log_2 D_1 D_2 \tag{1.64}$$

1.8. CONVOLUTION

The convolution of the input sequence and impulse response sequence in equations (1.1), or the two convolutions in system (1.2), may be obtained by direct multiplication and addition, as indicated by these equations. An alternative form of representation of these equations is that of matrix form.

1.8.1. Matrix Representation of Convolution

For a finite impulse response filter, equations (1.4) may be written as a matrix product. The input signal is written as a vector of infinite length (although in practice all signals will be limited in extent):

$$u = [u(0) u(1) u(2) \cdots]^{t} \tag{1.65}$$

and the output vector y is defined similarly. The matrix relationship may then be written as

$$y = hu \tag{1.66}$$

where $h = [h_{(i,j)}]$ in which

$$\left.\begin{aligned} h_{(i,j)} &= 0 && \text{for } i-j<0 \\ &= h(p) && \text{for } i-j=p \end{aligned}\right\}$$

Thus

$$h = \begin{bmatrix} h(0) & & & & 0 & \\ h(1) & h(0) & & & & \\ h(2) & h(1) & h(0) & & & \\ \vdots & & & \ddots & & \\ h(N) & \cdots & \cdots & h(1) & h(0) & \\ & h(N) & & & h(1) & \ddots \\ & & \ddots & & & \\ & & & \ddots & & \ddots \\ & 0 & & & h(N) & \end{bmatrix} \tag{1.67}$$

In two dimensions it is first necessary to rearrange the 2-D input data $u(m, n)$, $m = 0, 1, \ldots, P-1$ and $n = 0, 1, \ldots, Q-1$, in the form of a 1-D vector $u = (u_r)$, $r = 1, 2, \ldots, PQ$. This is done by ordering the 2-D signal lexicographically in the 1-D vector such that

$$u(r) = u(m, n)$$

where $r = mQ + n + 1$.

A similar rearrangement of the output signal y is made as a vector of length PQ.

The convolution given by expression (1.4b) may now be written as the matrix equation (1.66), where

$$h = \left[\begin{array}{llllll}
h(0,0) & & & & & \\
h(0,1) & h(0,0) & & & & \\
\vdots & \ddots & & & & \\
h(0,N) & & & & & \\
\vdots & h(0,N) & & & & \\
h(1,0) & & \ddots & h(0,0) & & \\
h(1,1) & h(1,0) & & h(0,1) & h(0,0) & \\
\vdots & \ddots & & \cdots & & \\
h(1,N) & & & h(0,N) & & \vdots \\
\vdots & h(1,N) & & \vdots & & \vdots \\
\vdots & & & \cdots & & h(N,N)
\end{array}\right] \qquad (1.68)$$

Such rearrangement of the convolution relationship will be found of value when attempting to reduce the number of computations necessary for evaluation. This will be considered in greater detail in Chapter 8.

1.8.2. Computation of Convolution

The computation of the convolution of two 1-D arrays, $f_1(n)$ of size N_1 and $f_2(n)$ of size N_2, will result in an array of size $(N_1 + N_2 - 1)$. It is therefore customary to increase both arrays to length $(N_1 + N_2 - 1)$ by adding extra zeros at the end. It is then possible to use the same location for storing the convolution product data as was used for one of the original data arrays.

In the case of 2-D arrays of size $M_1 \times N_1$ and $M_2 \times N_2$, the resultant convolution has a size $(M_1 + N_1 - 1) \times (M_2 + N_2 - 1)$. However, 2-D arrays—such as television pictures—frequently have a finite region of support which is identical for both the input and output arrays. The second of the two convolving arrays is usually the impulse response of some processing system and the dimensions $M_2 \times N_2$ are thus relatively small compared with the image size $M_1 \times N_1$. The output array will therefore be larger than the region of support and some picture elements at the boundaries will need to be discarded. As a consequence of this, the edges of a picture may frequently incorporate more distortion than the bulk; this may be reduced by padding the original image by values other than zero beyond the region of support.

1.8.2.1. Implementation of Convolution

The direct implementation of two 1-D arrays of size N_1 and N_2 will require a number of real multiplications approximately equal to $N_1 N_2$. This number may be reduced by employing the short convolution algorithm.[13]

As an illustration of this technique, consider a length-two convolution which may be written in matrix form as

$$\begin{bmatrix} y_0 \\ y_1 \end{bmatrix} = \begin{bmatrix} h_0 & h_{-1} \\ h_1 & h_0 \end{bmatrix} \begin{bmatrix} u_0 \\ u_1 \end{bmatrix} \tag{1.69}$$

This may be computed directly using four multiplications and two additions. However, we may modify this calculation by defining three intermediate variables

$$\begin{aligned} w_0 &= (h_0 + h_{-1})u_1 \\ w_1 &= h_0(u_0 - u_1) \\ w_2 &= (h_1 + h_0)u_0 \end{aligned} \tag{1.70}$$

requiring three additions and three multiplications. The desired result can now be obtained with two further additions from

$$\begin{aligned} y_0 &= w_0 + w_1 \\ y_1 &= w_2 - w_1 \end{aligned} \tag{1.71}$$

Since computation time is more critically dependent on the number of multiplications, this obviously results in approximately twenty-five percent reduction of computer time.

The application of this technique to much longer-length convolutions arises from partitioning the larger convolution matrices into a number of length-two convolutions, which may be evaluated as above. The economy in computation time may be further improved in many cases, since the convolution operators are frequently relatively sparse Toeplitz (or doubly Toeplitz) matrices, and thus only a limited number of length-two convolutions need to be evaluated.

REFERENCES

1. A. H. Kayran and R. A. King, Design of recursive and nonrecursive fan filters with complex transformations, *IEEE Trans. Circuits Syst.* **CAS-30,** 849–857 (1983).
2. H. Chang and J. K. Aggarwal, Design of two-dimensional recursive filters by interpolation, *IEEE Trans. Circuits Syst.* **CAS-24,** 281–291 (1977).
3. E. I. Jury, *Theory and Application of the Z-Transform Method,* John Wiley and Sons, New York (1964).
4. J. W. Cooley and J. W. Tukey, An algorithm for the machine calculation of complex Fourier series, *Math. Comput.* **19,** 297–301 (1965).
5. I. J. Good, The interaction algorithm and practical Fourier analysis, *J. R. Stat. Soc., B* **20,** 361–372 (1958); **22,** 372–375 (1960).
6. G. D. Bergland, The fast Fourier transform recursive equations for arbitrary length records, *Math. Comput.* **21,** 236–238 (1967).
7. G. D. Bergland, A fast Fourier transform algorithm using base 8 iterations, *Math. Comput.* **22,** 275–279 (1968).
8. R. C. Singleton, On computing the fast Fourier transform, *Commun. ACM* **10,** 647–654 (1967).
9. R. C. Singleton, An algorithm for computing the mixed radix fast Fourier transform, *IEEE Trans. Audio Electroacoust.* **AU-17,** 93–103 (1969).
10. W. M. Gentleman, Matrix multiplication and fast Fourier transforms, *Bell Syst. Tech. J.* **47,** 1099–1103 (1968).
11. W. M. Gentleman and G. Sande, Fast Fourier transform—for fun and profit, *AFIPS Conf. Proc.* **29,** Fall Joint Computer Conf., Nov. 7–10, San Francisco, 563–578 (1966).
12. M. R. Azimi-Sadjadi, *Block Implementation of Two-Dimensional Digital Filters,* PhD Thesis, University of London (1981).
13. R. C. Agarwal and C. S. Burrus, Fast one-dimensional digital convolution by multidimensional techniques, *IEEE Trans. Acoust., Speech, Signal Process* **ASSP-22,** 1–10 (1974).

2

Nonrecursive Filters

PART ONE. ONE-DIMENSIONAL FILTERS

2.1. INTRODUCTION

It was explained in Chapter 1 that digital filters can be divided into two classes, namely, nonrecursive and recursive. The z-transfer function of a causal nonrecursive digital filter is given by equation (1.31b),

$$H(z) = \sum_{n=0}^{N-1} h(nT)z^{-n} \tag{2.1}$$

Its frequency response is given by

$$H(e^{j\omega T}) = M(\omega)\,e^{j\phi(\omega)} = \sum_{n=0}^{N-1} h(nT)\,e^{-jn\omega T} \tag{2.2}$$

where

$$M(\omega) = |H(e^{j\omega T})| \tag{2.3}$$

and

$$\phi(\omega) = \arg H(e^{j\omega T}) \tag{2.4}$$

The phase and group delays of a filter are described by

$$\tau_p = -\frac{\phi(\omega)}{\omega} \qquad \text{and} \qquad \tau_g = -\frac{d\phi(\omega)}{d\omega}$$

respectively.

For constant group delay characteristic, the phase response must be linear and, assuming zero phase at zero frequency,

$$\phi(\omega) = -\tau\omega$$

Thus equations (2.2) and (2.4) yield

$$\phi(\omega) = -\tau\omega = \arctan \frac{-\sum_{n=0}^{N-1} h(nT)\sin\omega nT}{\sum_{n=0}^{N-1} h(nT)\cos\omega nT} \tag{2.5}$$

and taking the tangent of both sides of equation (2.5) gives

$$\tan\omega\tau = \frac{\sin\omega\tau}{\cos\omega\tau} = \frac{\sum_{n=0}^{N-1} h(nT)\sin\omega nT}{\sum_{n=0}^{N-1} h(nT)\cos\omega nT} \tag{2.6}$$

Cross multiplying both sides of equation (2.6) leads to

$$\sum_{n=0}^{N-1} h(nT)(\cos\omega nT\sin\omega\tau - \sin\omega nT\cos\omega\tau) = 0$$

or

$$\sum_{n=0}^{N-1} h(nT)\sin(\omega\tau - \omega nT) = 0 \tag{2.7}$$

The solution of equation (2.7) can be shown to be

$$\tau = (N-1)T/2 \tag{2.8}$$

and

$$h(nT) = h[(N-1-n)T] \qquad \text{for } 0 \le n \le N \tag{2.9}$$

Therefore, a nonrecursive filter is capable of producing linear phase (or constant group delay) over the entire pass band. To have a linear phase characteristic, it is only necessary for the impulse response to be symmetrical about the midpoint between samples $(N-2)/2$ and $N/2$ for even N or about sample $(N-2)/2$ for odd N.

In many applications it is only necessary for the group delay to be constant, in which case the phase response can have the form

$$\phi(\omega) = \phi_0 - \tau\omega \tag{2.10}$$

where ϕ_0 is a constant. On using the above procedure a second class of constant delay nonrecursive filters can be obtained. With $\phi_0 = \pm\pi/2$ the solution is

$$\tau = (N-1)T/2 \tag{2.11}$$

and

$$h(nT) = -h[(N-1-n)T] \tag{2.12}$$

In this case the impulse response is antisymmetrical about the midpoint between samples $(N-2)/2$ and $N/2$ for even N or about sample $(N-2)/2$ for odd N. It should be noted that this linear-phase property over the entire frequency band which is inherent in nonrecursive digital filters cannot be extended to recursive filters.

The desired frequency response for the FIR filter will be denoted by $\hat{H}(e^{j\omega T})$ with corresponding impulse response $\hat{h}(nT)$. Since $\hat{H}(e^{j\omega T})$ is simply the z transform of $\hat{h}(nT)$ evaluated on the unit circle, we have

$$\hat{H}(e^{j\omega T}) = \sum_{n=-\infty}^{\infty} \hat{h}(nT)\, e^{-j\omega nT} \tag{2.13}$$

and

$$\hat{h}(nT) = \frac{1}{\omega_s}\int_{-\omega_s/2}^{\omega_s/2} \hat{H}(e^{j\omega T})\, e^{j\omega nT}\, d\omega \tag{2.14}$$

where ω_s is the sampling frequency. This has the form of an infinite Fourier-series representation for $\hat{H}(e^{j\omega T})$, with $\hat{h}(nT)$ as the Fourier coefficients. The series is, in fact, infinite if $\hat{H}(e^{j\omega T})$ or any of its derivatives is discontinuous, which is the case for most filters of interest. On the other hand, the actual frequency response of the FIR filter is given, using equation (2.2), by

$$H(e^{j\omega T}) = \sum_{n=0}^{N-1} h(n)\, e^{-j\omega nT} \tag{2.15}$$

and thus $H(e^{j\omega T})$ corresponds to a finite Fourier-series approximation to $\hat{H}(e^{j\omega T})$. This will form the basis for some of the design problems considered in this chapter.

2.2. DESIGN OF FINITE-IMPULSE-RESPONSE FILTERS USING THE FOURIER-SERIES METHOD

A straightforward approach to the design of an FIR filter is just to truncate the ideal impulse response $\hat{h}(nT)$ outside the interval $0 \leq n \leq N-1$ to produce $h(nT)$, i.e.,

$$h(nT) = \begin{cases} \hat{h}(nT) & \text{for } 0 \leq n \leq N-1 \\ 0 & \text{otherwise} \end{cases} \tag{2.16}$$

This truncation of the Fourier series unfortunately produces the familiar Gibbs phenomenon (or Gibbs oscillation), which will be manifested in $H(e^{j\omega T})$, especially if $\hat{H}(e^{j\omega T})$ is discontinuous. Knowing that all frequency-selective filters are ideally discontinuous at their band edges (as depicted in Figure 2.1), simple truncation of the impulse response will often give an unacceptable FIR design.

The effect of simple truncation will now be shown in the following example.

EXAMPLE 2.1. Design a low-pass filter with a frequency response

$$\hat{H}(e^{j\omega T}) \simeq \begin{cases} 1 & \text{for } |\omega| \leq \omega_c \\ 0 & \text{for } \omega_c < |\omega| \leq \omega_s/2 \end{cases}$$

where ω_s is the sampling frequency.

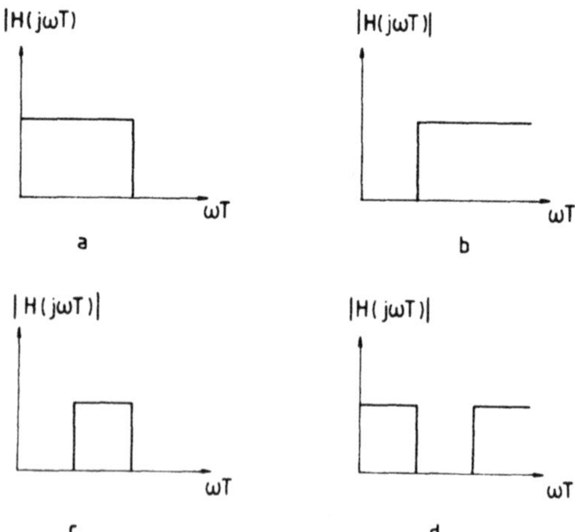

Figure 2.1. Ideal magnitude responses for (a) low-pass filter, (b) high-pass filter, (c) band-pass filter, and (d) band-stop filter.

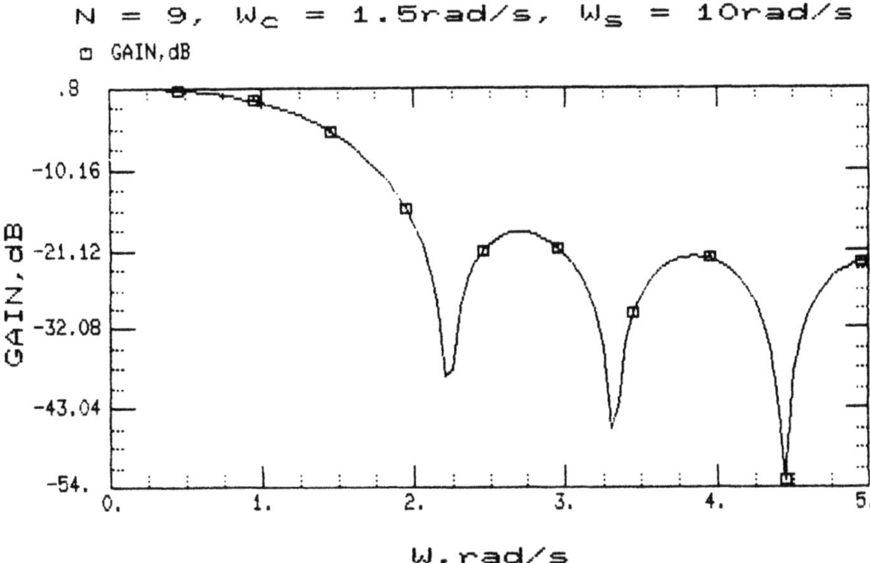

Figure 2.2a. Amplitude response of designed LP filter for $N = 9$, $\omega_c = 1.5$ rad/s, and $\omega_s = 10.0$ rad/s.

SOLUTION. From equation (2.14) we get

$$\hat{h}(nT) = \frac{1}{\omega_s}\int_{-\omega_c}^{\omega_c} e^{j\omega nT}\, d\omega = \frac{1}{\pi n}\sin \omega_c nT$$

Using the symmetry property of equation (2.9) and expression (2.15) we get

$$H(z) = z^{-(N-1)/2}\sum_{n=0}^{(N-1)/2} \frac{a_n}{2}(z^n + z^{-n})$$

where $a_0 = h(nT)$ and $a_n = 2h(nT)$.

The amplitude response of the preceding filter with ω_c and ω_s assumed to be 1.5 and 10 rad/s, respectively, is given in Figures 2.2a–c for $N = 9$, 19, and 29. The pass-band and stop-band oscillations observed are due to the slow convergence of the Fourier series, which in turn is caused by the discontinuity at the pass-band edge. As N is increased, the frequency of these oscillations is seen to increase, and at both low and high frequencies their amplitude is decreased. The amplitude of the last pass-band ripple, however, and that of the first stop-band ripple tend to remain virtually unchanged. This type of performance is often unacceptable in practice and ways must be sought to reduce Gibbs oscillations.

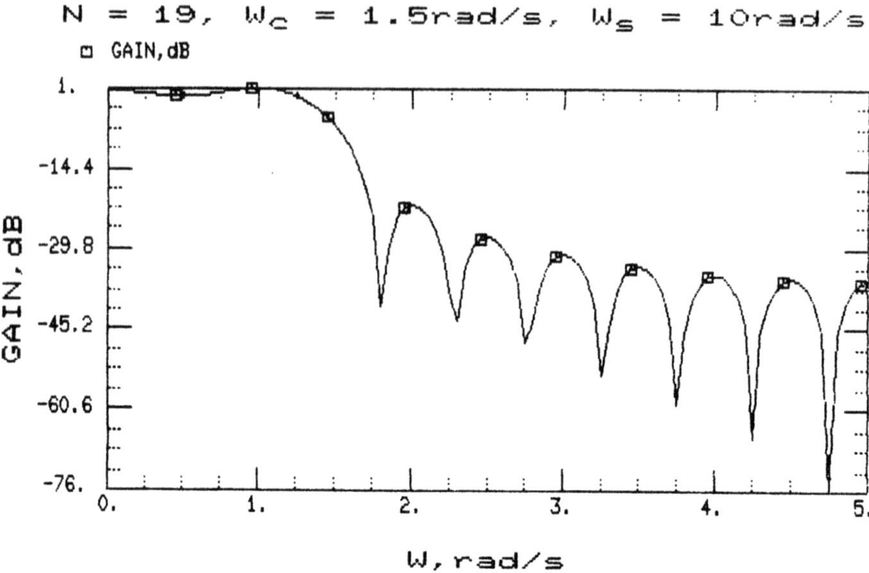

Figure 2.2b. Amplitude response of designed low-pass filter for $N = 19$, $\omega_c = 1.5$ rad/s, and $\omega_s = 10.0$ rad/s.

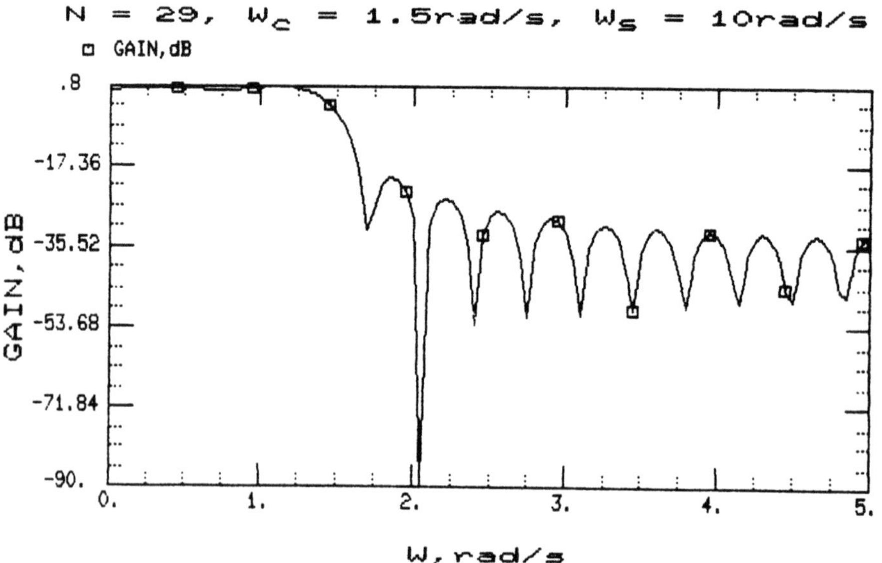

Figure 2.2c. Amplitude response of designed low-pass filter for $N = 29$, $\omega_c = 1.5$ rad/s, and $\omega_s = 10.0$ rad/s.

One way of avoiding discontinuities in the frequency response, which is the cause of Gibbs oscillations, is to introduce a transition band between the pass band and stop band.[1,6] For example, the response of the above low-pass filter could be redefined as

$$\hat{H}(e^{j\omega T}) = \begin{cases} 1 & \text{for } |\omega| \leq \omega_p \\ -\dfrac{\omega - \omega_a}{\omega_a - \omega_p} & \text{for } \omega_p < |\omega| < \omega_a \\ 0 & \text{for } \omega_a \leq |\omega| \leq \omega_s/2 \end{cases}$$

where ω_a and ω_p are the boundaries of the pass bands and stop bands, respectively. The other way of reducing Gibbs oscillations is by using window functions.

2.3. USE OF WINDOW FUNCTIONS

Another easy way of reducing the Gibbs oscillations is by preconditioning the impulse response $h(nT)$ in equation (2.14) using a class of time-domain functions known as window functions. It should be noted that truncation of the impulse response can be viewed as preconditioning the impulse response with a rectangular time wave form known as a rectangular window.

The effect of windowing the impulse response with a rectangular window can be presented more precisely by rewriting equation (2.16) in the form

$$h(nT) = w_R(nT) \cdot \hat{h}(nT) \tag{2.17}$$

where $w_R(nT)$ is the rectangular window function defined by

$$w_R(nT) = \begin{cases} 1 & \text{for } |n| \leq (N-1)/2 \\ 0 & \text{otherwise} \end{cases} \tag{2.18}$$

This corresponds to the direct truncation of the Fourier series, and its effect will be the convolution of the spectrum of the impulse response $\hat{h}(nT)$ and window function $w_R(nT)$ as follows:

$$H(e^{j\omega T}) = W_R(e^{j\omega T}) * \hat{H}(e^{j\omega T}) \tag{2.19}$$

where $*$ denotes convolution and $W_R(e^{j\omega T})$ is given by

$$W_R(e^{j\omega T}) = \sum_{n=-(N-1)/2}^{(N-1)/2} e^{-j\omega nT} = \frac{e^{j\omega(N-1)T/2} - e^{-j\omega(N+1)T/2}}{1 - e^{-j\omega T}}$$

$$= \frac{e^{j\omega NT/2} - e^{-j\omega NT/2}}{e^{j\omega T/2} - e^{-j\omega T/2}} = \frac{\sin(\omega NT/2)}{\sin(\omega T/2)} \tag{2.20}$$

2.3.1. Hann and Hamming Windows

The Hann and Hamming windows are given by

$$w_H(nT) = \begin{cases} \alpha + (1-\alpha)\cos\dfrac{2\pi n}{N-1} & \text{for } |n| \leq \dfrac{N-1}{2} \\ 0 & \text{otherwise} \end{cases} \tag{2.21}$$

with $\alpha = 0.5$ for the Hann window and $\alpha = 0.54$ in the Hamming window.

It can easily be shown from equation (2.21) that the spectra of these windows are given by the expression

$$W_H(e^{j\omega T}) = \frac{\alpha \sin(\omega NT/2)}{\sin(\omega T/2)} + \frac{1-\alpha \sin[\omega NT/2 - N\pi/(N-1)]}{2\sin[\omega T/2 - \pi(N-1)]}$$

$$+ \frac{1-\alpha \sin[\omega NT/2 + N\pi/(N-1)]}{2\sin[\omega T/2 + \pi(N-1)]} \tag{2.22}$$

The spectra of the Hann and Hamming windows are shown in Figures 2.3 and 2.4, respectively, for $N = 13$ and $\omega_s = 10$ rad/s.

2.3.2. Blackman Window

The Blackman window is given by the following expression:

$$w_B(nT) = \begin{cases} 0.42 + 0.5\cos\dfrac{2\pi n}{N-1} + 0.08\cos\dfrac{4\pi n}{N-1} & \text{for } |n| \leq \dfrac{N-1}{2} \\ 0 & \text{otherwise} \end{cases} \tag{2.23}$$

The additional cosine term leads to a further reduction in the amplitude of the Gibbs oscillations.

The following example demonstrates the usefulness of these window functions in one-dimensional (1-D) filter design.

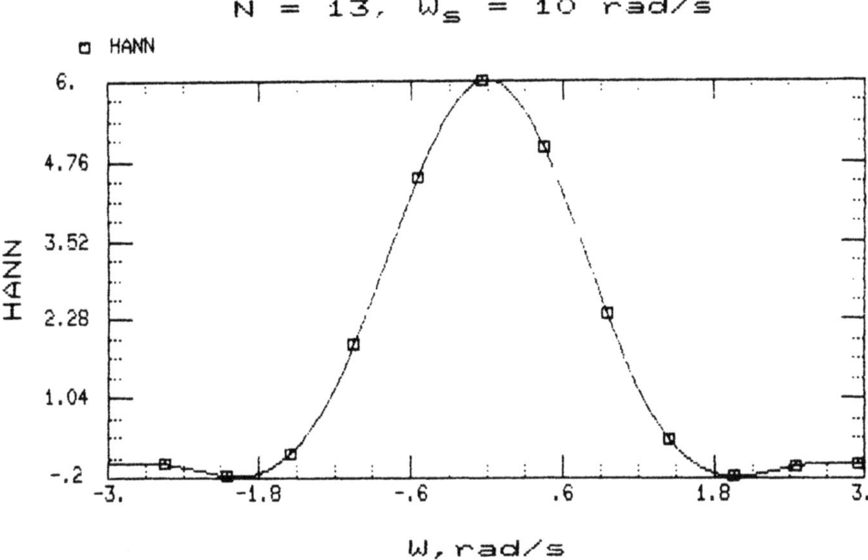

Figure 2.3. Spectrum of the Hann window for $N = 13$ and $\omega_s = 10.0$ rad/s.

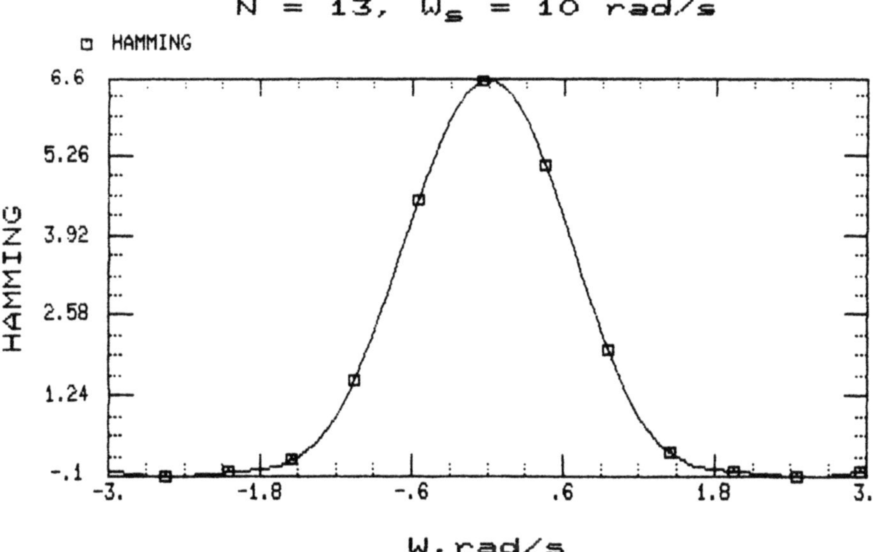

Figure 2.4. Spectrum of the Hamming window for $N = 13$ and $\omega_s = 10.0$ rad/s.

EXAMPLE 2.2. Design a low-pass filter satisfying the frequency specification

$$H(e^{j\omega T}) \simeq \begin{cases} 1 & \text{for } |\omega| \leq \omega_c \\ 0 & \text{for } \omega_c < |\omega| \leq \omega_s/2 \end{cases} \tag{2.24}$$

where ω_c is the cutoff frequency and its value is set equal to 1.5 rad/s, while ω_s is the sampling frequency and its value is set equal to 10 rad/s.

SOLUTION. Equation (2.14) yields the expression for the impulse response of the system in the form

$$\hat{h}(nT) = \frac{1}{\omega_s} \int_{-\omega_c}^{\omega_c} e^{j\omega nT}\, d\omega = \frac{1}{\pi n} \sin \omega_c nT \tag{2.25}$$

On multiplying function $\hat{h}(nT)$ by any of the (Hann, Hamming, or Blackman) window functions described earlier, we obtain

$$H_W(z) = z^{-(N-1)/2} \sum_{n=0}^{(N-1)/2} \frac{a_n}{2}(z^n + z^{-n}) \tag{2.26}$$

where

$$a_0 = w(0)\hat{h}(0) \qquad \text{and} \qquad a_n = 2w(nT)\hat{h}(nT) \tag{2.27}$$

The amplitude responses for the three filters are given by

$$M(\omega) = \left| \sum_{n=0}^{(N-1)/2} a_n \cos \omega nT \right| \tag{2.28}$$

These magnitude responses are plotted in Figures 2.5a–c for values of $N = 27$ and $\omega_s = 10$ rad/s. It is seen that the pass-band ripple is reduced, and the minimum stop-band attenuation as well as the transition width are increased progressively from the Hann to the Hamming to the Blackman window.

2.3.3. Kaiser Window

A very flexible family of window functions has been developed by Kaiser. These windows are nearly optimum in the sense of having the largest energy in the main lobe for a given peak side-lobe level. The Kaiser windows

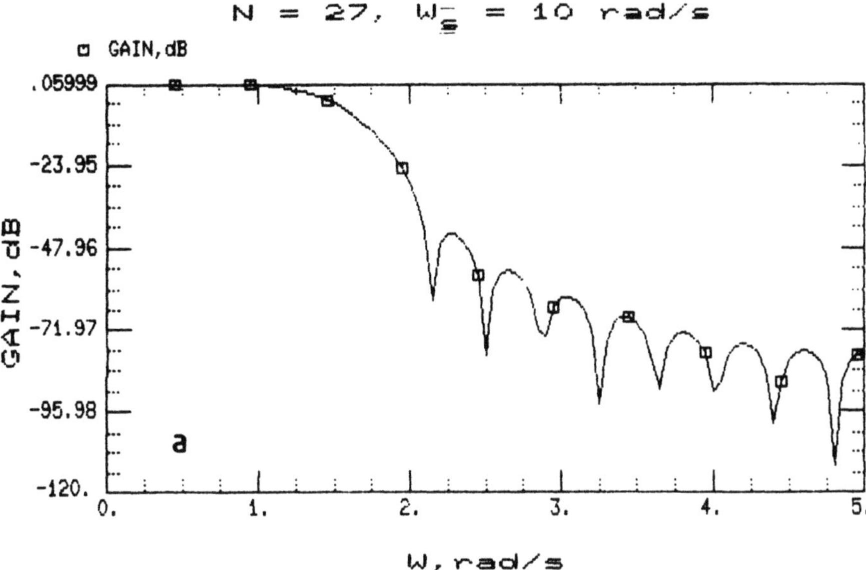

Figure 2.5a. Frequency response of a designed low-pass filter using the Hann window, $N = 27$, and $\omega_s = 10.0$ rad/s.

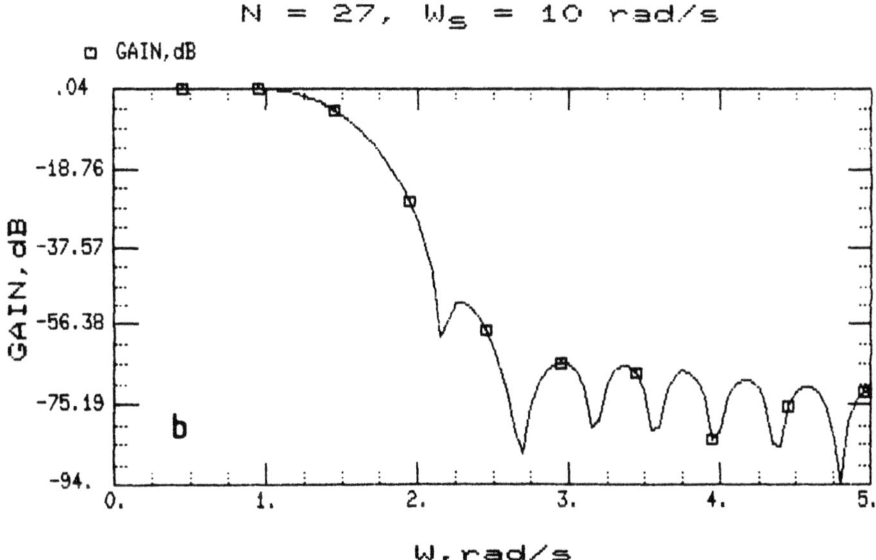

Figure 2.5b. Frequency response of a designed low-pass filter using the Hamming window, $N = 27$, and $\omega_s = 10.0$ rad/s.

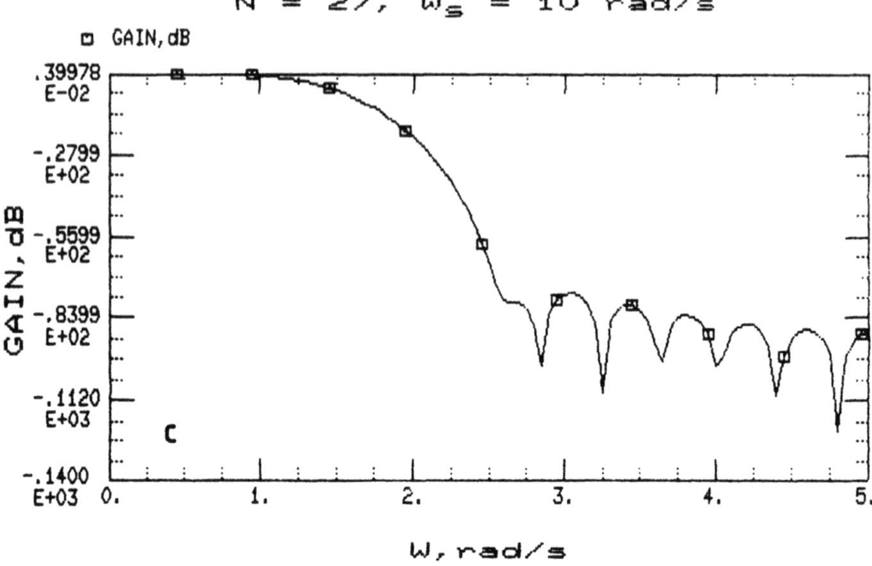

Figure 2.5c. Frequency response of a designed low-pass filter using the Blackman window, $N = 27$, and $\omega_s = 10.0$ rad/s.

are of the form

$$w_K(nT) = \begin{cases} \dfrac{I_0\{\beta[1-(2n/N-1)^2]^{1/2}\}}{I_0(\beta)} & \text{for } |n| \leq \dfrac{N-1}{2} \\ 0 & \text{otherwise} \end{cases} \tag{2.29}$$

where $I_0(\ )$ is the modified zeroth-order Bessel function of the first kind and β is a shape parameter determining the trade-off between the main-lobe width and the peak side-lobe level. The quantity $I_0(\ )$ can be computed to any desired degree of accuracy by using the rapidly converging series

$$I_0(x) = 1 + \sum_{m=1}^{\infty} \left[\frac{(x/2)^m}{m!}\right]^2 \tag{2.30}$$

with the first fifteen terms being sufficient for most applications.

It can easily be shown that the spectrum of $w_K(nT)$ is of the form

$$W_K(e^{j\omega T}) = w_K(0) + 2 \sum_{n=0}^{(N-1)/2} w_K(nT) \cos \omega nT \tag{2.31}$$

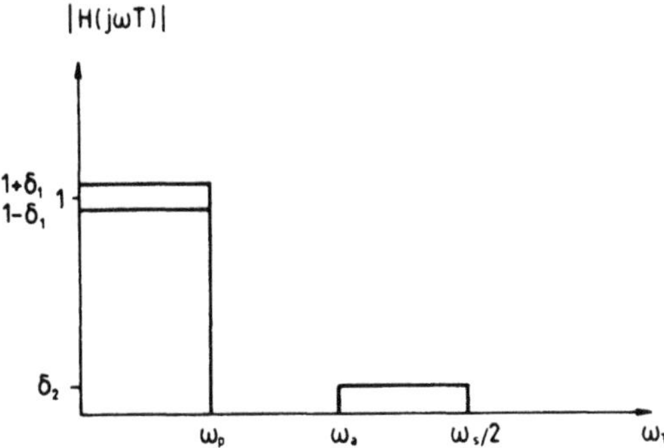

Figure 2.6. Magnitude-response specification for an idealized low-pass filter.

Empirical design formulas have also been derived by Kaiser for the parameters β and N, given the desired values for the transition bandwidth and the peak side-lobe level. Referring to Figure 2.6, the normalized transition bandwidth is defined by

$$\Delta\omega = (\omega_a - \omega_p)/\omega_s \tag{2.32}$$

and the stop-band attenuation is written as

$$A = -20 \log_{10} \delta \qquad \text{where } \delta = \min(\delta_1, \delta_2) \tag{2.33}$$

where δ_1 and δ_2 are defined in Figure 2.6, assuming that in our design technique $\delta_1 \simeq \delta_2$. The order of the FIR filter is then closely approximated by

$$N \simeq \frac{A - 7.95}{14.36\Delta\omega} \tag{2.34}$$

The shape parameter β can be determined from

$$\beta = \begin{cases} 0.1102(A - 8.7) & A \geq 50 \\ 0.5842(A - 21)^{0.4} + 0.07886(A - 21) & 21 < A < 50 \\ 0 & A \leq 21 \end{cases} \tag{2.35}$$

which is accurate to better than 1 percent over the useful range of A.

EXAMPLE 2.3. Design a low-pass filter using a Kaiser window function satisfying the following frequency specification: $\omega_p = 2000\,\pi$ rad/s, $\omega_a = 3000\pi$ rad/s, $\omega_s = 20{,}000\pi$ rad/s, and $A = 50$ dB.

SOLUTION. Expressions (2.32) and (2.34) yield

$$N = \frac{50 - 7.95}{14.36(0.05)} \cong 59$$

The order of the filter is thus determined. The parameter β is calculated using equation (2.35) as follows:

$$\beta = 0.1102(50 - 8.7) = 4.55$$

The coefficients of the filter can now readily be obtained from equation (2.27).

2.4. DESIGN OF NONRECURSIVE ONE-DIMENSIONAL FILTERS USING A MODIFIED KAISER WINDOW[7]

The design of a 1-D nonrecursive filter using a Kaiser window involves computing the parameter β and the order of the filter, N, via the empirical formulas of equations (2.34) and (2.35). These relationships, however, fail to produce the desired specification under certain conditions where the transition band is too narrow. The modified method to be presented here is aimed at alleviating this problem through an iterative approach for calculating the parameter β.

The transfer function of a low-pass nonrecursive, 1-D filter designed by the Kaiser method is given by

$$H(z) = \sum_{n=-r}^{r} w_K(nT)h(nT)z^{-n} \tag{2.36}$$

where

$$r = \frac{N-1}{2}, \qquad h(nT) = \frac{1}{n\pi}\sin(\omega_c nT), \qquad \omega_c = \tfrac{1}{2}(\omega_p + \omega_a)$$

$$w_K(nT) = \begin{cases} I_0(\alpha)/I_0(\beta) & \text{for } |n| \le r \\ 0 & \text{otherwise} \end{cases}$$

$\alpha = \beta[1 - (n/r)^2]^{1/2}$, N is the number of samples in the impulse response and is assumed to be odd, ω_p and ω_a are the respective pass-band and stop-band edges, and T is the sampling period.

The amplitude response of a filter designed by the preceding method depends on the independent parameter β and is given by

$$M(\beta, N, \omega) = |H(e^{j\omega T})| \tag{2.37}$$

The ideal amplitude response for a low-pass filter is given by

$$M_I(\omega) = \begin{cases} 1 & \text{for } 0 \le |\omega| \le \omega_p \\ 0 & \text{for } \omega_a \le |\omega| \le \omega_s/2 \end{cases} \tag{2.38}$$

Hence the error in the magnitude response is given by equations (2.37) and (2.38) as

$$E_{mag}(\beta, N, \omega) = M(\beta, N, \omega) - M_I(\omega) \tag{2.39}$$

The specifications of Figure 2.6 can be achieved by choosing β and N such that

$$L_p(\beta, N) = \max_{0 \le \omega \le \omega_p} |E_{mag}(\beta, N, \omega)| \le \delta_1$$

and

$$L_a(\beta, N) = \max_{\omega_a \le \omega \le \omega_s/2} |E_{mag}(\beta, N, \omega)| \le \delta_2$$

The value of N is required to be as low as possible, so as to keep the number of multiplications in the implementation of the filter to a minimum.

The approach of Kaiser for the determination of β and N is to use equations (2.32)–(2.35) as described in the previous section and, if the resulting value of N is even, set $N = N + 1$. An alternative approach for the determination of β and N, which is found to yield improved designs, is to use the 1-D minimax optimization technique of Antoniou and Charalambous.[8] The details of the improved approach are as follows:

1. Initialize β and N using equations (2.34) and (2.35).
2. If $\delta_2 < \delta_1$, replace $L_p(\beta, N)$ by $L_a(\beta, N)$ in steps 3 and 4 below.
3. Minimize $L_p(\beta, N)$ with respect to β until

$$|L_p^{(k)}(\beta_k, N) - L_p^{(k-1)}(\beta_{k-1}, N)| \le \varepsilon$$

where β_k and $L_p^{(k)}(\beta_k, N)$ are the values of β and $L_p(\beta, N)$, respectively, at the end of the kth stage of minimization, while ε is a minimization tolerance. Set $\beta = \beta^{(k)}$, where $\beta^{(k)}$ is the optimum value of β.

4a. If $L_p(\beta, N) = \delta_1$, go to step 5.
4b. If $L_p(\beta, N) > \delta_1$, set $N = N + 2$ and execute step 3 repeatedly until $L_p(\beta, N) < \delta_1$. Then go to step 5.
4c. If $L_p(\beta, N) < \delta_1$, set $N = N - 2$ and execute step 3 repeatedly until $L_p(\beta, N) > \delta_1$. Then go to step 5.

5. If step 4a or 4b was executed, use the current values of β and N to deduce the transfer function of the filter. Alternatively, if step 4c was executed, use the previous values of β and N for the determination of the transfer function.

In the above design procedure, the minimization of $L_p(\beta, N)$ or $L_a(\beta, N)$ can be performed over the frequency band with the tighter specification. This is permissible because

$$L_p(\beta, N) \simeq L_a(\beta, N)$$

and, as a result, $L_a(\beta, N)$ is minimized if $L_p(\beta, N)$ is minimized, and vice versa.

Through extensive computation it was found that, for a given value of β, the values of $L_p(\beta, N)$ and $L_a(\beta, N)$ occur in the narrow frequency ranges

$$\omega_p - \Delta\omega_s \le \omega \le \omega_p \qquad \text{and} \qquad \omega_a \le \omega \le \omega_a + \Delta\omega_s$$

respectively, where $\Delta\omega_s = \omega_s/(N-1)$. Consequently, the number of functions

$$f_i(\beta) = |E_{mag}(\beta, N, \omega_i)|$$

used in the minimization of $L_p(\beta, N)$ or $L_a(\beta, N)$, and hence the amount of computation, can be kept low.

If it is assumed that $T = 1$, the derivative of $f_i(\beta)$ with respect to β, which is necessary for minimization of $L_p(\beta, N)$, is given by

$$\frac{df_i(\beta)}{d\beta} = \begin{cases} \operatorname{sgn}\{|P(\beta)| - 1\} \operatorname{sgn}\{P(\beta)\} Q(\beta) & \text{for } 0 \le \omega \le \omega_p \\ \operatorname{sgn}\{P(\beta)\} Q(\beta) & \text{for } \omega_a \le \omega \le \omega_s/2 \end{cases} \tag{2.40}$$

where

$$P(\beta) = h(0) + 2 \sum_{n=1}^{r} w_K(n) h(n) \cos n\omega \tag{2.41}$$

$$Q(\beta) = 2 \sum_{n=1}^{r} \frac{dw_K(n)}{d\beta} h(n) \cos n\omega \tag{2.42}$$

$$\frac{dw_K(nT)}{d\beta} = \frac{\dfrac{dI_0(\alpha)}{d\alpha}\dfrac{d\alpha}{d\beta} I_0(\beta) - \dfrac{dI_0(\beta)}{d\beta} I_0(\alpha)}{[I_0(\beta)]^2} \tag{2.43}$$

$$\frac{d\alpha}{d\beta} = \left[1 - \left(\frac{n}{r}\right)^2\right]^{1/2} \tag{2.44}$$

$$I_0(x) = 1 + \sum_{m=1}^{\infty} \gamma_m \tag{2.45}$$

$$\gamma_m = \left[\frac{1}{m!}\left(\frac{x}{2}\right)^m\right]^{1/2} \tag{2.46}$$

and

$$\frac{dI_0(x)}{dx} = \begin{cases} 0 & \text{for } x = 0 \\ \dfrac{2}{x} \displaystyle\sum_{m=1}^{\infty} m\gamma_m & \text{otherwise} \end{cases} \tag{2.47}$$

2.5. FREQUENCY SAMPLING METHOD

A nonrecursive filter can be specified uniquely by giving either the impulse response coefficients $\{h(n)\}$ (assuming $T = 1$) or, equivalently, the DFT coefficients $\{H(k)\}$, because these sequences are related by the DFT relations

$$H(k) = \sum_{n=0}^{N-1} h(n)\, e^{-j(2\pi/N)nk} \tag{2.48}$$

and

$$h(n) = \frac{1}{N} \sum_{k=0}^{N-1} H(k)\, e^{j(2\pi/N)nk} \tag{2.49}$$

On the other hand, functions $H(k)$ for the nonrecursive filter can be regarded as samples of the filter's z transform, evaluated at N points equally spaced around the unit circle, i.e.,

$$H(k) = H(z)\big|_{z=e^{j(2\pi/N)k}} \tag{2.50}$$

Thus the z transform of an FIR filter can easily be expressed in terms of its DFT coefficients by substituting equation (2.49) into the z-transform equation to give

$$H(z) = \sum_{n=0}^{N-1} h(n) z^{-n} = \sum_{n=0}^{N-1} \left[\frac{1}{N} \sum_{k=0}^{N-1} H(k)\, e^{j(2\pi/N)nk}\right] z^{-n} \tag{2.51}$$

Interchanging the order of summation gives

$$H(z) = \sum_{k=0}^{N-1} \frac{H(k)}{N} \sum_{n=0}^{N-1} [e^{j(2\pi/N)k} z^{-1}]^n$$

$$= \sum_{k=0}^{N-1} \frac{H(k)}{N} \frac{(1 - e^{j2\pi k} z^{-N})}{[1 - e^{j(2\pi/N)k} z^{-1}]} \tag{2.52}$$

Since $e^{j2\pi k} = 1$, equation (2.52) reduces to

$$H(z) = \frac{(1 - z^{-N})}{N} \sum_{k=0}^{N-1} \frac{H(k)}{[1 - z^{-1} e^{j(2\pi/N)k}]} \tag{2.53}$$

which is the desired result.

The interpretation of equation (2.53) is that, to approximate any continuous frequency response, one could sample in frequency at N equispaced points around the unit circle (the frequency samples) and evaluate the continuous frequency response as an interpolation of the sampled frequency response. The approximation error would then be exactly zero at the sampling frequencies and finite between them. The greater the number of frequency points chosen, the smaller the error of interpolation between the sample points.

Although the procedure can be used directly as stated to design a nonrecursive filter, in order to improve the quality of the approximation by reducing the approximation error Rabiner and Gold[1] suggested that a number of frequency samples be made unconstrained variables. The values of these unconstrained variables are generally optimized by computer to minimize some cost function. Interested readers are encouraged to refer to their book[1] for details of the optimization algorithm using the frequency sampling method.

2.6. EQUIRIPPLE DESIGN OF A NONRECURSIVE FILTER

One of the best approaches for the design of nonrecursive filters is the one which utilizes the Remez exchange algorithm described by Rabiner *et al.*[9] and its improved version by Antoniou.[10] This algorithm, which is good for the design of an equiripple nonrecursive digital filter, is based on the following important theorem.

THEOREM 2.1. *If $P(\omega)$ is a linear combination of r cosine functions of the form*

$$P(\omega) = \sum_{n=0}^{r-1} \alpha(n) \cos(\omega n) \tag{2.54}$$

then a necessary and sufficient condition that $P(\omega)$ be the unique, best weighted Chebyshev approximation to a continuous function $D(\omega)$ on A, a compact subset of $[0, \pi]$, is that the weighted error function $E(\omega)$ exhibits at least $r+1$ extremal frequencies in A, i.e., there must exist $r+1$ points ω_i in A such that

$$\omega_0 < \omega_1 < \cdots < \omega_r \tag{2.55}$$

$$E(\omega_i) = -E(\omega_{i+1}) \tag{2.56}$$

and

$$|E(\omega_i)| = \max_{\omega \in A} |E(\omega)|, \qquad i = 0, 1, \ldots, r \tag{2.57}$$

2.6.1. Problem Formulation

For the sake of simplicity, we shall focus our attention on the case of low-pass filters with symmetrical impulse response and N odd, although the technique is also applicable for high-pass, band-pass, and band-stop filters.

The sampling frequency ω_s is assumed to be 2π (i.e., the sampling period is assumed to be unity). The frequency response of this filter is given by

$$H(e^{j\omega}) = e^{j\omega(r-1)} P(\omega) \tag{2.58}$$

where function $P(\omega)$ is given in equation (2.54):

$$P(\omega) = \sum_{n=0}^{r-1} \alpha(n) \cos(n\omega), \qquad r = (N+1)/2$$

The amplitude response of an equiripple low-pass filter is of the form illustrated in Figure 2.7, where δ_1 and δ_2 are the amplitudes of the pass-band

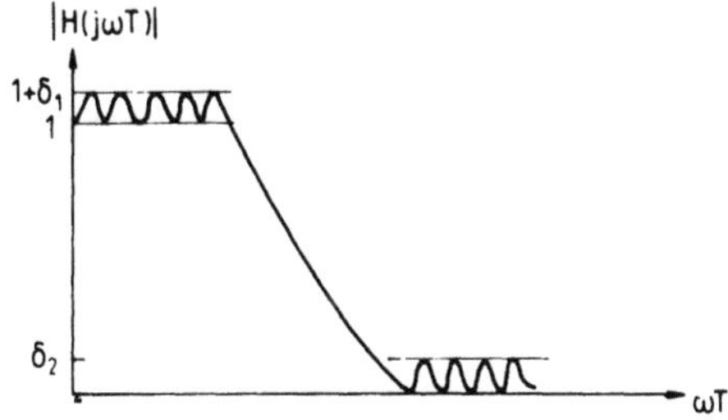

Figure 2.7. Magnitude response of an equiripple low-pass filter.

and stop-band ripple, while ω_p and ω_a are the pass-band and stop-band edges, respectively.

The desired value of $P(\omega)$ for a low-pass filter, designated $D(\omega)$, is given by

$$D(\omega) = \begin{cases} 1 & \text{for } 0 \le \omega \le \omega_p \\ 0 & \text{for } \omega_a \le \omega \le \pi \end{cases} \tag{2.59}$$

and hence an error function can be deduced in the form

$$E_0(\omega) = D(\omega) - P(\omega) \tag{2.60}$$

where

$$|E_0(\omega)| = \begin{cases} \delta_1 & \text{for } 0 \le \omega \le \omega_p \\ \delta_2 & \text{for } \omega_a \le \omega \le \pi \end{cases} \tag{2.61}$$

Alternatively, we can write

$$|E(\omega)| \le \delta_1 \tag{2.62}$$

where

$$E(\omega) = W(\omega)E_0(\omega) \tag{2.63}$$

where

$$W(\omega) = \begin{cases} 1 & \text{for } 0 \le \omega \le \omega_p \\ \delta_1/\delta_2 & \text{for } \omega_a \le \omega \le \pi \end{cases} \tag{2.64}$$

Equations (2.60) and (2.63) now yield

$$E(\omega) = W(\omega)[D(\omega) - P(\omega)] \tag{2.65}$$

From the alternation property cited in Theorem 2.1, a set of extremal frequencies $\{\omega_k\}$ and a cosine polynomial $P(\omega)$ exist such that

$$E(\omega_k) = W(\omega_k)[D(\omega_k) - P(\omega_k)] = (-1)^k\delta \tag{2.66}$$

for $k = 0, 1, \ldots, r$, where δ is a constant.

This relation can be expressed in the following matrix form:

$$\begin{bmatrix} 1 & \cos\omega_0 & \cos(2\omega_0) & \cdots & \cos[(r-1)\omega_0] & 1/W(\omega_0) \\ 1 & \cos\omega_1 & \cos(2\omega_1) & \cdots & & \\ \vdots & & & & & \\ 1 & \cos(\omega_r) & \cos(2\omega_r) & \cdots & & (-1)^r/W(\omega_r) \end{bmatrix} \begin{bmatrix} \alpha(0) \\ \alpha(1) \\ \vdots \\ \alpha(r-1) \\ \delta \end{bmatrix} = \begin{bmatrix} D(\omega_0) \\ D(\omega_1) \\ \vdots \\ \\ D(\omega_r) \end{bmatrix} \tag{2.67}$$

Therefore, if the extremal frequencies are known the coefficients $\alpha(n)$, and hence the frequency response of the filter, can be computed by inverting the above matrix. Invertibility of this matrix is guaranteed by the Haar condition on the basis functions.

2.6.2. Remez Exchange Algorithm

A powerful algorithm which can be used to deduce the unique low-pass filter for given values of ω_p, ω_a, and N is the so-called Remez exchange algorithm, which is given by Rabiner *et al.*[9] and can be described in the following steps.

Step 1. Select $r + 1$ extremal frequencies $\omega_0, \omega_1, \ldots, \omega_r$ such that $m + 1$ of them, namely $\omega_0, \omega_1, \ldots, \omega_m$, are uniformly spaced in the frequency range 0 to ω_p and $r - m$ of them, namely $\omega_{m+1}, \omega_{m+2}, \ldots, \omega_r$, are uniformly spaced in the frequency range ω_a to π. In addition, it is required that

$$\omega_m = \omega_p \qquad \text{and} \qquad \omega_{m+1} = \omega_a$$

Further, the frequency interval $\omega_i - \omega_{i-1}$ is required to be approximately the same in the two bands.

Step 2. Compute function $E(\omega)$ over a dense set of frequencies

$$\omega = 0, i_p, 2i_p, \ldots, \omega_p, \omega_a, \omega_a + i_a, \omega_a + 2i_a, \ldots, \pi,$$

where i_p and i_a are the pass-band and stop-band frequency increments, respectively.

Step 3. Let $\hat{\omega}_0, \hat{\omega}_1, \ldots, \hat{\omega}_{kp}, \ldots, \hat{\omega}_\mu, \hat{\omega}_{\mu+1}, \ldots, \hat{\omega}_{ka}, \ldots, \hat{\omega}_\zeta$ be potential extremal frequencies for the next iteration, and assume that $\hat{\omega}_0$ to $\hat{\omega}_\mu$ to $\hat{\omega}_\zeta$ as follows:

(a) If $[\{E(0) > 0 \text{ and } E(0) > E(i_p)\}$
  or $\{E(0) < 0 \text{ and } E(0) < E(i_p)\}]$ and
  $|E(0)| \geq |\delta|$, set $0 \to \hat{\omega}_0$.

(b) If $[\{E(\omega) > 0 \text{ and } E(\omega - i_p) < E(\omega) > E(\omega + i_p)\}$
  or $\{E(\omega) < 0 \text{ and } E(\omega - i_p) > E(\omega) < E(\omega + i_p)\}]$,
  set $\omega \to \hat{\omega}_{kp}$.

(c) If $[\{E(\omega_p) > 0 \text{ and } E(\omega_p) > E(\omega_p - i_p)\}$
  or $\{E(\omega_p) < 0 \text{ and } E(\omega_p) < E(\omega_p + i_p)\}]$ and
  $|E(\omega_p)| \geq |\delta|$, set $\omega_p \to \hat{\omega}_\mu$.

(d) If $[\{E(\omega_a) > 0 \text{ and } E(\omega_a) > E(\omega_a - i_a)\}$
  or $\{E(\omega_a) < 0 \text{ and } E(\omega_a) < E(\omega_a + i_a)\}]$ and
  $|E(\omega_a)| \geq |\delta|$, set $\omega_a \to \hat{\omega}_{\mu+1}$.

(e) If $[\{E(\omega) > 0$ and $E(\omega - i_a) < E(\omega) > E(\omega + i_a)\}$
or $\{E(\omega) < 0$ and $E(\omega - i_a) > E(\omega) < E(\omega - i_a)\}]$,
set $\omega \rightarrow \hat{\omega}_{ka}$.

(f) If $[\{E(\pi) > 0$ and $E(\pi) > E(\pi - i_a)\}$
or $\{E(\pi) < 0$ and $E(\pi) < E(\pi - i_a)\}]$ and
$|E(\pi)| \geq |\delta|$, set $\pi \rightarrow \hat{\omega}_{\zeta}$.

Step 4. Compute

$$Q = \frac{\max|E(\hat{\omega}_k)| - \min|E(\hat{\omega}_k)|}{\max|E(\hat{\omega}_k)|}$$

where $k = 0, 1, \ldots, \xi$.

Step 5. Reject the $\zeta - r$ frequencies $\hat{\omega}_k$ for which $|E(\hat{\omega}_k)|$ is lowest and renumber the remaining pass-band frequencies from 0 to m and the remaining stop-band frequencies from $m + 1$ to r. Then update extremal frequencies by letting $\omega_k = \hat{\omega}_k$, for $k = 0, 1, \ldots, r$.

Step 6. If $Q > 0.01$, repeat from step 2. Otherwise, continue to step 7.

Step 7. Compute $P(\omega)$ using the last set of extremal frequencies.

Step 8. Deduce the impulse response $h(n)$ of the filter.

The preceding algorithm converges very rapidly, usually within four to eight iterations.

2.6.3. Implementation of Algorithm

In step 1 of the preceding algorithm, it is convenient to set $\omega_0 = 0$ and $\omega_r = \pi$.

The pass-band and stop-band intervals, I_p and I_a, to achieve uniform distribution of extremal frequencies in the pass band as well as the stop band can be computed as follows: Let

$$I = \frac{\omega_p + \pi - \omega_a}{r},$$

$$m = \text{int}\left[\frac{\omega_p}{I} + 0.5\right], \qquad \text{and} \qquad m_1 = \text{int}\left[\frac{\pi - \omega_a}{I} + 0.5\right]$$

where int[ ] represents the closest integer by rounding up the value inside the bracket. Then compute

$$I_p = \frac{\omega_p}{m} \qquad \text{and} \qquad I_a = \frac{\pi - \omega_a}{m_1}$$

In step 2, a dense set of frequencies is required in order to achieve sufficient accuracy in the determination of frequencies $\hat{\omega}_k$ in step 3. A reasonable number of frequency points are $8(N-1)$. This corresponds to about 16 frequency points per ripple of $|E(\omega)|$. The frequency set should be chosen carefully to ensure that extremal frequencies belong to the set, in order to avoid numerical ill-conditioning in the evaluation of $E(\omega)$. Suitable pass-band and stop-band frequency intervals are $i_p = I_p/16$ and $i_a = I_a/16$, respectively.

The solution of equation (2.67) can be obtained only if precisely $r+1$ extremal frequencies are available. By differentiating $E(\omega)$ with respect to ω, one can show that the number of maxima in $|E(\omega)|$ (potential extremal frequencies) in steps 3(a), (b), (e), and (f) can be as high as $r+1$. Since two additional potential extremal frequencies can occur at $\omega = \omega_p$ and ω_a [see steps 3(c) and (d)], a maximum of two superfluous potential extremal frequencies can in principle occur in step 3. This problem is overcome by rejecting superfluous values of $\hat{\omega}_k$, if any, as described in step 5 of the algorithm.

In steps 2 and 7, $E(\omega)$ and $P(\omega)$ can be evaluated by determining coefficients $\alpha(n)$ through matrix inversion. However, this approach can slow down the computation, and possibly cause numerical ill-conditioning, in particular if δ is small and N is large. An alternative and more efficient approach is to deduce δ analytically and then to interpolate $P(\omega)$ on the r frequency points using the barycentric form of the Lagrange interpolation formula. The necessary formulation is as follows: parameter δ can be deduced as

$$\delta = \sum_{k=0}^{r} \alpha_k D(\omega_k) \Big/ \sum_{k=0}^{r} \frac{(-1)^k \alpha_k}{W(\omega_k)} \tag{2.68}$$

and function $P(\omega)$ is given by

$$P(\omega) = \begin{cases} C_k & \text{for } \omega = \omega_0, \omega_1, \ldots, \omega_{r-1} \\ \displaystyle\sum_{k=0}^{r-1} \frac{\beta_k}{x - x_k} C_k \Big/ \sum_{k=0}^{r-1} \frac{\beta_k}{x - x_k} & \text{otherwise} \end{cases} \tag{2.69}$$

where

$$\alpha_k = \prod_{\substack{i=0 \\ i \neq k}}^{r} \frac{1}{x_k - x_i} \tag{2.70}$$

$$C_k = D(\omega_k) - (-1)^k \frac{\delta}{W(\omega_k)} \tag{2.71}$$

$$\beta_k = \prod_{\substack{i=0 \\ i \neq k}}^{r-1} \frac{1}{x_k - x_i} \tag{2.72}$$

and

$$x = \cos \omega, \qquad x_k = \cos(\omega_k), \qquad k = 1, 2, \ldots, r-1 \tag{2.73}$$

The impulse response $h(n)$ in step 8 can be deduced by computing $P(k\Omega)$ for $k = 0, 1, \ldots, r$, where $\Omega = 2\pi/N$, and then using the inverse discrete Fourier transform. For a symmetrical impulse it can be shown that

$$h(n) = h(-n) = \frac{1}{N}\left\{P(0) + \sum_{k=1}^{r} 2P(k\Omega)\cos(nk\Omega)\right\} \tag{2.74}$$

PART TWO. TWO-DIMENSIONAL FILTERS

2.7. INTRODUCTION

A 2-D nonrecursive, digital filter is one whose impulse response possesses only a finite number of nonzero samples. For such a filter, the impulse response is always absolutely summable and thus nonrecursive filters are always stable.

A 2-D nonrecursive filter can be characterized by its transfer function as

$$H(z_1, z_2) = \sum_{m=0}^{M-1} \sum_{n=0}^{N-1} h(m, n) z_1^{-m} z_2^{-n}$$

One of the biggest advantages that nonrecursive filters enjoy over recursive filters is the fact that implementable nonrecursive filters can be designed to have purely real frequency responses. Since the imaginary component of the frequency response is equal to zero, the phase response of such a filter will also be equal to zero and therefore these filters are termed zero-phase filters. In this case we have

$$H(e^{j\omega_1 T}, e^{j\omega_2 T}) = H^*(e^{j\omega_1 T}, e^{j\omega_2 T}) \tag{2.75}$$

By taking the inverse Fourier transform of both sides of equation (2.75), we obtain the spatial domain symmetry requirement for the impulse response of a zero-phase filter:

$$h(m, n) = h^*(-m, -n) \tag{2.76}$$

Clearly, a nonrecursive filter can be constrained to satisfy this relation if its region of support is centered about the origin.

Most of the design techniques for 1-D nonrecursive filters can be extended to the design of 2-D nonrecursive filters.

2.8. DESIGN OF TWO-DIMENSIONAL NONRECURSIVE FILTERS USING WINDOWS

The window technique for the design of 2-D nonrecursive filters is a direct extension to two dimensions of the method for its one-dimension counterpart. It is a spatial domain method, in which we try to approximate an ideal-impulse response rather than the ideal-frequency response. We let $h_I(m, n)$ and $H_I(e^{j\omega_1 T}, e^{j\omega_2 T})$ denote the impulse response and frequency response of the ideal filter and let $h_D(m, n)$ and $H_D(e^{j\omega_1 T}, e^{j\omega_2 T})$ denote the impulse response and frequency response of the designed filter. With the window method the coefficients $h_D(m, n)$ are derived from

$$h_D(m, n) = h_I(m, n) \cdot w(m, n) \tag{2.77}$$

It should be noted that the region of support for $h_I(m, n)$ and $w(m, n)$ will be the same and this will also give the same region of support for $h_D(m, n)$. Since $h_D(m, n)$ is formed as the product of $h_I(m, n)$ and $w(m, n)$, the frequency response $H_D(e^{j\omega_1 T}, e^{j\omega_2 T})$ is related to $H_I(e^{j\omega_1 T}, e^{j\omega_2 T})$ by a frequency domain convolution as follows:

$$\begin{aligned} H_D(e^{j\omega_1 T}, e^{j\omega_2 T}) &= H_D(\omega_1, \omega_2) \\ &= \frac{1}{(2\pi)^2} \int_{-\pi}^{\pi} \int_{-\pi}^{\pi} H_I(u_1, u_2) W(\omega_1 - u_1, \omega_2 - u_2)\, du_1\, du_2 \end{aligned} \tag{2.78}$$

where $W(\omega_1, \omega_2)$, the smoothing function, is the Fourier transform of the window function. The frequency response $H_D(\omega_1, \omega_2)$ is a smoothed version of the ideal-frequency response.

For many filter design problems, the desired filter behavior is given in terms of the frequency specification $H_I(\omega_1, \omega_2)$ (magnitude, phase, or both) rather than the impulse response $h_I(m, n)$. In this case $h_I(m, n)$ can be computed either analytically or by sampling $H_I(\omega_1, \omega_2)$ and then taking its inverse discrete Fourier transform. Knowing that the support of $h_I(m, n)$ is generally of infinite extent, the result of sampling and truncation of the ideal-impulse response to a finite-size array is a spatially aliased approximation to $h_I(m, n)$. This aliasing error can be minimized by increasing the extent of support of the inverse discrete Fourier transform to several times larger than its original size.

2.8.1. Two-Dimensional Window Function

According to Dudgeon and Mersereau[11] the 2-D window function must satisfy the following three requirements.

1. It must have the correct region of support.
2. For $H_D(\omega_1, \omega_2)$ to approximate $H_I(\omega_1, \omega_2)$, $W(\omega_1, \omega_2)$ should behave like a 2-D impulse function.
3. If $h_D(m, n)$ is to be zero-phase, the window function should satisfy the zero-phase relation

$$w(m, n) = w^*(-m, -n) \tag{2.79}$$

It has been shown[11] that since the properties of 1-D and 2-D window functions are identical, a 1-D window function can be used as the basis for generating a 2-D window function. There are two approaches to the generation of 2-D windows from 1-D windows. The first method forms a 2-D window with a square or rectangular region of support by cascading two 1-D windows, as follows:

$$w_R(m, n) = w_1(m) \cdot w_2(n) \tag{2.80}$$

The second method, which is due to Huang,[12] forms a 2-D window by sampling a circularly rotated, continuous window function,

$$w_C(m, n) = w(\sqrt{m^2 + n^2}) \tag{2.81}$$

The resulting 2-D windows in this case have nearly circular regions of support.

The Fourier transform of $w_R(m, n)$ is the outer product of the Fourier transforms of $w_1(m)$ and $w_2(n)$. Thus

$$W_R(\omega_1, \omega_2) = W_1(\omega_1)\, W_2(\omega_2) \tag{2.82}$$

The Fourier transform of $w_C(m, n)$ resembles, but differs in detail from, a circularly rotated version of the 1-D Fourier transform of $w(t)$.

If $w(t)$, $w_1(m)$, and $w_2(n)$ are good 1-D windows (i.e., satisfy the above three conditions), then w_C and w_R will be good 2-D windows. Therefore any of a number of 1-D windows which are explained in Section 2.3 can be used. The design approach then follows that of a 1-D filter.

2.9. OPTIMAL DESIGN OF TWO-DIMENSIONAL NONRECURSIVE FILTERS USING KAISER WINDOWS[11]

In the method to be presented here the 2-D nonrecursive filter is characterized by the transfer function

$$H(z_1, z_2) = \sum_{m=-M}^{M} \sum_{n=-M}^{M} h(m, n) z_1^{-m} z_2^{-n} \tag{2.83}$$

where $h(m, n)$ are the values of the impulse response.

If

$$z_i = e^{j\omega_i T_i} \qquad \text{where } T_i = 2\pi/\omega_{si} \tag{2.84}$$

assuming $T_1 = T_2 = T$ are the sampling periods and ω_{si} are sampling frequencies, then we have

$$H(e^{j\omega_1 T}, e^{j\omega_2 T}) = M(\omega_1, \omega_2)\, e^{j\theta(\omega_1, \omega_2)} \tag{2.85}$$

and

$$\theta(\omega_1, \omega_2) = \arg H(e^{j\omega_1 T}, e^{j\omega_2 T}) \tag{2.86}$$

as the gain and phase shift of the filter. A low-pass 2-D filter with a circular cutoff boundary is a filter in which

$$M_I(\omega_1, \omega_2) = \begin{cases} 1 & \text{for } (\omega_1^2 + \omega_2^2)^{1/2} \le \omega_c \\ 0 & \text{otherwise} \end{cases} \tag{2.87}$$

A practical 2-D filter of this class, on the other hand, is a filter in which

$$\begin{aligned} E(\omega_1, \omega_2) &= |M_D(\omega_1, \omega_2) - M_I(\omega_1, \omega_2)| \\ &\le \begin{cases} \Delta_p & \text{for } (\omega_1^2 + \omega_2^2)^{1/2} \le \omega_p \\ \Delta_a & \omega_s/2 \ge (\omega_1^2 + \omega_2^2)^{1/2} \ge \omega_a \end{cases} \end{aligned} \tag{2.88}$$

where M_I and M_D are ideal and designed magnitude response, respectively; ω_p and ω_a are referred to as the pass-band and stop-band edge, respectively; Δ_p and Δ_a are the maximum pass-band and stop-band error, respectively.

Function $h_D(m, n)$, the impulse response of a 2-D low-pass nonrecursive digital filter, can be obtained using the Kaiser window as follows:

$$h_D(m, n) = h_I(m, n) \cdot w(m, n) \tag{2.89}$$

where $h_I(m, n)$ is the ideal-impulse response of the filter and is given by

$$h_I(m, n) = \frac{\omega_C J_1(\omega_C \sqrt{m^2 + n^2})}{2\pi\sqrt{m^2 + n^2}} \tag{2.90}$$

where $J_1(\cdot)$ is the first-order Bessel function and $\omega_C = (\omega_p + \omega_a)/2$ is the cutoff frequency of the filter.

In addition, function $w(m, n)$ is given by

$$w(m, n) = \begin{cases} \dfrac{I_0(\beta\sqrt{1 - [m^2 + n^2]/M^2})}{I_0(\beta)} & \text{for } \sqrt{m^2 + n^2} \leq M \\ 0 & \text{for } \sqrt{m^2 + n^2} > M \end{cases} \tag{2.91}$$

where M is the radius, in samples, of the desired filter and is odd; $I_0(\cdot)$ is the modified Bessel function of the first kind of zeroth order and β is an adjustable parameter that specifies a frequency domain trade-off between main-lobe width and side-lobe ripple.

The maximum pass-band and stop-band errors in such a filter are given by

$$L_p(\beta, M) = \max|E(\omega_1, \omega_2)|, \qquad 0 \leq (\omega_1^2 + \omega_2^2)^{1/2} \leq \omega_p \tag{2.92}$$

and

$$L_a(\beta, M) = \max|E(\omega_1, \omega_2)|, \qquad \omega_a \leq (\omega_1^2 + \omega_2^2)^{1/2} \leq \omega_s/2 \tag{2.93}$$

where $E(\omega_1, \omega_2)$ is given by equation (2.88).

Prescribed maximum pass-band and stop-band errors, Δ_p and Δ_a, can be achieved if

$$L(\beta, M) \leq \Delta_p \tag{2.94}$$

where

$$L(\beta, M) = \max\left[L_p(\beta, M), \frac{\Delta_p}{\Delta_a} L_a(\beta, M)\right] \tag{2.95}$$

under the conditions

$$L_p(\beta, M) \leq \Delta_p \tag{2.96}$$

and

$$L_a(\beta, M) \leq \Delta_a \tag{2.97}$$

Using the above, the design of a 2-D low-pass nonrecursive filter with circular cutoff boundary can now be obtained by the following steps.

1. Initialize β and M.
2. Minimize $L(\beta, M)$ with respect to β until

$$|L^{(k)}(\beta, M) - L^{(k-1)}(\beta, M)| \leq \varepsilon \tag{2.98}$$

where $L^{(k)}(\beta, M)$ is the value of $L(\beta, M)$ at the end of the kth stage of minimization and ε is the tolerance of the optimization and is a positive number.
3. If $L^k(\beta, M) > \Delta_p$
Set $\beta = \beta^{(k)}$, increase M by 2, and repeat step 2 until $L^k(\beta, M) \leq \Delta_p$.
The last values of β and M are desired parameters.
4. If $L^{(k)}(\beta, M) < \Delta_p$
Set $\beta = \beta^{(k)}$, decrease M by 2, and repeat step 2 until $L^{(k)}(\beta, M) > \Delta_p$.
In this case, the last but one set of values for β and M are the desired parameters.

If one uses the values of β and M obtained in step 3 or 4 of the above design, then function $H(z_1, z_2)$ is determined.

EXAMPLE 2.4. Design a 2-D low-pass filter with the following specification using the technique described earlier: $\omega_s = 10$ rad/s, $\omega_p = 1.0$ rad/s, $\omega_a = 2.5$ rad/s, $\Delta_p = 0.05$, and $\Delta_a = 0.05$.

Table 2.1 shows the optimum value of M and β and the resulting maximum pass-band and stop-band errors for the designed filter, while Figure 2.8 shows the amplitude plot for the same filter.

2.10. McCLELLAN TRANSFORMATION[14,15]

The McClellan transformation is a transformation of a 1-D zero-phase nonrecursive filter into a 2-D zero-phase nonrecursive filter by a substitution of variables. It can be applied to 1-D filters of odd length and, in one special case, also to filters of even length.

Table 2.1. Parameter Values for a Designed Filter

	M	β	Maximum error in stop band and pass band
Using minimax technique of Ahmadi and Chottera[13]	13	3.08656	0.5×10^{-1}
Using Kaiser	15	3.3953	0.84×10^{-1}
empirical formulas	17	3.5035	0.31×10^{-1}

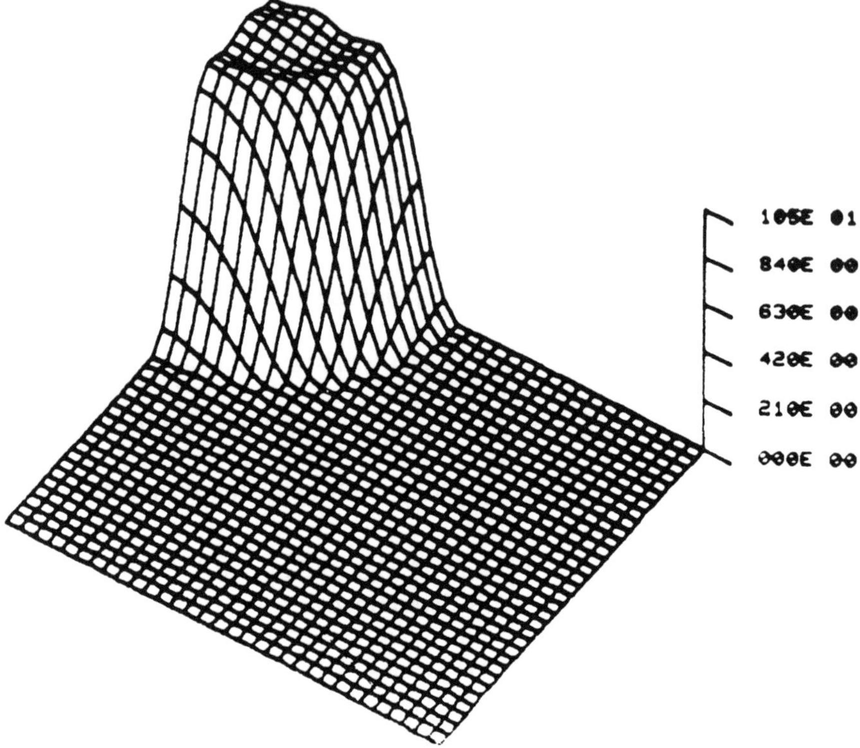

Figure 2.8. Three-dimensional plot of the amplitude response of the two-dimensional low-pass filter using the modified Kaiser window technique.

For a 1-D filter of length $2M + 1$ to be zero-phase, its impulse response $h(nT)$ must have Hermitian symmetric coefficients. Thus if $h(nT)$ is real, it must also be even. The frequency response $H(e^{j\omega T})$, assuming $T = 1$, can thus be expressed in the form

$$\begin{aligned} H(e^{j\omega}) &= h(0) + \sum_{n=1}^{M} h(n)[e^{-j\omega n} + e^{j\omega n}] \\ &= h(0) + \sum_{n=1}^{M} 2h(n) \cos \omega n \end{aligned} \tag{2.99}$$

Two-dimensional digital filters which have frequency responses of the form

$$H(e^{j\omega_1}, e^{j\omega_2}) = \sum_{m=0}^{M} \sum_{n=0}^{N} h(m, n) \cos m\omega_1 \cos n\omega_2 \tag{2.100}$$

are examples of 2-D zero-phase filters. Although equation (2.100) does not represent the most general class of such filters, many useful filters are in this class and it is the only class which will be considered for implementation by this technique. The impulse response of such a system $h(m, n)$ is a real sequence, which is an even function of its arguments.

The McClellan transformation converts 1-D filters of the form (2.99) into 2-D filters of the form (2.100) by means of the substitution

$$\cos \omega = \sum_{k=0}^{1} \sum_{l=0}^{1} t(k, l) \cos k\omega_1 \cos l\omega_2 \tag{2.101}$$

The relation between functions $h(n)$ and $h(m, n)$ can be seen by rewriting equation (2.99) in the form

$$H(e^{j\omega}) = \sum_{n=0}^{M} b(n)(\cos \omega)^n \tag{2.102}$$

Substitution into this summation from equation (2.101) yields

$$H(e^{j\omega_1}, e^{j\omega_2}) = \sum_{n=0}^{M} b(n) \left[\sum_{k=0}^{1} \sum_{l=0}^{1} t(k, l) \cos k\omega_1 \cos l\omega_2 \right]^n \tag{2.103}$$

Application of the recurrence formula for Chebyshev polynomials enables this latter relation to be expressed in the form of equation (2.100). The resulting 2-D filter is of size $(2M + 1) \times (2M + 1)$.

It follows implicitly from equation (2.101) that points in the frequency response of the 1-D filter are mapped to contours in the (ω_1, ω_2) plane. Furthermore, the shape of these contours is determined only by the parameters t. The variation of the frequency response from contour to contour, on the other hand, is controlled by the impulse response of the 1-D filter $h(n)$.

The problem of designing nonrecursive filters using this technique thus splits into two parts:

1. Design of the contour parameters t such that the contours (ω = constant) produced by the transformation of equation (2.101) in the (ω_1, ω_2) plane have some desired shape.
2. Design of the 1-D prototype filter $h(n)$.

Techniques discussed earlier in this chapter for 1-D filters can be employed to design $h(n)$ of the prototype filter, so we shall only be concerned here with the first problem, namely, calculation of the parameters t.

2.10.1. The Contour Approximation Problem

The transformation parameters $t(k, l)$ need to be chosen so that the contours (ω = constant) produced by the transformation

$$\cos \omega = F(\omega_1, \omega_2) = \sum_{k=0}^{K} \sum_{l=0}^{L} t(k, l) \cos k\omega_1 \cos l\omega_2 \tag{2.104}$$

in the (ω_1, ω_2) plane have some desired shape. As a working example, the design of a 2-D low-pass filter whose pass band is in the shape of a circle is considered. The contour in the frequency plane to which the pass-band edge of the 1-D filter should map is then described by the relation

$$\omega_1^2 + \omega_2^2 = R^2 \tag{2.105}$$

For simplicity, the transformation considered here is the first-order one having the form

$$\begin{aligned} \cos \omega = F(\omega_1, \omega_2) &= t(0, 0) + t(1, 0) \cos \omega_1 \\ &\quad + t(0, 1) \cos \omega_2 + t(1, 1) \cos \omega_1 \cos \omega_2 \end{aligned} \tag{2.106}$$

If the prototype is low-pass and the 2-D filter is to be low-pass, the 1-D origin will usually map to the 2-D origin. This gives the constraint equation

$$t(0, 0) + t(1, 0) + t(0, 1) + t(1, 1) = 1 \tag{2.107}$$

One variable, say $t(1, 1)$, is thus constrained to be a function of the other three, while those three are still unconstrained. To find values for these free variables the following equation should be solved for ω_2 in terms of ω and ω_1 and the free mapping parameters

$$\cos \omega = F(\omega_1, \omega_2) \tag{2.108}$$

This yields the equation

$$\begin{aligned} \omega_2 &= G[\omega, \omega_1, t(0, 0), t(0, 1), t(1, 0), t(1, 1)] = G(\omega, \omega_1, t) \\ &= \arccos\left[\frac{\cos \omega - t(0, 0) - t(1, 0) \cos \omega_1}{t(0, 1) + t(1, 1) \cos \omega_1}\right] \end{aligned} \tag{2.109}$$

An error function at the cutoff frequency can be defined as

$$E(\omega_1) = G(\omega_c, \omega_1, t) - (R^2 - \omega_1^2)^{1/2} \tag{2.110}$$

where ω_c is the cutoff frequency of the prototype. The parameters t can then be chosen to minimize some function of $E(\omega_1)$, such as its integral square value (l_2 approximation) or its maximum absolute value (l_∞ or the Chebyshev approximation). There are two difficulties with this formulation. First, the error function is not a linear function of known parameters; thus nonlinear optimization routines must be used for the minimization. Second, for transformations other than of first order, an explicit relation for G cannot be found.

A suboptimum approach reformulates the problem as a linear approximation problem. If the mapping were exact, then as the circular contour was traversed the values of $F(\omega_1, \omega_2)$ would be constant. This would result in

$$\cos \omega_0 = t(0,0) + t(1,0)\cos\omega_1 + t(0,1)\cos(R^2 - \omega_1^2)^{1/2} + [1 - t(0,0) - t(1,0) - t(0,1)]\cos\omega_1 \cos(R^2 - \omega_1^2)^{1/2} \quad (2.111)$$

where ω_0 can be any desired frequency, including the pass-band cutoff of the prototype. If the mapping is not exact, however, the equality in equation (2.111) will only be approximate and an error function can be defined by

$$E_1(\omega_1) = \cos\omega_0 - t(0,0) - t(1,0)\cos\omega_1 - t(0,1)\cos(R^2 - \omega_1^2)^{1/2} - [1 - t(0,0) - t(0,1) - t(1,0)]\cos\omega_1 \cos(R^2 - \omega_1^2)^{1/2} \quad (2.112)$$

This error is now a linear function of $t(0,0)$, $t(0,1)$, and $t(1,0)$, and thus linear programming optimization routines can be used to minimize the integral square error or the maximum absolute error. The former problem can be solved trivially using a classical least-squares formulation, and the parameters for minimizing the maximum absolute error can be found by linear programming.

REFERENCES

1. L. R. Rabiner and B. Gold, *Theory and Application of Digital Signal Processing*, Prentice-Hall, Englewood Cliffs, NJ (1978).
2. A. Antoniou, *Digital Filters: Analysis and Design*, McGraw-Hill, New York (1979).
3. L. C. Ludeman, *Fundamentals of Digital Signal Processing*, Harper and Row, New York (1986).
4. L. B. Jackson, *Digital Filters and Signal Processing*, Kluwer Academic Publishers, Boston (1986).
5. A. V. Oppenheim and R. W. Schafer, *Digital Signal Processing*, Prentice-Hall, Englewood Cliffs, NJ (1975).
6. F. F. Kuo and J. F. Kaiser, *System Analysis by Digital Computer*, John Wiley and Sons, New York (1966).

7. A. Antoniou, M. Ahmadi, and C. Charalambous, Design of factorable lowpass 2-dimensional digital filters satisfying prescribed specifications, *Proc. IEE* **128,** Part G, No. 2, 53–60 (1981).
8. A. Antoniou and C. Charalambous, An Efficient Approach for the Design of Digital Differentiators based on the Kaiser Window Function, *Proc. IEEE Int. Conf. on Circuits and Computers,* 1166–1171 (1980).
9. L. R. Rabiner, J. H. McClellan, and T. W. Parks, FIR digital filter design techniques using weighted Chebyshev approximation, *Proc. IEEE* **63,** 595–610 (1975).
10. A. Antoniou, Accelerated procedure for the design of equiripple nonrecursive digital filters, *Proc. IEE* **129,** Part G, No. 1, 1–10 (1982).
11. D. E. Dudgeon and R. M. Mersereau, *Multi-Dimensional Digital Signal Processing,* Prentice-Hall, Englewood Cliffs, NJ (1984).
12. T. S. Huang, Two-dimensional windows, *IEEE Trans. Audio Electroacoust.* **AU-20,** 88–89 (1972).
13. M. Ahmadi and A. Chottera, An improved method for the design of 2-D FIR digital filters with circular and rectangular cut-off boundary using Kaiser window, *Can. Electron. Eng. J.* **8,** 3–8 (1983).
14. J. H. McClellan, *On the Design of One-Dimensional and Two-Dimensional FIR Digital Filters,* Ph.D. Dissertation, Department of Electrical Engineering, Rice University, Houston, Texas, USA (1973).
15. R. M. Mersereau, W. F. G. Mecklenbräuker, and T. F. Quatieri, Jr., McClellan transformations for two-dimensional digital filtering: I—Design, *IEEE Trans. Circuits Syst.* **CAS-23,** 405–414 (1976).

3

Stability and Stabilization Techniques

PART ONE. ONE-DIMENSIONAL SYSTEM

3.1. INTRODUCTION

The term stability is generally used to describe a filter in which the convolution of the impulse response with some bounded input sequence will always yield a bounded output. We examine a one-dimensional (1-D) recursive digital filter whose transfer function is described in equation (1.33a) as follows:

$$H(z) = \frac{Y(z)}{U(z)} = \frac{\sum_{i=0}^{M} a_i z^{-i}}{\sum_{i=0}^{N} b_i z^{-i}} = \frac{N(z)}{D(z)} \tag{3.1}$$

(with $N \geq M$) where $Y(\cdot)$ and $U(\cdot)$ are output and input functions, respectively. This can be expressed alternatively in terms of the difference equation (1.2a) as follows:

$$y(n) = \sum_{i=0}^{M} a_i u(n-i) - \sum_{i=1}^{N} b_i y(n-i) \tag{3.2}$$

Without loss of generality we can assume $N = M$ in equations (3.1) and (3.2).

It can be seen from equation (3.2) that in recursive filters, past output values are used by the recursion algorithm in calculating the present output. This value can become arbitrarily large independent of the size of the input values. Recursive digital filters can therefore be unstable.

In order to understand the various known conditions and tests for stability of 1-D recursive digital filters, some definitions relating to signal-representing sequences must first be stated.

A function $u(m)$ in variable m is absolutely bounded when

$$|u(m)| \leq L < \infty \tag{3.3}$$

and is absolutely summable when

$$\sum_{m} |u(m)| \leq K < \infty \tag{3.4}$$

for all values of integer variable m, where L and K are positive real numbers.

One kind of stability to consider is bounded-input, bounded-output (BIBO) stability. With the input sequence $u(m)$ absolutely bounded, there arises the question as to which restrictions are to be placed on the filter's impulse response $h(m)$ to ensure that the output $y(m)$ is also absolutely bounded. We recall that the output $y(m)$ of a system may be expressed as the convolution of the input $u(m)$ and the impulse response of the network function [see equation (1.3a)]:

$$y(m) = \sum_{n=0}^{\infty} h(n)u(m-n) \tag{3.5}$$

Application of Schwarz's inequality yields

$$|y(m)| \leq \sum_{n=0}^{\infty} |h(n)||u(m-n)| \tag{3.6}$$

but since $|u(\cdot)| \leq L$ for all integer values of its argument, the inequality reduces to

$$|y(m)| \leq L \sum_{n=0}^{\infty} |h(n)| \tag{3.7}$$

This restricts the impulse response to an absolutely summable function. Hence if a system is BIBO stable, the output $y(m)$ is absolutely bounded if the impulse response $h(m)$ is absolutely summable and the input array $u(m)$ is absolutely bounded; the converse is also true.

Another kind of stability is summable-input, summable-output (SISO) stability. Here again the restrictions on the filter's impulse response $h(m)$ that ensure summability of the output $y(m)$ with summable input $u(m)$ are derived by applying Schwarz's inequality to the convolution sum of equation (3.5):

$$y(m) = \sum_{n=0}^{\infty} h(n)u(m-n)$$

or

$$\sum_{m=0}^{\infty} |y(m)| = \sum_{m=0}^{\infty} \left| \sum_{n=0}^{\infty} h(n)u(m-n) \right|$$

$$\leq \sum_{n=0}^{\infty} |h(n)| \sum_{m=0}^{\infty} |u(m-n)|$$

$$\leq K \sum_{n=0}^{\infty} |h(n)| \tag{3.8}$$

Therefore a sufficient condition for the filter to be considered stable is for its impulse response to be absolutely summable, although this is not a necessary condition. It is required that polynomial $D(z)$ in equation (3.1) should not possess any poles outside the unit circle, namely

$$D(z) \neq 0 \qquad \text{for } |z| \geq 1 \tag{3.9}$$

3.2. TESTING THE STABILITY OF ONE-DIMENSIONAL RECURSIVE FILTERS

There are many ways to test the stability of a 1-D recursive digital filter. One such approach is to find the zeros of polynomial $D(z)$ and to check whether any of the zeros of $D(z)$ has magnitude greater than unity, which indicates instability of the filter. This method is tedious and time consuming.

A second approach is based on generation of Jury's table.[1] We consider a recursive digital filter which is described by

$$H(z) = \frac{N(z)}{D(z)} = \frac{\sum_{i=0}^{M} a_i z^{-i}}{\sum_{i=0}^{M} b_i z^{-i}} \tag{3.10}$$

assuming that the order of polynomials $N(z)$ and $D(z)$ are both equal to M. Equation (3.10) can be expressed in the alternative form

$$H(z) = \frac{N(z)}{D(z)} = \frac{\sum_{i=0}^{M} a_i z^{M-i}}{\sum_{i=0}^{M} b_i z^{M-i}} \tag{3.11}$$

Table 3.1. Jury's Table

Row	Coefficients						
1	b_0^0	b_1^0	b_2^0	$\cdots$	b_{M-2}^0	b_{M-1}^0	b_M^0
2	b_M^0	b_{M-1}^0	b_{M-2}^0	$\cdots$	b_2^0	b_1^0	b_0^0
3	b_0^1	b_1^1	b_2^1	$\cdots$	b_{M-2}^1	b_{M-1}^1	
4	b_{M-1}^1	b_{M-2}^1	b_{M-3}^1	$\cdots$	b_1^1	b_0^1	
5	b_0^2	b_1^2	b_2^2	$\cdots$	b_{M-2}^2		
6	b_{M-2}^2	b_{M-3}^2	b_{M-4}^2	$\cdots$	b_0^2		
$\vdots$	$\vdots$						
$2M-3$	b_0^{M-2}	b_1^{M-2}	b_2^{M-2}				

and we may assume that $b_0 > 0$ without loss of generality. Jury's table can now be constructed using the coefficients of polynomial $D(z)$ as in Table 3.1. The first two rows of the table can be formed by using the coefficients of $D(z)$ directly, where $b_i^0 \equiv b_i$. The third and fourth rows of the table are calculated using the first two rows as follows:

$$b_i^1 = \begin{vmatrix} b_0^0 & b_{M-i}^0 \\ b_M^0 & b_i^0 \end{vmatrix} \qquad \text{for } i = 0, 1, \ldots, M-1$$

In a similar manner the fifth and sixth rows are calculated as

$$b_i^2 = \begin{vmatrix} b_0^1 & b_{M-1-i}^1 \\ b_{M-1}^1 & b_i^1 \end{vmatrix} \qquad \text{for } i = 0, 1, \ldots, M-2$$

This process is continued until $2M-3$ rows of the table are formed. The last row will have only three elements, namely, b_0^{M-2}, b_1^{M-2}, and b_2^{M-2}. The recursive filter of equation (3.11) is stable if and only if

1. $D(1) > 0$.
2. $(-1)^M D(-1) > 0$.
3. $b_0^0 > |b_M^0|$.

   $|b_0^1| > |b_{M-1}^1|$.

   $|b_0^2| > |b_{M-2}^2|$.

   $\vdots$

   $|b_0^j| > |b_2^j|$.

EXAMPLE 3.1. Determine whether the following transfer functions represent stable recursive filters:

(a) $$H(z) = \frac{0.01z^4 + 2.3z^3 + 1.7z^2 - 2.9z + 1}{z^4 + 2.6z^3 + 3.15z^2 + 2.15z + 0.55}$$

(b) $$H(z) = \frac{0.01z^4 + 2.3z^3 + 1.7z^2 + 2.9z + 1}{5z^4 - 0.26z^3 - 1.134z^2 + 1.02z + 1}$$

SOLUTION. (a) Apply Jury's three conditions; we obtain

(i) $D(1) = 9.45 > 0$,

(ii) $(-1)^4 D(-1) = -0.05 < 0$.

So the filter is unstable. In fact, it can be found that $D(z)$ has a pole with $|z| = 1.1$, which justifies Jury's stability criterion.

(b) Application of the three Jury conditions would give

(i) $D(1) = 5.626 > 0$,

(ii) $(-1)^4 D(-1) = 4.106 > 0$.

Row					
1	5	−0.26	−1.134	1.02	1
2	1	1.02	−1.134	−0.26	5
3	24	−2.32	−4.536	5.36	
4	5.36	−4.536	−2.32	24	
5	547.270	−31.367	−96.429		

Since $|b_0^0| > |b_4^0|$, $|b_0^1| > |b_3^1|$, and $|b_0^2| > |b_2^2|$, the filter is stable.

3.3. GENERATION OF ONE-DIMENSIONAL STABLE POLYNOMIALS

It has been shown by Schüssler[2] that for a polynomial of degree m possessing real coefficients as shown below,

$$D(z) = d_m z^m + d_{m-1} z^{m-1} + \cdots + d_1 z + d_0$$
$$= \sum_{i=0}^{m} d_i z^i \qquad (3.12)$$

(with d_m positive), function $D(z)$ can be decomposed as the sum of the mirror-image polynomial

$$F_1(z) = [D(z) + z^m D(z^{-1})]/2 \qquad (3.13)$$

and the antimirror-image polynomial

$$F_2(z) = [D(z) - z^m D(z^{-1})]/2 \tag{3.14}$$

The necessary and sufficient conditions for $D(z)$ to have zeros inside the unit circle are:

1. The zeros of $F_1(z)$ and $F_2(z)$ are located on the unit circle.
2. They are simple.
3. They separate each other, and $|d_0/d_m| < 1$.

Ramachandran and Gargour[3] have used these results to generate a 1-D polynomial of order m which has all its zeros inside the unit circle. They used the following relationship:

For m even

$$D(z) = K_e \prod_{i=1}^{n} (z^2 - 2\alpha_i z + 1) + (z^2 - 1) \prod_{i=1}^{n-1} (z^2 - 2\beta_i z + 1) \tag{3.15}$$

where $n = m/2$, with $K_e > 1$ and

$$1 > \alpha_1 > \beta_1 > \alpha_2 > \beta_2 > \cdots > \beta_{n-1} > \alpha_n > -1 \tag{3.16}$$

For m odd

$$D(z) = K_o(z + 1) \prod_{i=1}^{n} (z^2 - 2\alpha_i z + 1) + (z - 1) \prod_{i=1}^{n} (z^2 - 2\beta_i z + 1) \tag{3.17}$$

where $n = (m - 1)/2$, with $K_o > 1$ and

$$1 > \alpha_1 > \beta_1 > \alpha_2 > \beta_2 > \cdots > \alpha_n > \beta_n > -1 \tag{3.18}$$

This procedure also requires $|d_0/d_m| < 1$ in both cases.

3.4. STABILIZATION OF ONE-DIMENSIONAL RECURSIVE DIGITAL FILTERS

3.4.1. Pole-Zero Cancellation Technique

In the design of a recursive digital filter, it may happen that the filter's transfer function will have some of its poles located outside the unit circle despite a satisfactory magnitude response associated with the filter's transfer function. In this case it is desirable to stabilize the filter's transfer function without affecting its magnitude characteristics. Let us assume that the designed filter which has the desired magnitude response is expressed as follows:

$$H(z) = \frac{\sum_{i=0}^{M} a_i z^i}{\prod_{i=1}^{K} (z - r_i\, e^{j\theta_i}) \prod_{i=K+1}^{M} \left(z - \frac{1}{r_i} e^{j\theta_i}\right)} \tag{3.19}$$

where $|r_i| < 1$.

Equation (3.19) represents the transfer function of a filter with K of its poles inside the unit circle and the remaining $(M - K)$ poles located outside the unit circle. If the transfer function of equation (3.19) is cascaded with the all-pass transfer function of the form

$$H_1(z) = \prod_{i=K+1}^{M} \left[\left(z - \frac{1}{r_i} e^{j\theta_i}\right) \Big/ (z - r_i\, e^{j\theta_i})\right] \tag{3.20}$$

the destabilizing effect will be removed without changing the magnitude response of the resultant filter. This may be seen from

$$H_{\text{Res}}(z) = H_1(z) \cdot H(z) \tag{3.21}$$

and since $|H_1(e^{j\omega T})| = 1$, we have

$$|H_{\text{Res}}(e^{j\omega T})| = |H(e^{j\omega T})| \tag{3.22}$$

The above procedure can be summarized by stating that all poles outside the same unit circle are changed into their mirror image with respect to the unit circle. This stabilization procedure, despite its simplicity, has a drawback, that is, if a filter with a prespecified phase and magnitude response is desired through this stabilization method, the phase response will be distorted and will no longer resemble the original phase response of the unstable transfer function; furthermore, exact cancellation of the poles is not possible in practice.

3.4.2. Stabilization via a Discrete Hilbert Transform

The development of this stabilization procedure is due to Read and Treitel,[4] Gold and Rader,[5] and Cizek.[6] The discrete Fourier transform (DFT) plays an important role in this presentation.

It is assumed that the discrete time sequence $d(n)$ is of finite length N, where N is an even integer. Thus $d(n) = d(n + kN)$ for any integer k.

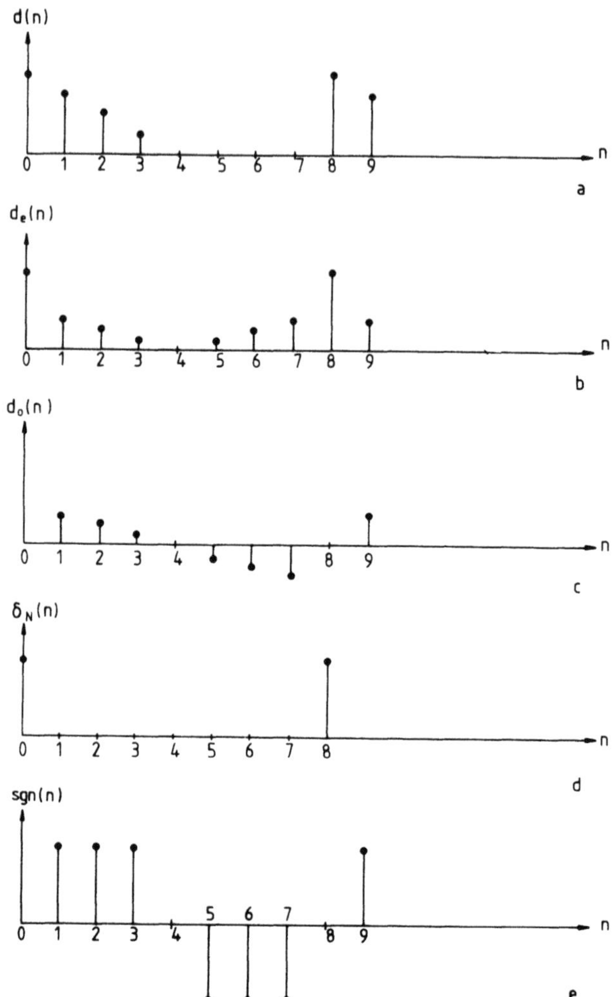

Figure 3.1. Example of one-dimensional stabilization by a Hilbert transform: (a) original sequence, (b) even part of sequence, (c) odd part of sequence, (d) modulo N Kronecker delta, (e) signum function.

Furthermore, we also assume that $d(n)$ $(n = 0, 1, \ldots, N-1)$ is a causal sequence over one period of length N. This would require that

$$d(n) = 0 \qquad \text{for } n \geq N/2 \tag{3.23}$$

It should be noted that any finite discrete time sequence can be made causal by choosing an appropriate N and appending zeros. An example of the sequence $d(n)$ is given in Figure 3.1a for the case $N = 8$.

The even and odd parts of a finite discrete sequence are given by

$$d_e(n) = \tfrac{1}{2}[d(n) + d(N - n)] \tag{3.24a}$$

and

$$d_o(n) = \tfrac{1}{2}[d(n) - d(N - n)] \tag{3.24b}$$

The even and odd parts of the pulse sequence in Figure 3.1a are shown in Figures 3.1b and 3.1c, respectively. Thus for a finite discrete causal pulse train one has

$$d_e(n) = \begin{cases} d(n), & n = 0 \\ \tfrac{1}{2}d(n), & 0 < n < N/2 \text{ and } N/2 < n < N \\ 0, & n = N/2 \end{cases} \tag{3.25}$$

with even symmetry, i.e., $d_e(n) = d_e(N - n)$. Similarly,

$$d_o(n) = \begin{cases} 0, & n = 0 \\ \tfrac{1}{2}d(n), & 0 < n < N/2 \\ 0 & n = N/2 \\ -\tfrac{1}{2}d(n), & N/2 < n < N \end{cases} \tag{3.26}$$

and this pulse train has odd symmetry, i.e., $d_o(n) = -d_o(N - n)$. From these equations it follows that the odd and even parts of the pulse sequence $d(n)$ are related by

$$d_e(n) = \operatorname{sgn}(n)d_o(n) + d(n)\delta_N(n) \tag{3.27a}$$

and

$$d_o(n) = \operatorname{sgn}(n)d_e(n) \tag{3.27b}$$

where $\delta_N(n)$ is the modulo N Kronecker delta given by

$$\delta_N(n) = \begin{cases} 1, & \text{for } n = kN,\ k \text{ any integer} \\ 0, & \text{otherwise} \end{cases} \tag{3.28}$$

and $\operatorname{sgn}(\cdot)$ denotes the finite discrete signum function given by

$$\operatorname{sgn}(n) = \begin{cases} 0, & n = 0,\ N/2 \\ 1, & 0 < n < N/2 \\ -1, & N/2 < n < N \end{cases} \tag{3.29}$$

Table 3.2. Transform Symmetry Relations

$d(n)$	DFT[$d(n)$]
Even	Even
Odd	Odd
Even and real	Even and real
Odd and real	Odd and imaginary
Real	Real part even, imaginary part odd
Imaginary	Real part odd, imaginary part even
Even and imaginary	Even and imaginary
Odd and imaginary	Odd and real

Functions $\delta_N(n)$ and $\text{sgn}(n)$ are presented in Figures 3.1d and 3.1e. Equations (3.27a) and (3.27b) follow logically by inspection of Figure 3.1.

The above equations can be used to relate the real and imaginary parts of the Fourier transform of the finite discrete causal pulse sequence. In order to do so we require the symmetry relations given in Table 3.2, where the terms "odd" and "even" are to be construed in the sense of equations (3.25) and (3.26).

Table 3.2 and equations (3.27a) and (3.27b) allow us to relate the real and imaginary parts of the discrete Fourier transform of the finite discrete causal pulse train, $d(n)$. We have

$$D(k) = \text{DFT}[d(n)] = \sum_{n=0}^{N-1} d(n)\, W^{kn} \tag{3.30}$$

where

$$W = e^{-j2\pi/N}, \qquad j = (-1)^{1/2}$$

and

$$d(n) = \text{IDFT}[D(k)] = \frac{1}{N} \sum_{k=0}^{N-1} D(k)\, W^{-kn} \tag{3.31}$$

Knowing that

$$d(n) = d_o(n) + d_e(n) \tag{3.32}$$

and by taking the DFT of both sides of equation (3.32), we derive

$$\text{DFT}[d(n)] = \text{DFT}[d_o(n)] + \text{DFT}[d_e(n)] \tag{3.33}$$

Both $d_o(n)$ and $d_e(n)$ are real. From Table 3.2 we see that DFT$[d_o(n)]$ is odd and imaginary, while DFT$[d_e(n)]$ is even and real. Thus

$$\mathrm{PR}(k) = \mathrm{Re}[D(k)] = \mathrm{DFT}[d_e(n)] \tag{3.34a}$$

and

$$\mathrm{PI}(k) = \mathrm{Im}[D(k)] = -\mathrm{j}\,\mathrm{DFT}[d_o(n)] \tag{3.34b}$$

where PR(k) and PI(k) are the real and imaginary parts of DFT$[d(n)]$, respectively. Since the result of transforming a real and odd pulse $d_o(n)$ is purely imaginary (see Table 3.2), multiplication by $-\mathrm{j}$ in equation (3.34b) results in a real quantity PI(k). Inverse transformation of these equations yields

$$d_e(n) = \mathrm{IDFT}[\mathrm{PR}(k)] \tag{3.35a}$$

and

$$d_o(n) = \mathrm{IDFT}[\mathrm{j}\,\mathrm{PI}(k)] \tag{3.35b}$$

Substitution of equations (3.35a) and (3.35b) into the right-hand sides of equations (3.27a) and (3.27b), respectively, yields

$$d_e(n) = \mathrm{sgn}(n)\,\mathrm{IDFT}[\mathrm{j}\,\mathrm{PI}(k)] + d(n)\delta_N(n) \tag{3.36a}$$

and

$$d_o(n) = \mathrm{sgn}(n)\,\mathrm{IDFT}[\mathrm{PR}(k)] \tag{3.36b}$$

Direct transformation of both sides of equation (3.36a) together with the use of equations (3.34a) and (3.34b) gives

$$\mathrm{PR}(k) = \mathrm{DFT}\{\mathrm{sgn}(n)\,\mathrm{IDFT}[\mathrm{j}\,\mathrm{PI}(k)] + d(n)\delta_N(n)\} \tag{3.37a}$$

and

$$\mathrm{PI}(k) = -\mathrm{j}\,\mathrm{DFT}\{\mathrm{sgn}(n)\,\mathrm{IDFT}[\mathrm{PR}(k)]\} \tag{3.37b}$$

The latter two equations relate the real and imaginary parts of the DFT of a finite causal sequence. In this respect, these equations are similar to the integral Hilbert transform equations that relate the real and imaginary parts of the integral Fourier transform of a causal function. Cizek[6] shows that equation (3.37b) can be written as

$$\mathrm{PI}(k) = \frac{1}{N}\sum_{n=0}^{N-1} \mathrm{PR}(n)[1-(-1)^{k-n}]\,\mathrm{cotan}\,\frac{\pi}{N}(k-n) \tag{3.38}$$

This equation can be expressed for the even and odd values of n as follows:

$$\mathrm{PI}(k) = \frac{2}{N} \sum_{n=1,3,5,\ldots}^{N-1} \mathrm{PR}(n) \cot \frac{\pi}{N}(k-n), \qquad k \text{ even} \tag{3.39a}$$

and

$$\mathrm{PI}(k) = \frac{2}{N} \sum_{n=0,2,4,\ldots}^{N-1} \mathrm{PR}(n) \cot \frac{\pi}{N}(k-n), \qquad k \text{ odd} \tag{3.39b}$$

Equations (3.39a) and (3.39b) are approximations of the integral form of the Hilbert transform of a continuous periodic function.

The discrete Hilbert transform described above forms the core of a computational procedure that yields an approximation to the minimum-phase (a minimum-phase sequence is one whose z transform has no zeros outside the unit circle) version of a mixed or maximum-phase sequence.

It is well known that for a minimum-phase sequence $d(n)$ the phase $\theta(e^{j\omega})$ and the log magnitude of the amplitude spectrum, $\log|D(e^{j\omega})|$, are related through the Hilbert transform,

$$\theta(e^{j\omega}) = -\frac{1}{2\pi} \int_0^{2\pi} \log|D(e^{j\Omega})| \cos \frac{\omega - \Omega}{2} \, d\Omega \tag{3.40}$$

which results from the fact that $d(n)$ $(n = 0, 1, \ldots, N-1)$ is minimum phase if and only if the inverse z transform of $\log|D(e^{j\omega})|$ is causal. Since the Hilbert transform relates the real and imaginary parts of causal functions, application of the Hilbert transform to $\log|D(e^{j\omega})|$ in equation (3.40) yields the imaginary part of $\log[D(e^{j\omega})]$, which is the phase of $D(z)$.

It can easily be shown that equations (3.39a) and (3.39b) are approximations of equation (3.40), and the procedure for obtaining the approximate minimum-phase spectrum from knowledge of the amplitude spectrum of a mixed or maximum-phase finite sequence $d(n)$ can be given by the equation

$$\theta(k) = -j\, \mathrm{DFT}\{\mathrm{sgn}(n)\, \mathrm{IDFT}[\log|D(k)|]\} \tag{3.41}$$

Here, $|D(k)|$ is the kth component of the amplitude spectrum and $\theta(k)$ is the kth component of the phase spectrum. A block diagram summarizing the procedure required to obtain the minimum-phase version of a mixed or maximum-phase pulse is shown in Figure 3.2.

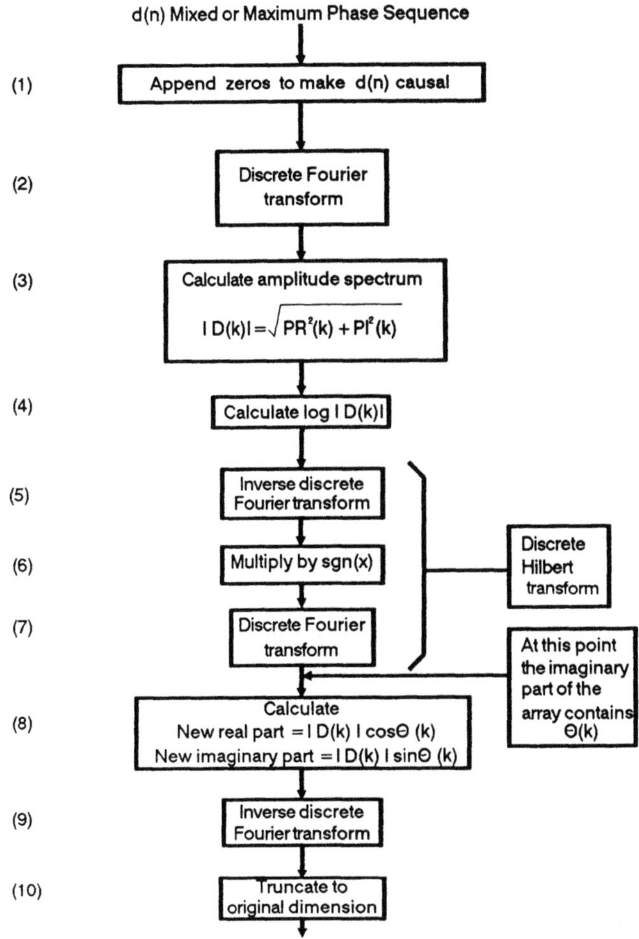

Figure 3.2. Flow chart for obtaining the minimum-phase version of a function.

PART TWO. TWO-DIMENSIONAL SYSTEM

3.5. INTRODUCTION

As in a 1-D system, the stability of a two-dimensional (2-D) system depends on the impulse response of the system in question, i.e., if the input is a bounded signal, the result of convolving the input sequence with the filter's impulse response should yield a bounded output sequence. We consider the transfer function of a 2-D recursive digital filter which can be written as follows:

$$H(z_1, z_2) = \frac{Y(z_1, z_2)}{U(z_1, z_2)} = \frac{N(z_1, z_2)}{D(z_1, z_2)} = \frac{\sum_{i=0}^{M} \sum_{j=0}^{N} a(i,j) z_1^{-i} z_2^{-j}}{\sum_{i=0}^{M} \sum_{j=0}^{N} b(i,j) z_1^{-i} z_2^{-j}} \tag{3.42}$$

where $Y(\cdot,\cdot)$ and $U(\cdot,\cdot)$ are the output and input functions, respectively, while a and b are the sets of coefficients of the system. Equation (1.2b) enables this to be expressed in the spatial domain in terms of the linear difference equation

$$y(m,n) = \sum_{i=0}^{M} \sum_{j=0}^{N} a(i,j) u(m-i, n-j) - \sum_{\substack{i=0 \\ i+j\neq 0}}^{M} \sum_{j=0}^{N} b(i,j) y(m-i, n-j), \qquad b_{00} = 1 \tag{3.43}$$

The latter equation shows that the present value of the output in recursive filters is calculated using present and past input as well as past output samples through the recursion formula. Such values can become arbitrarily large, independent of the size of the input values, resulting in instability of the 2-D recursive filter.

In order to understand the various known conditions and tests for stability of 2-D recursive filters, some definitions relating to signal-representing sequences (as in the 1-D case) must first be provided.

A function $u(m, n)$ in two variables m and n is absolutely bounded when

$$|u(m,n)| \leq L < \infty \tag{3.44}$$

and is absolutely summable when

$$\sum_{m} \sum_{n} |u(m,n)| \leq K < \infty \tag{3.45}$$

for all nonnegative integer pairs (m, n), where L and K are positive real numbers.

One kind of stability to consider is BIBO stability. With the input array $u(m, n)$ absolutely bounded, there arises the question as to which restrictions are to be placed on the filter's impulse response $h(m, n)$ to ensure that the output $y(m, n)$ is also absolutely bounded. From equation (1.3b) the output $y(m, n)$ is expressed as the convolution of the input array $u(m, n)$ and the impulse response of the system $h(m, n)$ as follows:

$$y(m, n) = \sum_k \sum_l h(k, l) u(m - k, n - l) \tag{3.46}$$

Application of Schwarz's inequality gives

$$|y(m, n)| \leq \sum_k \sum_l |h(k, l)| |u(m - k, n - l)| \tag{3.47}$$

but since $|u(\cdot)| \leq L$ for all integer values of its arguments, the inequality reduces to

$$|y(m, n)| \leq L \sum_k \sum_l |h(k, l)| \tag{3.48}$$

which restricts the impulse response to an absolutely summable function. This has been shown to be a necessary and sufficient condition for stability.[7]

Another kind of stability is SISO stability. Here again the restriction on the filter's impulse response $h(m, n)$ that ensures the summability of the output $y(m, n)$ with summable input $u(m, n)$ is derived by applying Schwarz's inequality to the convolutional sum of equation (3.46). Hence

$$\begin{aligned} \sum_m \sum_n |y(m, n)| &= \sum_m \sum_n \left| \sum_k \sum_l h(k, l) u(m - k, n - l) \right| \\ &\leq \sum_m \sum_n \sum_k \sum_l |h(k, l)| |u(m - k, n - l)| \\ &\leq \sum_k \sum_l |h(k, l)| \sum_m \sum_n |u(m - k, n - l)| \\ &\leq K \sum_k \sum_l |h(k, l)| \end{aligned} \tag{3.49}$$

and therefore a sufficient condition for the filter to be considered SISO stable is for its impulse response to be absolutely summable.

Generally speaking, the stability of a 2-D recursive digital filter is determined by the coefficients of the denominator $D(z_1, z_2)$ of the z-transfer function of the filter when such a filter is expressed in the form of equation (3.42). However, testing the stability is difficult because the fundamental theorem of algebra is not applicable to two-variable functions, namely, denominator factorization is not always possible. Testing the stability by finding the poles of the z-transfer function, as in the 1-D case, is hence not possible, nor is filter stabilization by replacement of the poles in the instability region by poles in conjugate reciprocal positions with respect to the unit circle. A direct extension of the condition for stability in the 1-D case to the 2-D case was first proposed by Shanks *et al.*[8] and can be stated as follows:

THEOREM 3.1. *Given that $D(z_1, z_2)$ is a polynomial in z_1 and z_2, a necessary and sufficient condition for the coefficients of the expansion of $H(z_1, z_2) = 1/D(z_1, z_2)$ in negative powers of z_1 and z_2 to converge absolutely, and hence for $h(m, n)$ to be absolutely summable, is*

$$D(z_1, z_2) \neq 0 \qquad \text{for} \bigcap_{i=1}^{2} |z_i| \geq 1 \tag{3.50}$$

The above condition suggests a test procedure for checking the stability of the filter by finding the continuum of (z_1, z_2) values for which $D(z_1, z_2) = 0$. This is done by assigning values to the variable z_1 and finding the roots of $D(z_1, z_2) = 0$ as a function of z_2. For stability, it follows from Theorem 3.1 that all roots of z_2 must be less than unity in magnitude when $|z_1|$ is greater than one. The testing for stability as stated above is very tedious to apply, since it involves mapping an infinite number of points from the z_1 plane into the z_2 plane. However, a considerably simplified version has been achieved by Huang[9] in terms of the following theorem:

THEOREM 3.2. *A causal 2-D recursive filter with a z-transfer function*

$$H(z_1, z_2) = \frac{N(z_1, z_2)}{D(z_1, z_2)}$$

is stable if and only if:

1. *the map of $R_1 \equiv (z_1, |z_1| = 1)$ in the z_2 plane according to $D(z_1, z_2) = 0$ lies outside $d_2 \equiv (z_2, |z_2| \geq 1)$,*
2. *no point in $d_1 \equiv (z_1, |z_1| \geq 1)$ maps into the point $z_2 = 0$ by the relation $D(z_1, z_2) = 0$.*

To check the stability using Theorem 3.2, R_1 is mapped into the z_2 plane according to $D(z_1, z_2) = 0$ and the resulting image tested to determine whether it lies outside d_2. Also, $D(z_1, 0) = 0$ must be solved to find whether there are any roots with magnitude greater than one.

3.6. ANSELL STABILITY THEORY[10]

Theorem 3.2 can be reduced to a stability test involving only a finite number of steps. However, it will be shown that this can still be very tedious.

Let us consider the following change of variables applied to the filter z-transfer function $H(z_1, z_2)$:

$$s_1 = \frac{1 - z_1^{-1}}{1 + z_1^{-1}} \tag{3.51}$$

and

$$s_2 = \frac{1 - z_2^{-1}}{1 + z_2^{-1}} \tag{3.52}$$

We set

$$H(z_1, z_2) = \frac{N(z_1, z_2)}{D(z_1, z_2)} = \frac{A(s_1, s_2)}{B(s_1, s_2)} \qquad \text{where } s_i = \frac{1 - z_i^{-1}}{1 + z_i^{-1}}, \qquad i = 1, 2 \tag{3.53}$$

where $A(\cdot)$ and $B(\cdot)$ are polynomials in s_1 and s_2. Theorem 3.2 can now be restated as follows:

THEOREM 3.3. *The causal recursive filter $H(z_1, z_2)$ is stable if and only if:*

1. *in all real finite ω_1, the complex polynomial in s_2, $B(j\omega_1, s_2)$, has no zeros in* $\text{Re}[s_2] \geq 0$, *and*
2. *the real polynomial in s_1, $B(s_1, 1)$, has no zeros in* $\text{Re}[s_1] \geq 0$.

Moreover, condition (1) of Theorem 3.3 can be derived in another form. We express $B(j\omega_1, j\omega_2)$ for real ω_1 and ω_2 as

$$\begin{aligned} B(j\omega_1, j\omega_2) = {} & \beta_0(\omega_1)\omega_2^n + \beta_1(\omega_1)\omega_2^{n-1} + \cdots + \beta_n(\omega_1) \\ & + j[\alpha_0(\omega_1)\omega_2^n + \alpha_1(\omega_1)\omega_2^{n-1} + \cdots + \alpha_n(\omega_1)] \end{aligned} \tag{3.54}$$

where $\alpha_i(\omega_1)$ and $\beta_i(\omega_1)$ for $i = 1, \ldots, n$ are real polynomials in ω_1 and where neither $\alpha_0(\omega_1)$ nor $\beta_0(\omega_1)$ is identically zero. Also, we define $\gamma_{k,l}(\omega_1)$ as

$$\gamma_{k,l} = \alpha_k \beta_l - \alpha_l \beta_k \tag{3.55}$$

for $0 \leq l,\ k \leq n$, equating to zero αs and βs not present in function $B(j\omega_1, j\omega_2)$. Quantity $D(\omega_1)$ denotes the $n \times n$ symmetrical polynomial matrix whose typical element $D_{ij}(\omega_1)$ $(1 \leq i, j \leq n)$ is the sum of all those $\gamma_{k,l}(\omega_1)$ $(0 \leq k, l \leq n)$ for which both

$$k + l = i + j - 1 \qquad \text{and} \qquad |l - k| > |i - j| \tag{3.56}$$

are satisfied. Then the n successive principal minors of $D(\omega_1)$ must be positive for all real ω_1.

In the above, Sturm's method[11] can be used to test whether each minor of $D(\omega_1)$ is positive for all real ω_1.

It can thus be seen that while Ansell's test requires only a finite number of steps, the mathematics involved are still tedious.

3.7. THE ANDERSON AND JURY STABILITY TEST[12]

Huang's simplification of the stability test rests on the fact that the denominator in the z-transfer function $D(z_1, z_2) \neq 0$ for $|z_1| \geq 1 \cap |z_2| \geq 1$ if the following two conditions hold:

$$D(z_1, 0) \neq 0, \qquad |z_1| \geq 1 \tag{3.57}$$

and

$$D(z_1, z_2) \neq 0, \qquad |z_1| = 1 \cap |z_2| \geq 1 \tag{3.58}$$

The Anderson and Jury stability test is divided into two parts: first, condition (3.57) is checked for the values of D with z_1 restricted, and second, condition (3.58) is checked for the values of D with both z_1 and z_2 restricted.

3.7.1. Denominator Examination with z_1 Limited

Two methods for checking condition (3.57) have been proposed by Anderson and Jury.[12] The first is based on the use of the Schur-Cohn matrix.[13] This matrix is square, Hermitian, of size equal to the degree of $D(z_1, 0)$, and with elements which are simple functions of the coefficients of $D(z_1, 0)$. The matrix is negative definite if $D(z_1, 0)$ has all its zeros in $|z_1| < 1$. The negative definiteness can be established by examining the sign of the leading principal minors of the matrix.

The Schur-Cohn criterion for checking condition (3.57) will now be discussed in detail. We set

$$f(z) = \sum_{i=0}^{n} a_i z^i \qquad (a_n \neq 0) \tag{3.59}$$

where the associated $n \times n$ Hermitian matrix $C = (\gamma_{ij})$ is defined by

$$\gamma_{ij} = \sum_{p=1}^{i} (a_{n-i+p} a^*_{n-j+p} - a^*_{i-p} a_{j-p}), \qquad i \leq j \tag{3.60}$$

Then:

1. The number of zeros z_i of $f(z)$ for which $|z_i| > 1$ and for which z_i^{-1} is not also a zero is the number of positive eigenvalues of C.
2. The number of zeros z_i for which $|z_i| < 1$ and for which z_i^{-1} is not also a zero is the number of negative eigenvalues of C.
3. The number of zeros z_i for which either $|z_i| = 1$ or z_i^{-1} is also a zero (or both) is the nullity of C.

For condition (3.57) to hold, with $D(z_1, 0)$ a real polynomial and C a real matrix, all eigenvalues of C must be negative definite. This is so if and only if all odd-order leading principal minors are negative and all even-order leading principal minors are positive.

An alternative procedure for checking condition (3.57) can be derived from the theorem of Jury,[(14)] which is based on forming a sequence of polynomials.

THEOREM 3.4. *If*

$$f(z) = \sum_{i=0}^{n} a_i z^i \tag{3.61}$$

then a sequence of polynomials

$$F_0(z) = f(z) = \sum_{i=0}^{n} a_i^{(0)} z^i \qquad \textit{where } a_i^{(0)} \triangleq a_i$$

$$F_1(z) = \sum_{i=0}^{n-1} a_i^{(1)} z^i$$

$$F_2(z) = \sum_{i=0}^{n-2} a_i^{(2)} z^i$$

$$\vdots$$

$$F_j(z) = \sum_{i=0}^{n-j} a_i^{(j)} z^i \tag{3.62}$$

can be defined by

$$F_{j+1}(z) = a_0^{*(j)} F_j(z) - a_{n-j}^{(j)} F_j^*(z) \tag{3.63}$$

where $a_0^{*(j)}$ is the complex conjugate of $a_0^{(j)}$ and

$$F_j^*(z) = a_0^{*(j)} z^{n-j} + a_1^{*(j)} z^{n-j-1} + \cdots + a_{n-j}^*$$
$$= \sum_{i=0}^{n-1} a_i^{*(j)} z^{n-j-i} \qquad (3.64)$$

(Thus $F_j^*(z)$ is obtained from $F_j(z)$ by coefficient reversal and conjugation.)

If we set $\delta_j = F_j(0)$ and $P_j = \delta_1 \delta_2 \cdots \delta_j$, then:

1. All zeros of $f(z)$ lie inside $|z| \geq 1$ iff $P_j < 0$ for all j.
2. All zeros of $f(z)$ lie outside $|z| \geq 1$ iff $P_j > 0$ for all j, or equivalently, $\delta_j > 0$ for all j.
3. If all P_j are nonzero, the number of negative P_j is the number of zeros of $f(z)$ inside $|z| \geq 1$ and the number of positive P_j is the number of zeros outside $|z| \geq 1$.

It should be noted that (2) above covers the requirement for condition (3.57).

3.7.2. Denominator Examination with both z_1 and z_2 Limited

Broadly speaking, checking condition (3.58) comprises two distinct steps. First, a Schur-Cohn test will be applied. Second, the positiveness of a number of polynomials on $|z_1| = 1$ will be checked.

Step 1: Applying the Schur-Cohn Test. Condition (3.58) is checked by replacing $f(z)$ in expression (3.59) by $D(z_1, z_2)$ written as a polynomial in z_2:

$$D(z_1, z_2) = \sum_{j=0}^{N} \left(\sum_{i=0}^{M} b_{ij} z_1^i \right) z_2^j \qquad (3.65)$$

The coefficient a_j is therefore $\sum_{i=0}^{M} b_{ij} z_1^i$ and the matrix C is $N \times N$ with entries which, from equation (3.60), are seen to be polynomials in z_1 and z_1^* with real coefficients (matrix C is Hermitian).

The condition $D(z_1, z_2) \neq 0$ for $|z_1| = 1$ and $|z_2| \geq 1$ holds if and only if, for all $|z_1| = 1$, C is negative definite, i.e., if and only if the leading principal minors of C have appropriate signs. These principal minors, being linear combinations of products of quantities γ_{ij}, are themselves polynomials in z_1 and z_1^* with real coefficients; they are also real, since C is Hermitian. By setting $z_1^* = z_1^{-1}$, on $|z_1| = 1$, the polynomials will have the form $\sum_{j=0}^{M} C_j(z_1^j + z_1^{-j})$.

Such polynomials are termed self-inverse, i.e., if $z_1 = z_{1\alpha}$ is a zero, then so is $z_{1\alpha}^{-1}$.

Step 2. This step involves checking the positiveness of a self-inverse polynomial on $|z_1| = 1$. We consider the self-inverse polynomial

$$f(z_1) = \sum_{j=0}^{M} C_j(z_1^j + z_1^{-j}) \tag{3.66}$$

where coefficients C_j are real constants; some immediate necessary conditions are obtained by putting $z_1 = 1, -1$, namely,

$$\sum_{j=0}^{M} C_j > 0 \qquad \text{and} \qquad \sum_{j=0}^{M} (-1)^j C_j > 0 \tag{3.67}$$

There are two techniques for checking sign definiteness of a self-inverse polynomial. The first approach substitutes $z_i = \exp(j\theta)$, $|z_i| = 1$, in expression (3.66). This yields

$$f(\cos\theta) = 2\sum_{j=0}^{M} C_j \cos j\theta \tag{3.68}$$

Positiveness is to be checked for $0 \leq \theta < 2\pi$ (or $-\pi \leq \theta < \pi$). But since $\cos j\theta = \cos(-j\theta)$, it would be sufficient to examine the range 0 to π.

The following change of variable will be introduced:

$$x = \cos\theta \qquad \text{and} \qquad T_k(x) = \cos k\theta \tag{3.69}$$

where $T_k(x)$ is the kth Chebyshev polynomial of the first kind, defined recursively by

$$T_{k+1}(x) = 2xT_k(x) - T_{k-1}(x), \qquad T_1(x) = x$$
$$T_0(x) = 1 \tag{3.70}$$

Then the requirement will be

$$g(x) = \sum_{j=0}^{M} C_j T_j(x) > 0 \tag{3.71}$$

for $-1 \leq x \leq 1$. Function $T_j(x)$ is a polynomial in x of degree j; therefore $g(x)$ has the form

$$g(x) = \sum_{j=0}^{M} d_j x^j \tag{3.72}$$

for some real d_j. Positiveness can be checked by forming a Sturm chain. An example of this can be found elsewhere.[11]

An alternative approach for checking positiveness is based on the determination of the zero distribution of $f(z_1)$ in equation (3.66). Because $f(z_1)$ is self-inverse, there are as many zeros of $z^M f(z_1)$ inside $|z_1|<1$ as outside. Therefore, $f(z_1)$ is positive on $|z_1|=1$ if and only if $f(1)>0$ [or $f(z_1)$ is positive at any one point of $|z_1|=1$ and $z_1^M f(z_1)$ has M zeros inside $|z_1|<1$, for then $z_1^M f(z_1)$ has M zeros outside, and thus no zeros on $|z_1|=1$].

Unfortunately, the Schur-Cohn matrix for a self-inverse polynomial is zero, and the other previously mentioned procedure based on setting up a sequence of polynomials leads to a zero polynomial at the first recursion. Hence neither procedure, as it stands, is of help. However, the following result is of assistance.

THEOREM 3.5.[13,14] *Let $f(z_1)$ be as in equation* (3.66). *The number of zeros of $g(z_1)=z_1^M f(z_1)$ in $|z_1|<1$ is the same as the number of zeros of $z_1^{2M-1}g'(1/z_1)$ in $|z_1|<1$, where $g'(1/z_1)$ is obtained by differentiating $g(z_1)$ with respect to z_1 and substituting z_1^{-1} for z_1.*

Assuming $z_1^{2M-1}g'(1/z_1)$ has neither zero reciprocal with respect to $|z_1|=1$, nor zeros on $|z_1|=1$, the Schur-Cohn criterion will yield the number of zeros inside $|z_1|<1$; so we may use the procedure based on generating a recursive set of polynomials. If, on the other hand, $z_1^{2M-1}g'(1/z_1)$ does have reciprocal zeros or zeros on the unit circle, the situation must be analyzed further. In view of the easily established relation

$$z_1 g'(z_1)+z_1^{2M-1}g'(1/z_1)=f(z_1) \tag{3.73}$$

it follows that such zeros of $z_1^{2M-1}g'(1/z_1)$ are also zeros of $f(z_1)$.

Accordingly, the highest common factor of $z^{2M-1}g'(1/z_1)$ and $f(z_1)$ can be found, and then its zero properties can in turn be studied, it too being self-inverse. Proceeding in this way will result in determining the number of zeros of $z_1^{2M-1}g'(1/z_1)$ and therefore of $z_1^M f(z_1)$ inside $|z_1|<1$, to conclude whether or not $f(z_1)$ is positive.

The Anderson and Jury method for testing the stability can be summarized as follows:

1. For checking condition (3.57):
   (i) Use the bilinear transformation of $D(z_1, 0)=0$ and apply the Hurwitz method to the transformed polynomial.
   (ii) An alternative test involves forming the Schur-Cohn matrix from $f(z_1, 0)=0$ and testing this for positiveness.

2. For checking condition (3.58) two successive tests as follows are needed:
   (i) Apply a Schur-Cohn test to $D(z_1, z_2)=0$ to get the self-inverse polynomials and check the positiveness of these polynomials.
   (ii) Second, the positiveness of a number of polynomials of $D(z_1, z_2)=0$ on $|z_1|=1$ should be checked.

3.8. THE MARIA AND FAHMY METHOD FOR TESTING STABILITY

A method has been introduced by Maria and Fahmy[15] for checking condition (3.57) of Huang's criterion. This method, based on modification of Jury's table,[1] is as stated below:

THEOREM 3.6. *Let $f(z)$ be the nth degree polynomial given by*

$$f(z) = a_0 + a_1 z + a_2 z^2 + \cdots + a_n z^n \tag{3.74}$$

where coefficients a_i, $i = 0, 1, \ldots, n$ are complex numbers. The roots of $f(z)$ are inside the unit circle if and only if

$$b_0 < 0, \qquad c_0 > 0, \qquad d_0 > 0, \ldots, g_0 > 0, \ldots, t_0 > 0 \tag{3.75}$$

where $b_0, c_0, \ldots, t_0$ are obtained from the modified Jury's table formed as follows:

$$\begin{array}{cccccc}
z^0 & z^1 & z^2 & \cdots & z^{n-2} & z^{n-1} & z^n \\
a_0 & a_1 & a_2 & \cdots & a_{n-2} & a_{n-1} & a_n \\
a_n^* & a_{n-1}^* & a_{n-2}^* & \cdots & a_2^* & a_1^* & a_0^* \\
b_0 & b_1 & b_2 & & b_{n-2} & b_{n-1} & \\
b_{n-1}^* & b_{n-2}^* & b_{n-3}^* & & b_1^* & b_0^* & \\
c_0 & c_1 & c_2 & & c_{n-2} & & \\
c_{n-2}^* & c_{n-3}^* & & & c_0^* & & \\
\vdots & & & & & & \\
r_0 & r_1 & & & & & \\
r_1^* & r_0^* & & & & & \\
t_0 & & & & & &
\end{array} \tag{3.76}$$

where

$$b_k = \begin{vmatrix} a_0 & a_{n-k} \\ a_n^* & a_k^* \end{vmatrix} \quad \textit{and} \quad c_k = \begin{vmatrix} b_0 & b_{n-1-k} \\ b_{n-1}^* & b_k^* \end{vmatrix} \tag{3.77}$$

and a_k^* *is the complex conjugate of* a_k.

To check the first condition (3.57) of Huang's theory using the above results, it should be noted that the condition is satisfied if and only if the roots of z_2 in $D(z_1, z_2)|_{|z_1|=1} = 0$ are inside the unit circle $|z_2| = 1$. To test this equivalent condition, $D(z_1, z_2)$ should be written in the form

$$D(z_1, z_2) = [a_N(z_1)]z_2^N + [a_{N-1}(z_1)]z_2^{N-1} + \cdots + [a_j(z_1)]z_2^j + \cdots + a_0(z_1) \tag{3.78}$$

where

$$a_j(z_1) = \sum_{i=0}^{M} b_{ij} z_1^i \tag{3.79}$$

In such a form $D(z_1, z_2)$ is viewed as a polynomial in the single variable z_2 with the coefficients being functions in z_1. The above table should be constructed for this $D(z_1, z_2) = 0$.

Computation of the entries of the modified Jury's table is considerably simplified because:

1. all the b_{ij} are real, and thus $a_j^*(z_1) = a_j(z_1^*)$,
2. z_1 is restricted to the boundary $|z_1| = 1$, and thus $(z_1 z_1^*)^l = 1$ for all l.

All the entries of the table will be functions of z_1 with $b_0, c_0, \ldots, g_0, \ldots, t_0$ assuming the form

$$g_0 = g_{00} + g_{01}(z_1^* + z_1) + g_{02}(z_1^{*2} + z_1^2) + \cdots \tag{3.80}$$

By setting $z_1 = X + jY$ and noting that $|z_1| = 1$, the coefficients $b_0, c_0, \ldots, g_0, \ldots, t_0$ can be expressed as functions of the real variable X by employing the following substitutions for $(z_1^{*k} + z_1^k)$, $k = 1, 2, \ldots$:

$$\begin{aligned}
(z_1^* + z_1) &= 2X \\
(z_1^{*2} + z_1^2) &= 4X^2 - 2 \\
(z_1^{*3} + z_1^3) &= 8X^3 - 6X \\
(z_1^{*4} + z_1^4) &= 16X^4 - 16X^2 + 2
\end{aligned} \tag{3.81}$$

Thus the test for condition (3.58) may be represented as a modified Jury's criterion, which takes the following form:

For $-1 \leq X \leq +1$,

$$b_0(X) < 0$$

$$c_0(X) > 0, \qquad d_0(X) > 0, \ldots, t_0(x) > 0$$

This criterion can be further simplified because

(i) $b_0(0) < 0$, $c_0(0), \ldots, t_0(0) > 0$, and

(ii) the polynomials $b_0(X), c_0(X), \ldots, t_0(X)$ have no real roots in the interval $|X| \leq 1$.

Sturm's test can be used to check condition (ii).

A perusal of all the above stability tests indicates that the techniques mentioned suffer from the high cost of computing and complexity in terms of handling for a general class of higher-order 2-D polynomials. However, Huang[9] in his paper has derived conditions on the coefficients for a special class of second-order polynomial to have all its zeros inside the unit circles in the z_1 and z_2 planes. Let us consider

$$D(z_1, z_2) = 1 + b_{10}z_1 + b_{01}z_2 + b_{11}z_1z_2 \tag{3.82}$$

in which case $D(z_1, z_2)$ has all its zeros inside the unit circles in the z_1 and z_2 planes iff:

(i) $$\left|\frac{1 - b_{10}}{b_{01} + b_{11}}\right| > 1 \tag{3.83a}$$

(ii) $$\left|\frac{1 + b_{10}}{b_{01} + b_{11}}\right| > 1 \tag{3.83b}$$

(iii) $$|b_{10}| < 1 \tag{3.83c}$$

3.9. STABILIZATION OF TWO-DIMENSIONAL RECURSIVE DIGITAL FILTERS

In the previous sections we have examined several methods for testing the stability of 2-D recursive digital filters. We will now focus our attention on techniques for the stabilization of an unstable 2-D recursive digital filter.

Unlike the 1-D case, where the stabilization is a straightforward procedure, in the 2-D case the problem is as yet unsolved due to the following reasons.

1. Factorization of polynomials of two variables is not in general possible, therefore the pole cancellation technique of the 1-D case cannot be extended to two dimensions.
2. Unlike the 1-D case, where the stability of a filter depends only on the location of its poles, in 2-D filters the numerator of the filter's transfer function can affect the stability of the filter. This problem has been treated thoroughly by Goodman.[16] We consider the following three transfer functions:

$$H_1(z_1, z_2) = \frac{1}{1 - 0.5z_1^{-1} - 0.5z_2^{-1}} \tag{3.84}$$

$$H_2(z_1, z_2) = \frac{(1 - z_1^{-1})^8(1 - z_2^{-1})^8}{1 - 0.5z_1^{-1} - 0.5z_2^{-1}} \tag{3.85}$$

and

$$H_3(z_1, z_2) = \frac{(1 - z_1^{-1})(1 - z_2^{-1})}{1 - 0.5z_1^{-1} - 0.5z_2^{-1}} \tag{3.86}$$

It is seen that in equation (3.84) function H_1 possesses a pole at $z_1 = z_2 = 1$ (this is called a nonessential singularity of the first kind) and its zeros at $z_1 = z_2 = \infty$. Functions H_2 and H_3, however, possess nonessential singularities of the second kind at $z_1 = z_2 = 1$ (meaning that for the same values of z_1 and z_2, both the numerator and denominator become identically zero).

Goodman has shown that $H_2(z_1, z_2)$ is stable while $H_3(z_1, z_2)$ is not, despite the fact that both functions have identical poles.
3. Nonessential singularities of the second kind in 1-D filters can be removed by pole and zero cancellation, while this is not possible in 2-D filters as can be easily seen from equation (3.86).

The above factors have stalled any further development in the area of stabilization of unstable 2-D filters, as many of the extensions of well-known stabilization techniques from one to two dimensions have been invalidated through counterexamples. Some of the techniques for stabilization of unstable 2-D filter will now be presented despite their invalidity, which has been proven by several researchers.

3.10. DOUBLE PLANAR LEAST-SQUARES INVERSE METHOD

This method is based on a conjecture of Shanks *et al.*[8] and is a direct extension of a well-established 1-D method[17] based on some properties of the planar least-squares inverse of a matrix. Before reviewing the method in detail, some useful definitions and preliminary theorems relevant to its understanding are given.

3.10.1. Definitions

1. A minimum-phase 1-D discrete sequence is one with a z transform having no zeros outside the unit circle in the z plane.
2. A minimum-phase 2-D discrete array $a(m, n)$ is defined as one that would satisfy both of the following two conditions:
   - When the spectrum $A(\omega_1, \omega_2)$ of the 2-D array $a(m, n)$ is evaluated at any real frequency ω_1, the resulting 1-D function in frequency ω_2 is minimum-phase.
   - The same spectrum $A(\omega_1, \omega_2)$ when evaluated at any real frequency ω_2 forms a 1-D minimum-phase function in ω_1.

The spectrum $A(\omega_1, \omega_2)$ is the 2-D discrete Fourier transform of the array $a(m, n)$. The above two conditions can be restated as the following theorem.

THEOREM 3.7. *A 2-D array $a(m, n)$, with z transform $A(z_1, z_2)$, is minimum-phase iff*

$$A(z_1, z_2) \neq 0 \qquad \textit{for} \bigcap_{i=1}^{2} |z_i| \geq 1$$

From the above theorem it can be concluded that the filter $F(z_1, z_2) = 1/A(z_1, z_2)$ is stable if $A(z_1, z_2)$ is minimum-phase. In 1-D filter theory it is known that the planar least-squares inverse of a filter is minimum-phase.[17] The object here is to find how to make use of this property for stabilizing an unstable 2-D recursive digital filter without changing its amplitude spectrum.

Given an array C, an array P can be found such that the convolution of C and P is approximately equal to the unit impulse array δ, that is,

$$C * P \cong \delta \tag{3.87}$$

where the symbol $*$ denotes (2-D) convolution. In general, it is not possible to make $C * P$ exactly equal to δ. Let $C * P = G$; if P is now chosen such that the sum of the squares of the elements of $\delta - G$ is minimized, then P is called a planar least-squares inverse (PLSI) of C.

The size of the array P is arbitrary. However, δ and G must have the same size, which depends on the size of the arrays C and P. If C is an array of dimension (m, n) and P of (k, l), then G and δ must be $(m + k - 1) \times (n + l - 1)$. Since the size of P is arbitrary, there are many PLSI of the array C, one for each possible set of dimensions of P. However, once the dimensions k and l of the matrix P are fixed, there is one and only one array P that minimizes the mean-square difference between G and δ. This array P is determined from the matrices δ and C by the (2-D) Wiener technique.[18]

As a group PLSIs have some interesting properties. One particular such property, which has not yet been proved, is described by the following conjecture.

CONJECTURE 3.1. Given an arbitrary real finite array C, any planar least-squares inverse of C is minimum-phase.

This is an important conjecture, because it implies that the filter $F(z_1, z_2) = 1/D(z_1, z_2)$ is stable when $D(z_1, z_2)$ is PLSI.

When a (2-D) recursive digital filter $F(z_1, z_2) = 1/D(z_1, z_2)$ is found to be unstable after completing the design, the question arises as to whether the coefficients of the filter's denominator $D(z_1, z_2)$ could be altered in order to produce a final stable filter. The denominator array of the unstable filter is denoted by D. Quantity D', a planar least-squares inverse of D, can be formed. Further, a planar least-squares inverse of D' can also be formed and called D''. Now, D'' is the inverse of the inverse, or the "double inverse" of D. Intuitively, D'' and D will have some characteristics in common. Moreover, D'' is itself a PLSI. Hence it is minimum-phase, which D is not. Therefore, the filter $F(z_1, z_2) = 1/D''(z_1, z_2)$ is conjectured to be stable.

The validity of assuming some degree of similarity between D'' and D is, of course, questionable. It has been so far taken for granted that D'', being the double PLSI of D, is an approximate minimum-phase version of the latter. Thus the amplitude spectra of D'' and D would be expected to be roughly equal.

One of the factors governing the quality of the approximation is the size of the intermediate array D'. The larger the size of D', the better is the resemblance of D'' to the minimum-phase-version array D. Several examples, where the stabilized filters so obtained were a good approximation to the original unstable ones, can be found.[8] They cover a number of cases involving filters of different classes and degrees. However, Genin and Kamp,[19] among others, have given a counterexample that invalidated the method.

3.11. STABILIZATION VIA TWO-DIMENSIONAL DISCRETE HILBERT TRANSFORMS

Read and Treitel[4] have extended the one-dimensional stabilization technique, which was based on the discrete Hilbert transform relation, to two dimensions. Further treatment of the technique necessitates the reader to be familiar with some alternative definitions of a few, already defined functions, as given below.

A finite discrete impulse response function $d(m, n)$ is causal if

$$d(m, n) = 0 \qquad \text{for} \begin{cases} n \geq M/2 \\ n \geq N/2 \end{cases} \tag{3.88}$$

where m varies over the discrete set $\{0, 1, \ldots, M-1\}$ and n over the set $\{0, 1, \ldots, N-1\}$. The even and odd parts of such a sequence are defined as

$$d_e(m, n) = \tfrac{1}{2}[d(m, n) + d(M-m, N-n)] \tag{3.89}$$

and

$$d_o(m, n) = \tfrac{1}{2}[d(m, n) - d(M-m, N-n)] \tag{3.90}$$

respectively, and are related by

$$d_o(m, n) = [\text{sgn}(m, n) + \text{bdy}(m, n)]d_e(m, n) \tag{3.91}$$

where the sgn function is a finite 2-D version of the 1-D signum function and is given by

$$\text{sgn}(m, n) = \begin{cases} 1 & 0 < m < M/2 \text{ and } 0 < n < N/2 \\ -1 & M/2 < m < M \text{ and } N/2 < n < N \\ 0 & \text{elsewhere} \end{cases} \tag{3.92}$$

The bdy function makes boundary adjustments and is defined by

$$\text{bdy}(m, n) = \begin{cases} 1 & n = 0 \text{ and } 0 < m < M/2 \\ -1 & n = 0 \text{ and } M/2 < m < M \\ 1 & m = 0 \text{ and } 0 < n < N/2 \\ -1 & m = 0 \text{ and } N/2 < n < N \\ 0 & \text{elsewhere} \end{cases} \tag{3.93}$$

The sequence $d(m, n)$ is the sum of its even and odd parts:

$$d(m, n) = d_e(m, n) + d_o(m, n) \tag{3.94}$$

Taking the DFT of both sides yields

$$\text{DFT}[d(m, n)] = \text{DFT}[d_e(m, n)] + \text{DFT}[d_o(m, n)] \tag{3.95}$$

and, as is well known, the DFT of a real and even function is real and even, and that of a real and odd function is odd and imaginary (see Table 3.2 above). Thus if we write

$$\mathrm{DFT}[d(m, n)] = \mathrm{PR}(m, n) + \mathrm{j}\,\mathrm{PI}(m, n)$$

we have

$$\mathrm{PR}(m, n) = \mathrm{DFT}[d_e(m, n)] \tag{3.96}$$

and

$$\mathrm{PI}(m, n) = -\mathrm{j}\,\mathrm{DFT}[d_o(m, n)] \tag{3.97}$$

On taking the IDFT of both sides of equation (3.96), substituting into expression (3.91), and using equation (3.97), we obtain

$$\mathrm{PI}(m, n) = -\mathrm{j}\,\mathrm{DFT}\{[\mathrm{sgn}(m, n) + \mathrm{bdy}(m, n)]\mathrm{IDFT}[\mathrm{PR}(m, n)]\} \tag{3.98}$$

This relation defines the 2-D discrete Hilbert transform.

It will now be shown how to use the discrete Hilbert transform of equation (3.98) to obtain a minimum-phase version of a 2-D array. The z transform of such an array could be the polynomial $D(z_1, z_2)$ in the equation $H(z_1, z_2) = N(z_1, z_2)/D(z_1, z_2)$.

Application of the scheme leads to a rational filter transfer function $H(z_1, z_2)$ with very nearly the same amplitude spectrum as the original unstable filter. To arrive at a stable result, the phase of the 2-D denominator polynomial $D(z_1, z_2)$ should be minimized. The 2-D discrete Hilbert transform can be applied in the same way as in the 1-D case to obtain the minimum-phase version of a given array. Thus, from the 2-D amplitude spectrum $D(z_1, z_2)$ of a causal sequence, the minimum-phase spectrum $\theta(m, n)$ is calculated from the equation

$$\theta(m, n) = -\mathrm{j}\,\mathrm{DFT}\{[\mathrm{sgn}(m, n) + \mathrm{bdy}(m, n)]\mathrm{IDFT}[\log|D(z_1, z_2)|]\} \tag{3.99}$$

The derivation of the above expression can be found elsewhere.[4] The formation of a minimum-phase version of an array by employing equation (3.99) can be summarized by the steps below:

Step 1. Given a finite discrete 2-D array, the coefficient array should be augmented with zeros to satisfy the condition for causality, equation (3.88). The added zeros increase the size of the array, so that it becomes amenable to fast Fourier transform analysis.

Step 2. The natural logarithm of the amplitude spectrum of the augmented (2-D) array should be calculated.

Step 3. The 2-D discrete Hilbert transform must be applied to this 2-D array. Thus the log of the magnitude is treated as the real part and the discrete Hilbert transform then yields the imaginary part.

Step 4. The imaginary part is used as the phase spectrum corresponding to the given amplitude spectrum. These two spectral characteristics completely describe the transform of the minimum-phase array.

Step 5. After conversion from amplitude and phase to real and imaginary parts, the inverse transform is determined and truncated to obtain the same dimensions as the original array. This yields the minimum-phase version of the original array.

Although the discrete Hilbert transform procedure works for most examples, it has been shown[20] that there exist some cases where it proves to be of no value.

3.12. STABILIZATION OF TWO-DIMENSIONAL ZERO-PHASE FILTERS

The stabilization techniques studied so far cause a modification of the transfer function of the network in such a manner that the amplitude response is kept approximately unchanged while the phase response is adjusted to ensure stability of the modified transfer function. Such procedures are thus not applicable to 2-D zero-phase functions which do not permit phase modification in any manner that would improve the stability.

In one dimension, zero-phase functions can only represent (apart from a trivial case) noncausal sequences. It has been shown that such functions may be realized by processing data, first in the positive sense along the axis and cascading this with a processor working in the negative sense. As noncausal sequences do not exist in the time domain, this presents no problem; if the data are distributed in a single spatial dimension, they may be stored and processed recursively (or nonrecursively) if required in any manner demanded.

With this in mind, the approach used by Pistor[21] in two dimensions is to decompose the array representing the impulse response of the 2-D filter into four single quadrant arrays, ${}^{1}f$, ${}^{2}f$, ${}^{3}f$, and ${}^{4}f$, each recursing in a different direction as discussed in Section 1.2.1. The output array for an arbitrary input array may be obtained by convolving the input successively with the four quadrant arrays in the appropriate directions as shown in Figure 3.3.

The technique may be applied to any unstable filter function but in its simplest form, and probably most useful application, it is used with zero-phase functions.

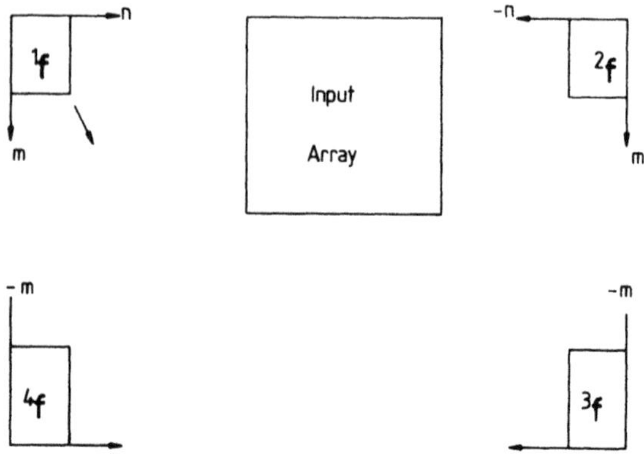

Figure 3.3. Two-dimensional factorization into four single quadrant arrays.

Let us consider a real-valued discrete function $d(m, n)$ with a limited number of sample points, in which

$$d = \{d(m, n)\} \qquad \begin{cases} |m| \leq 2M = \alpha_1 \\ |n| \leq 2N = \alpha_2 \end{cases} \tag{3.100}$$

the z transform of which is zero-phase and nonnegative for all $(z_1, z_2) \in R_2$ where

$$R_2 = \{(z_1, z_2): |z_1| = 1 \cap |z_2| = 1\} \tag{3.101}$$

This satisfies

$$\text{Im}[D'(\omega_1, \omega_2)] = 0 \tag{3.102}$$

and

$$\text{Re}[D'(\omega_1, \omega_2)] > 0 \tag{3.103}$$

where, assuming frequency is normalized to $\omega_s = 2\pi$,

$$\begin{aligned} D'(\omega_1, \omega_2) &= \sum_{m=-\alpha_1}^{\alpha_1} \sum_{n=-\alpha_2}^{\alpha_2} d(m, n) \exp[-\text{j}2\pi(m\omega_1 + n\omega_2)] \\ &= D(z_1, z_2) \qquad \text{for } (z_1, z_2) \in R_2 \end{aligned} \tag{3.104}$$

Equations (3.102) and (3.104) imply central symmetry of d:

$$d(m, n) = d(-m, -n) \tag{3.105}$$

Owing to this symmetry, d is not a one-quadrant function. However, because d has a finite number of sample points, it can be transformed by translation to any single quadrant function. Thus, the term $1/D(z_1, z_2)$ could be associated with four different recursive filters. None, however, would be stable. It is desired that the unstable $1/D(z_1, z_2)$ be decomposed into four stable filters that recurse in four different directions:

$$\frac{1}{D(z_1, z_2)} = \prod_{i=1}^{4} 1/k_i(z_1, z_2) \qquad (3.106)$$

This is done by transforming equation (3.106) into the cepstrum domain:

$$\hat{d}(m, n) = \sum_{i=1}^{4} {}^{i}\hat{k}(m, n) \qquad (3.107)$$

where $\hat{d}$ is the cepstrum of d, defined as a function whose z transform $\hat{D}(z_1, z_2)$ is derived by taking the logarithm of $D(z_1, z_2)$, the z transform of $d(m, n)$.

The z transform of the cepstrum $d(m, n)$ is readily obtained by evaluating the z transform of $d(m, n)$ on R_2 as given by equation (3.101) and taking the logarithm:

$$\hat{D}'(\omega_1, \omega_2) = \ln\left\{\sum_{m=-2M}^{2M} \sum_{n=-2N}^{2N} d(m, n) \exp[-j2\pi(m\omega_1 + n\omega_2)]\right\} \qquad (3.108)$$

Since D' is real and positive as in equations (3.102) and (3.103), the logarithm in equation (3.108) presents no problem. It is easily verified that all partial derivatives of $\hat{D}'(\omega_1, \omega_2)$ can be given by the Fourier series expansion

$$\hat{D}'(\omega_1, \omega_2) = \sum_{m=-\infty}^{\infty} \sum_{n=-\infty}^{\infty} \hat{d}(m, n) \exp - [j2\pi(m\omega_1 + n\omega_2)] \qquad (3.109)$$

The coefficients $\hat{d}(m, n)$ that make up the cepstrum $\hat{d}$ are obtained from the inverse Fourier transform

$$\hat{d}(m, n) = \int_0^1 \int_0^1 \hat{D}'(\omega_1, \omega_2) \exp[j2\pi(m\omega_1 + n\omega_2)]\, d\omega_1\, d\omega_2 \qquad (3.110)$$

From the properties of $\hat{D}'(\omega_1, \omega_2)$ and from equation (3.110), it follows that

$$\hat{d}(m, n) = \hat{d}(-m, -n) \qquad (3.111)$$

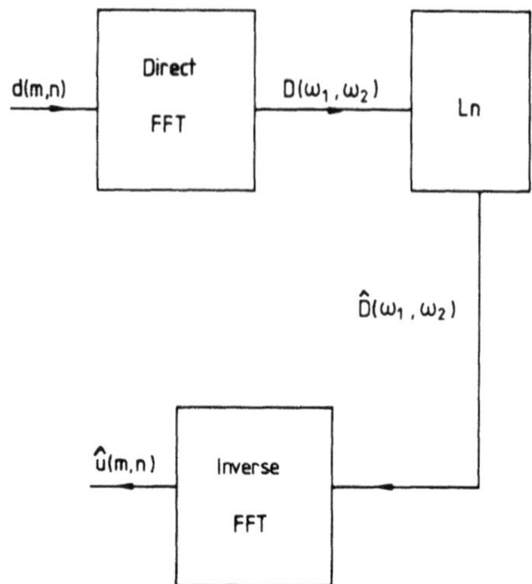

Figure 3.4. Implementation of the Pistor stabilization method.

The procedure can be implemented by determining $d(m, n)$, from equations (3.104), (3.108), and (3.110). A flow chart of this procedure is shown in Figure 3.4, from which it is seen that functions $d(m, n)$ and $\hat{d}(m, n)$ are IDFTs of a sampled version of $D'(\omega_1, \omega_2)$ and $\hat{D}'(\omega_1, \omega_2)$. Function $\hat{u}(m, n)$ is defined as the aliased version of the desired result $d(m, n)$, the degree of aliasing being controlled by the rate at which $D'(\omega_1, \omega_2)$ is sampled.

The function $\hat{u}(m, n)$, truncated at the Nyquist subscripts I_m and I_n, is decomposed according to

$$\hat{u}_{\text{tr}} = {}^1\hat{k}' + {}^2\hat{k}' + {}^3\hat{k}' + {}^4\hat{k}' \tag{3.112}$$

where quantities $\hat{k}'$ are approximations to the function $\hat{k}$ in equation (3.107).

3.13. GENERATION OF TWO-DIMENSIONAL DIGITAL FUNCTIONS WITHOUT NONESSENTIAL SINGULARITIES OF THE SECOND KIND

In the previous sections, different procedures for testing the stability of the 2-D filters were discussed. Most of these test procedures were found to be complex and cumbersome for computer implementation besides lacking a solid theoretical background and procedure for stabilizing an unstable 2-D filter. This has inhibited the widespread application of recursive filters, despite their significant advantages over nonrecursive filters.

A straightforward and successful method of generating a 2-D stable digital function has been proposed in which a double bilinear transformation is applied to a two-variable analogue function possessing a strictly Hurwitz polynomial denominator.[22]

However, it has recently been shown by Goodman[23] that not all two-variable analogue functions with strictly Hurwitz polynomial denominators yield stable 2-D digital transfer functions upon bilinear transformation. There exist analogue functions which, on bilinear transformation, result in digital functions possessing singularities of the second kind on the distinguished boundary of the unit circles in the (z_1, z_2) plane. In this section we present several methods of generating 2-D strictly Hurwitz polynomials free of nonessential singularities of the second kind. But first we explain several definitions and certain mathematical preliminaries.

3.13.1. Value of a Function at Infinity

The 2-D biplane (s_1, s_2) consists of two complex planes s_1 and s_2 and an infinite distant point which can have infinite coordinates in any one or both of these planes, and so there exist an infinite number of points at infinity. They can be classified into three categories as follows:

Category 1. $s_1 = \infty$ and $s_2 =$ finite.
Category 2. $s_1 =$ finite and $s_2 = \infty$.
Category 3. $s_1 = \infty$ and $s_2 = \infty$.

By applying the transformation method to each variable, the value of the function at each of the above points is defined as

$$f(\infty, s_2) = \operatorname*{Lim}_{s_1 \to 0} f(1/s_1, s_2) \qquad \text{Category 1}$$

$$f(s_1, \infty) = \operatorname*{Lim}_{s_2 \to 0} f(s_1, 1/s_2) \qquad \text{Category 2}$$

$$f(\infty, \infty) = \operatorname*{Lim}_{\substack{s_1 \to 0 \\ s_2 \to 0}} f(1/s_1, 1/s_2) \qquad \text{Category 3}$$

3.13.2. Singularities

It is well known that a two-variable rational function $H(s_1, s_2) = A(s_1, s_2)/B(s_1, s_2)$ may possess two types of singularities and they may be defined as follows (essential singularities cannot occur in rational functions):

1. *Nonessential singularities of the first kind.* $H(s_1, s_2)$ is said to possess a nonessential singularity of the first kind at (s_1^*, s_2^*) if $B(s_1^*, s_2^*) = 0$ and $A(s_1^*, s_2^*) \neq 0$.

2. *Nonessential singularities of the second kind.* $H(s_1, s_2)$ is said to possess a nonessential singularity of the second kind at (s_1^*, s_2^*) if $A(s_1^*, s_2^*) = B(s_1^*, s_2^*) = 0$.

Goodman[23] has pointed out some difficulties with the generation of stable 2-D digital functions using double bilinear transformation. However, it may be verified that this maps the entire s_1 and s_2 planes on the entire z_1 and z_2 planes on a one-to-one basis and so the behavior of the function is unaltered by the application of the double bilinear transformation. For instance, if $H(z_1, z_2)$ is the transformed function of $H(s_1, s_2)$ and $H(z_1, z_2)$ has a singularity at (z_1^*, z_2^*), then $H(s_1, s_2)$ ought to have the same type of singularity at (s_1^*, s_2^*) where (s_1^*, s_2^*) is the point corresponding to (z_1^*, z_2^*). This simply means that an analogue function with strict Hurwitz polynomial denominators also possesses the second-kind singularities. This is not surprising in view of the fact that a two-variable polynomial may possess, in addition to zeros and poles, singularities of the second kind at infinite-distance points. In the definition of strict Hurwitz polynomials these second-kind singularities are not taken into account. Naturally, one obvious remedy for this problem is then to define a Hurwitz polynomial which avoids the nonessential singularities of the second kind. This we do in the next section after considering the various types of two-variable Hurwitz polynomial discussed in the literature.

3.13.3. Two-Variable Hurwitz Polynomials

There are four types of Hurwitz polynomials and their definitions are stated below in a slightly different form in terms of singularities rather than zeros, as has been the common practice. This has been done so as to facilitate a uniform definition for all the four types of polynomials, differing only in the region of analyticity. In the following definitions $D(s_1, s_2)$ is a polynomial in s_1 and s_2 and $\text{Re}(s)$ refers to the real part of s.

Definition 1. $D(s_1, s_2)$ is a broad sense Hurwitz polynomial (BHP) if $1/D(s_1, s_2)$ does not possess any singularities in the region

$$\{(s_1, s_2) \mid \text{Re}(s_1) > 0, \text{Re}(s_2) > 0, |s_1| < \infty, \text{ and } |s_2| < \infty\}$$

Definition 2. $D(s_1, s_2)$ is a narrow sense Hurwitz polynomial (NHP) if $1/D(s_1, s_2)$ does not possess any singularities in the region

$$\begin{aligned}&\{(s_1, s_2) \mid \text{Re}(s_1) > 0, \text{Re}(s_2) > 0, |s_1| < \infty, \text{ and } |s_2| < \infty\}\\ &\cup \{(s_1, s_2) \mid \text{Re}(s_1) = 0, \text{Re}(s_2) > 0, |s_1| \le \infty \text{ and } |s_2| < \infty\}\\ &\cup \{(s_1, s_2) \mid \text{Re}(s_1) > 0, \text{Re}(s_2) = 0, |s_1| < \infty \text{ and } |s_2| \le \infty\}\end{aligned}$$

DEFINITION 3. $D(s_1, s_2)$ is a strict Hurwitz polynomial (SHP) if $1/D(s_1, s_2)$ does not possess any singularities in the region

$$\{(s_1, s_2) \mid \text{Re}(s_1) \geq 0, \text{Re}(s_2) \geq 0, |s_1| < \infty, \text{ and } |s_2| < \infty\}$$

DEFINITION 4. $D(s_1, s_2)$ is a very strict Hurwitz polynomial (VSHP) if $1/D(s_1, s_2)$ does not possess any singularities in the region

$$\{(s_1, s_2) \mid \text{Re}(s_1) \geq 0 \text{ and } \text{Re}(s_2) \geq 0\}$$

It may be noted that each successive condition above includes the previous conditions, and so each successive class of Hurwitz polynomials is a subclass of the earlier classes. From experience of 2-D digital filters we may state that a denominator polynomial must be SHP as a necessary and VSHP as a sufficient condition to ensure stability.

We next give a few examples[24] to distinguish the various classes.

EXAMPLE 3.2. $D_1(s_1, s_2) = s_1 + s_1 s_2$. It can be verified that $1/D_1(s_1, s_2) = 1/(s_1 + s_1 s_2)$ does not have any singularity in the region specified in Definition 1. However, it has a first-kind singularity at $s_1 = 0$ for any s_2. Hence $s_1 + s_1 s_2$ is a BHP.

EXAMPLE 3.3. $D_2(s_1, s_2) = 1 + s_1 s_2$. $D_2(s_1, s_2)$ is a NHP.

EXAMPLE 3.4. $D_3(s_1, s_2) = 1 + s_1 + s_2$ is a SHP. However, it is not a VSHP because at $s_1 = s_2 = \infty$ we have $D_3(\infty, \infty) = 0/0$ and $D_3(s_1, s_2)$ has a second-kind singularity at (∞, ∞). In a similar way one can show that $1 + s_1 + s_2 s_1$ and $1 + s_2 + s_1 s_2$ are SHPs but not VSHPs.

EXAMPLE 3.5. $D_4(s_1, s_2) = 1 + s_1 + s_2 + s_1 s_2$ can be shown to be VSHP.

Procedures to test whether a given two-variable polynomial is a BHP, NHP, or SHP have been well documented.[9,10] A test to ascertain whether a SHP is also a VSHP is given elsewhere,[24] as follows:

Step 1. Test $D_1(s_1, s_2)$ to see whether it is a SHP. For this purpose, the procedure suggested by Ansell[10] and modified by Huang[9] can be used. If the given polynomial is a SHP, then proceed to Step 2. Otherwise $D(s_1, s_2)$ is not a VSHP.

Step 2. Check whether the following conditions are satisfied:

- $\lim_{s_1 \to 0} D(1/s_1, s_2) \neq 0/0$ for $\text{Re}(s_2) = 0$.
- $\lim_{s_2 \to 0} D(s_1, 1/s_2) \neq 0/0$ for $\text{Re}(s_1) = 0$.
- $\lim_{\substack{s_1 \to 0 \\ s_2 \to 0}} D(1/s_1, 1/s_2) \neq 0/0$.

If these three conditions are satisfied, $D(s_1, s_2)$ is a VSHP. Otherwise it is not.

3.13.4. Generation of a Very Strict Hurwitz Polynomial using Properties of the Derivatives of Even or Odd Parts of Hurwitz Polynomials[25]

The method consists of the following steps:

1. A suitable even or odd part of an n-variable Hurwitz polynomial is generated.
2. The corresponding derivative giving the odd or even part is associated with it.
3. The resulting n-variable Hurwitz polynomial is converted to a two-variable VSHP.[24]

It will be seen later that a large number of possibilities exist. Two cases are discussed below:

Case A: *The even part of a Hurwitz polynomial as the starting point*
Consider the polynomial M_{2n} given by

$$M_{2n} = \det|\mu I_{2n} + A_{2n}| \tag{3.113}$$

where μ is a diagonal matrix of order $2n$ given by

$$\mu = \text{diag}[\mu_1, \mu_2, \mu_3, \ldots, \mu_{2n}] \tag{3.114}$$

and A_{2n} is a skew-symmetric matrix of order $2n$ given by

$$A_{2n} \triangleq A = \begin{bmatrix} 0 & a_{12} & a_{13} & \cdots & a_{1,2n} \\ -a_{12} & 0 & a_{23} & \cdots & a_{2,2n} \\ -a_{13} & -a_{23} & 0 & \cdots & \\ \vdots & & & & \\ -a_{1,2n} & -a_{2,2n} & \cdots & \cdots & 0 \end{bmatrix} \tag{3.115}$$

From the diagonal expansion of the determinant of a matrix, M_{2n} can be written as

$$M_{2n} = \det A + \sum_{1\le i_1<i_2\le 2n} \mu_{i_1}\mu_{i_2}A_{i_1 i_2} + \sum_{1\le i_1<i_2<i_3<i_4\le 2n} \mu_{i_1}\mu_{i_2}\mu_{i_3}\mu_{i_4}A_{i_1 i_2 i_3 i_4} + \cdots + \mu_1\mu_2\mu_3\cdots\mu_{2n} \tag{3.116}$$

where $A_{i_1 i_2}$ is the determinant of the submatrix of A obtained by deleting both the i_1th and i_2th rows and columns and is of order $2n-2$; $A_{i_1 i_2 i_3 i_4}$ is the determinant of the submatrix of A obtained by deleting the i_1th, i_2th, i_3th, and i_4th rows and columns and is of order $2n-4$; and so on. The following properties should be noted:

1. All odd-order terms of the type μ_{i_1} and $\mu_{i_1}\mu_{i_2}\mu_{i_3}$, etc., are absent, since the determinant of an odd-order skew-symmetric matrix is zero.
2. The degree of any μ_i $(i = 1, 2, \ldots, 2n)$ is unity.
3. The quantities $\det A$, $A_{i_1 i_2}$, $A_{i_1 i_2 i_3 i_4}$, etc., are nonnegative numbers, since the determinant of an even-order skew-symmetric matrix is a perfect square.

Since the matrix $[\mu I_{2n} + A_{2n}]$ is always physically realizable, M_{2n} represents the even part of a $2n$-variable Hurwitz polynomial. Therefore, $(\partial M_{2n}/\partial \mu_i)/M_{2n}$ is a reactance function. As a consequence

$$M' = M_{2n} + \sum_{j=1}^{2n} K_j \frac{\partial M_{2n}}{\partial \mu_j} \tag{3.117}$$

is a $2n$-variable Hurwitz polynomial, where K_j is a real nonnegative constant.

From equation (3.117), a two-variable VSHP can be generated by equating some of the μ_is to s_1 and the rest to s_2, while ensuring that the conditions of a two-variable VSHP are satisfied. Also, there will be a large number of possibilities, which can be seen from the following example:

EXAMPLE 3.6. We consider

$$M_4 = |\mu I_4 + A_4| = \begin{vmatrix} \mu_1 & a_{12} & a_{13} & a_{14} \\ -a_{12} & \mu_2 & a_{23} & a_{24} \\ -a_{13} & -a_{23} & \mu_3 & a_{34} \\ -a_{14} & -a_{24} & -a_{34} & \mu_4 \end{vmatrix} \tag{3.118}$$

With the aid of equation (3.116) we obtain

$$\begin{aligned} M_4 = {} & \begin{vmatrix} 0 & a_{12} & a_{13} & a_{14} \\ -a_{12} & 0 & a_{23} & a_{24} \\ -a_{13} & -a_{23} & 0 & a_{34} \\ -a_{14} & -a_{24} & -a_{34} & 0 \end{vmatrix} + \mu_1\mu_2 \begin{vmatrix} 0 & a_{34} \\ -a_{34} & 0 \end{vmatrix} \\ & + \mu_1\mu_3 \begin{vmatrix} 0 & a_{24} \\ -a_{24} & 0 \end{vmatrix} + \mu_1\mu_4 \begin{vmatrix} 0 & a_{23} \\ -a_{23} & 0 \end{vmatrix} + \mu_2\mu_3 \begin{vmatrix} 0 & a_{14} \\ -a_{14} & 0 \end{vmatrix} \\ & + \mu_2\mu_4 \begin{vmatrix} 0 & a_{13} \\ -a_{13} & 0 \end{vmatrix} + \mu_3\mu_4 \begin{vmatrix} 0 & a_{12} \\ -a_{12} & 0 \end{vmatrix} + \mu_1\mu_2\mu_3\mu_4 \end{aligned} \tag{3.119}$$

Expansion of this equation yields

$$M_4 = \det A_4 + a_{34}^2\mu_1\mu_2 + a_{24}^2\mu_1\mu_3 + a_{23}^2\mu_1\mu_4 + a_{14}^2\mu_2\mu_3 + a_{13}^2\mu_2\mu_4 + a_{12}^2\mu_3\mu_4 + \mu_1\mu_2\mu_3\mu_4 \tag{3.120}$$

where

$$\det A_4 = (a_{12}a_{34} - a_{13}a_{24} + a_{14}a_{23})^2 \tag{3.121}$$

Now, relations (3.117) and (3.120) enable us to obtain the four-variable Hurwitz polynomial in the form

$$\begin{aligned} M' = {} & [\det A_4 + a_{34}^2\mu_1\mu_2 + a_{24}^2\mu_1\mu_3 + a_{23}^2\mu_1\mu_4 + a_{14}^2\mu_2\mu_3 \\ & + a_{13}^2\mu_2\mu_4 + a_{12}^2\mu_3\mu_4 + \mu_1\mu_2\mu_3\mu_4] \\ & + K_1[a_{34}^2\mu_2 + a_{24}^2\mu_3 + a_{23}^2\mu_4 + \mu_2\mu_3\mu_4] \\ & + K_2[a_{34}^2\mu_1 + a_{14}^2\mu_3 + a_{13}^2\mu_4 + \mu_1\mu_3\mu_4] \\ & + K_3[a_{24}^2\mu_1 + a_{14}^2\mu_2 + a_{12}^2\mu_4 + \mu_1\mu_2\mu_4] \\ & + K_4[a_{23}^2\mu_1 + a_{13}^2\mu_2 + a_{12}^2\mu_3 + \mu_1\mu_2\mu_3] \end{aligned} \tag{3.122}$$

On setting $\mu_1 = \mu_2 = s_1$ and $\mu_3 = \mu_4 = s_2$, we get finally

$$\begin{aligned} D(s_1, s_2) = {} & s_1^2[s_2^2 + s_2(K_3 + K_4) + a_{34}^2] \\ & + s_1[s_2^2(K_1 + K_2) + s_2(a_{24}^2 + a_{23}^2 + a_{13}^2 + a_{14}^2) \\ & \quad + a_{34}^2(K_1 + K_2) + K_3(a_{24}^2 + a_{14}^2) + K_4(a_{23}^2 + a_{13}^2)] \\ & + [a_{12}^2 s_2^2 + s_2[K_1(a_{24}^2 + a_{23}^2) + K_2(a_{14}^2 + a_{13}^2) + a_{12}^2(K_3 + K_4)] \\ & \quad + (a_{12}a_{34} - a_{13}a_{24} + a_{14}a_{23})^2] \end{aligned} \tag{3.123}$$

This is a VSHP. However, it is seen that the VSHP property is not destroyed if $a_{24} = 0$ or $a_{13} = 0$ or $a_{23} = a_{24} = 0$. Many other combinations are possible. It is also noteworthy that certain variables like a_{12} and a_{34} cannot be equated to zero.

A similar treatment can be applied to any K-variable VSHP, and it can be shown that a large number of variations are possible as the order of M increases.

Case B: The odd part of a Hurwitz polynomial as the starting point
Consider the polynomial N_{2n+1} given by

$$N_{2n+1} = \det|\mu I_{2n+1} + A_{2n+1}| \tag{3.124}$$

where μ is a diagonal matrix of order $2n+1$ given by

$$\mu = \text{diag}[\mu_1, \mu_2, \ldots, \mu_{2n+1}] \tag{3.125}$$

and A_{2n+1} is a skew-symmetric matrix of order $2n+1$ given by

$$A_{2n+1} \triangleq A = \begin{bmatrix} 0 & a_{12} & a_{13} & \cdots & a_{1,2n+1} \\ -a_{12} & 0 & a_{23} & \cdots & a_{2,2n+1} \\ -a_{13} & -a_{23} & 0 & \cdots & a_{3,2n+1} \\ \vdots & & & & \\ -a_{1,2n+1} & -a_{2,2n+1} & -a_{3,2n+1} & \cdots & 0 \end{bmatrix} \tag{3.126}$$

From the diagonal expansion of the determinant of a matrix, N_{2n+1} can be written as

$$N_{2n+1} = \sum_{1 \le i_1 \le 2n+1} \mu_{i_1} A_{i_1} + \sum_{1 \le i_1 < i_2 < i_3 \le 2n+1} \mu_{i_1}\mu_{i_2}\mu_{i_3} A_{i_1 i_2 i_3} + \cdots + \mu_1\mu_2\mu_3 \cdots \mu_{2n+1} \tag{3.127}$$

where A_{i_1} is the determinant of the submatrix of A obtained by deleting the i_1th row and column and is of order $2n$; $A_{i_1 i_2 i_3}$ is the determinant of the submatrix of A obtained by deleting the i_1th, i_2th, and i_3th rows and columns and is of order $2n-2$; and so on. The following properties should be noted:

1. All even-order terms of the type $\mu_{i_1}\mu_{i_2}$ and $\mu_{i_1}\mu_{i_2}\mu_{i_3}\mu_{i_4}$, etc., are absent, since the determinant of an odd-order skew-symmetric matrix is zero.
2. The degree of any μ_i $(i = 1, 2, \ldots, 2n+1)$ is unity.
3. The quantities A_{i_1}, $A_{i_1 i_2 i_3}$, etc., are nonnegative numbers, since the determinant of an even-order skew-symmetric matrix is a perfect square.

Since the matrix $[\mu I_{2n+1} + A_{2n+1}]$ is always physically realizable, N_{2n+1} represents the odd part of a $(2n+1)$-variable Hurwitz polynomial. Therefore, $(\partial N_{2n+1}/\partial \mu_i)/N_{2n+1}$ is a reactance function. As a consequence

$$N' = N_{2n+1} + \sum_{j=1}^{2n+1} K_j \frac{\partial N_{2n+1}}{\partial \mu_j} \tag{3.128}$$

is a $(2n+1)$-variable Hurwitz polynomial.

From equation (3.128), a two-variable VSHP can be obtained by equating some of the μ_is to s_1 and the rest to s_2, while ensuring that the conditions of a VSHP are satisfied. As can be anticipated, there will be a large number of possibilities, which can be seen from the following example:

EXAMPLE 3.7. We consider

$$N_5 = |\mu I_5 + A_5| = \begin{vmatrix} \mu_1 & a_{12} & a_{13} & a_{14} & a_{15} \\ -a_{12} & \mu_2 & a_{23} & a_{24} & a_{25} \\ -a_{13} & -a_{23} & \mu_3 & a_{34} & a_{35} \\ -a_{14} & -a_{24} & -a_{34} & \mu_4 & a_{45} \\ -a_{15} & -a_{25} & -a_{35} & -a_{45} & \mu_5 \end{vmatrix}$$

$$\begin{aligned} &= \mu_1\mu_2\mu_3\mu_4\mu_5 + \mu_1\mu_2\mu_3 a_{45}^2 + \mu_1\mu_2\mu_4 a_{35}^2 + \mu_1\mu_2\mu_5 a_{34}^2 \\ &\quad + \mu_1\mu_3\mu_4 a_{25}^2 + \mu_1\mu_3\mu_5 a_{24}^2 + \mu_1\mu_4\mu_5 a_{23}^2 \\ &\quad + \mu_2\mu_3\mu_4 a_{15}^2 + \mu_2\mu_3\mu_5 a_{14}^2 + \mu_2\mu_4\mu_5 a_{13}^2 + \mu_3\mu_4\mu_5 a_{12}^2 \\ &\quad + \mu_1(a_{23}a_{45} - a_{24}a_{35} + a_{25}a_{34})^2 + \mu_2(a_{13}a_{45} - a_{14}a_{35} + a_{15}a_{34})^2 \\ &\quad + \mu_3(a_{12}a_{45} - a_{14}a_{25} + a_{15}a_{24})^2 + \mu_4(a_{12}a_{35} - a_{13}a_{25} + a_{15}a_{23})^2 \\ &\quad + \mu_5(a_{12}a_{34} - a_{13}a_{24} + a_{14}a_{23})^2 \end{aligned} \tag{3.129}$$

On setting $\mu_1 = \mu_2 = s_1$ and $\mu_3 = \mu_4 = \mu_5 = s_2$, we obtain finally from equation (3.129)

$$\begin{aligned} D(s_1, s_2) = s_1^2[&s_2^3 + s_2^2(K_3 + K_4 + K_5) + s_2(a_{34}^2 + a_{35}^2 + a_{45}^2) \\ &+ (K_3 a_{45}^2 + K_4 a_{35}^2 + K_5 a_{34}^2)] \\ &+ s_1\{s_2^3(K_1 + K_2) + s_2^2(a_{13}^2 + a_{14}^2 + a_{15}^2 + a_{23}^2 + a_{24}^2 + a_{25}^2) \\ &\quad + s_2[K_1(a_{34}^2 + a_{35}^2 + a_{45}^2) + K_2(a_{34}^2 + a_{35}^2 + a_{45}^2) \\ &\qquad + K_3(a_{14}^2 + a_{15}^2 + a_{24}^2 + a_{25}^2) \\ &\qquad + K_4(a_{13}^2 + a_{15}^2 + a_{23}^2 + a_{25}^2) \\ &\qquad + K_5(a_{13}^2 + a_{14}^2 + a_{23}^2 + a_{24}^2)] \\ &\quad + [(a_{23}a_{45} - a_{24}a_{35} + a_{25}a_{34})^2 \\ &\qquad + (a_{13}a_{45} - a_{14}a_{35} + a_{15}a_{34})^2]\} \\ &+ \{s_2^3 a_{12}^2 + s_2^2[K_1(a_{23}^2 + a_{24}^2 + a_{25}^2) + K_2(a_{13}^2 + a_{14}^2 + a_{15}^2) \\ &\qquad + (K_3 + K_4 + K_5)a_{12}^2] \\ &\quad + s_2[(a_{12}a_{45} - a_{14}a_{25} + a_{15}a_{24})^2 \\ &\qquad + (a_{12}a_{35} - a_{13}a_{25} + a_{15}a_{23})^2 \\ &\qquad + (a_{12}a_{34} - a_{13}a_{24} + a_{14}a_{23})^2] \end{aligned}$$

$$+ [K_1(a_{23}a_{45} - a_{24}a_{35} + a_{25}a_{34})^2$$
$$+ K_2(a_{13}a_{45} - a_{14}a_{35} + a_{15}a_{34})^2$$
$$+ K_3(a_{12}a_{45} - a_{14}a_{25} + a_{15}a_{24})^2$$
$$+ K_4(a_{12}a_{35} - a_{13}a_{25} + a_{15}a_{23})^2$$
$$+ K_5(a_{12}a_{34} - a_{13}a_{24} + a_{14}a_{23})^2]\} \qquad (3.130)$$

This is a VSHP. However, it is seen that the VSHP property is destroyed if $a_{12} = 0$ while the VSHP property is preserved if some parameters (like a_{24} and a_{35}) are equated to zero.

A large number of variations are possible as the order of N increases. In addition, it is noteworthy that the degrees of s_1 and s_2 will never be equal; they can be made equal only when two sections are properly cascaded. We consider, as an alternative, substituting in equation (3.129) $\mu_1 = \mu_2 = \mu_3 = s_1$ and $\mu_4 = \mu_5 = s_2$ to give

$$N_5 = s_1^3[s_2^2 + (K_4 + K_5)s_2 + a_{45}^2]$$
$$+ s_1^2\{s_2^2(K_1 + K_2 + K_3) + s_2(a_{14}^2 + a_{15}^2 + a_{24}^2 + a_{25}^2 + a_{34}^2 + a_{35}^2)$$
$$+ [a_{45}^2(K_1 + K_2 + K_3) + K_4(a_{15}^2 + a_{25}^2 + a_{35}^2) + K_5(a_{24}^2 + a_{34}^2)]\}$$
$$+ s_1\{s_2^2(a_{12}^2 + a_{13}^2 + a_{23}^2)$$
$$+ s_2[K_1(a_{24}^2 + a_{25}^2 + a_{34}^2 + a_{35}^2) + K_2(a_{14}^2 + a_{15}^2 + a_{34}^2 + a_{35}^2)$$
$$+ K_3(a_{14}^2 + a_{25}^2) + (K_4 + K_5)(a_{12}^2 + a_{13}^2 + a_{23}^2)]$$
$$+ [(a_{23}a_{45} - a_{24}a_{35} + a_{25}a_{34})^2 + (a_{13}a_{45} - a_{14}a_{35} + a_{25}a_{34})^2$$
$$+ (a_{12}a_{45} - a_{14}a_{25} + a_{15}a_{24})^2]\}$$
$$+ \{s_2^2(K_1a_{23}^2 + K_2a_{13}^2 + K_3a_{12}^2 + K_5a_{14}^2)$$
$$+ s_2[(a_{12}a_{35} - a_{13}a_{25} + a_{15}a_{23})^2 + (a_{12}a_{34} - a_{13}a_{24} + a_{14}a_{23})^2]$$
$$+ [K_1(a_{23}a_{45} - a_{24}a_{35} + a_{25}a_{34})^2 + K_2(a_{13}a_{45} - a_{14}a_{35} + a_{15}a_{34})^2$$
$$+ K_3(a_{12}a_{45} - a_{14}a_{25} + a_{15}a_{24})^2$$
$$+ K_4(a_{12}a_{35} - a_{13}a_{25} + a_{15}a_{23})^2$$
$$+ K_5(a_{12}a_{34} - a_{13}a_{24} + a_{14}a_{23})^2]\} \qquad (3.131)$$

It is seen that cascading equations (3.130) and (3.131) will give a polynomial with equal degrees in s_1 and s_2. This is not the case for a matrix of even order.

3.13.5. Generation of Two-Variable Very Strict Hurwitz Polynomials using Matrix Theory

It is well known that a symmetric positive-definite (or positive-semidefinite) matrix is always physically realizable. It is further known that any positive-definite matrix P can always be decomposed as a product of two matrices, Q and Q^t, where Q is either an upper-triangular or a lower-triangular matrix. We define the matrix[26]

$$D = A\Gamma A^t s_1 + B\Delta B^t s_2 + G \tag{3.132}$$

where A and B are upper-triangular matrices, Γ and Δ are arbitrary diagonal nonnegative matrices, and G is a skew-symmetric matrix. These matrices have the forms

$$A = \begin{bmatrix} 1 & a_{12} & a_{13} & \cdots & a_{1n} \\ 0 & 1 & a_{23} & \cdots & a_{2n} \\ 0 & 0 & 1 & \cdots & a_{3n} \\ \vdots & & & \ddots & \vdots \\ 0 & 0 & 0 & \cdots & 1 \end{bmatrix} \tag{3.133a}$$

$$\Gamma = \begin{bmatrix} \gamma_1^2 & 0 & 0 & \cdots & 0 \\ 0 & \gamma_2^2 & 0 & \cdots & 0 \\ 0 & 0 & \gamma_3^2 & \cdots & 0 \\ \vdots & & & \ddots & \vdots \\ & & & & \gamma_n^2 \end{bmatrix} \tag{3.133b}$$

$$B = \begin{bmatrix} 1 & b_{12} & b_{13} & \cdots & b_{1n} \\ 0 & 1 & b_{23} & \cdots & b_{2n} \\ 0 & 0 & 1 & \cdots & b_{3n} \\ \vdots & & & \ddots & \vdots \\ 0 & 0 & 0 & \cdots & 1 \end{bmatrix} \tag{3.133c}$$

$$\Delta = \begin{bmatrix} \delta_1^2 & 0 & 0 & 0 & \cdots & 0 \\ 0 & \delta_2^2 & 0 & 0 & \cdots & 0 \\ 0 & 0 & \delta_3^2 & 0 & \cdots & 0 \\ \vdots & & & & \ddots & \vdots \\ 0 & 0 & 0 & 0 & \cdots & \delta_n^2 \end{bmatrix} \tag{3.133d}$$

and

$$G = \begin{bmatrix} 0 & g_{12} & g_{13} & \cdots & g_{1n} \\ -g_{12} & 0 & g_{23} & \cdots & g_{2n} \\ \vdots & & & \ddots & \vdots \\ -g_{1n} & -g_{2n} & -g_{3n} & \cdots & 0 \end{bmatrix} \tag{3.133e}$$

Matrix D is realizable as a two-variable reactance network. Therefore $\det D$ constitutes either the even part or the odd part of a two-variable Hurwitz polynomial (HP), depending on whether the order is even or odd. It should be noted that some of the elements of Γ and Δ can be zero. Since D is a physically realizable matrix, we derive a two-variable HP

$$D(s_1, s_2) = \det D + k_1 \frac{\partial(\det D)}{\partial s_1} + k_2 \frac{\partial(\det D)}{\partial s_2} \tag{3.134}$$

where k_1 and k_2 are nonnegative constants. Similarly, other HPs can be formed using higher-order derivatives. The required VSHP is generated using higher-order partial polynomial derivatives. The various steps are given below, assuming that $\det D = M$:

1. Form

$$M_1 = M + k_{11} \frac{\partial M}{\partial s_1} + k_{21} \frac{\partial M}{\partial s_2}$$

   This is known to be a two-variable HP.

2. Now form

$$M_2 = M_1 + k_{12} \frac{\partial M_1}{\partial s_1} + k_{22} \frac{\partial M_1}{\partial s_2} \tag{3.135}$$

   which involves higher-order partial derivatives of M. This will also be a two-variable HP.

3. This process can be continued until we form

$$M_n = M_{n-1} + k_{1n} \frac{\partial M_{n-1}}{\partial s_1} + k_{2n} \frac{\partial M_{n-1}}{\partial s_2} \tag{3.136}$$

   which can easily be shown to be a VSHP. Other VSHPs can be formed using other derivatives of M, hence the choice is not restrictive.

EXAMPLE 3.8. As an example, consider

$$D_1 = \left\| \begin{bmatrix} 1 & a_1 \\ 0 & 1 \end{bmatrix} \begin{bmatrix} 0 & 0 \\ 0 & \alpha_2^2 \end{bmatrix} \begin{bmatrix} 1 & 0 \\ a_1 & 1 \end{bmatrix} s_1 + \begin{bmatrix} 1 & b_1 \\ 0 & 1 \end{bmatrix} \begin{bmatrix} 0 & 0 \\ 0 & \beta_2^2 \end{bmatrix} \begin{bmatrix} 1 & 0 \\ b_1 & 1 \end{bmatrix} s_2 + \begin{bmatrix} 0 & g_1 \\ -g_1 & 0 \end{bmatrix} \right\|$$

$$= \left\| \begin{bmatrix} a_1^2\alpha_2^2 s_1 + b_1^2\beta_2^2 s_2 & a_1\alpha_2^2 s_1 + b_1\beta_2^2 s_2 + g_1 \\ a_1\alpha_2^2 s_1 + b_1\beta_2^2 s_2 - g_1 & \alpha_2^2 s_1 + \beta_2^2 s_2 \end{bmatrix} \right\|$$

$$= (a_1 - b_1)^2\alpha_2^2\beta_2^2 s_1 s_2 + g_1^2 \tag{3.137}$$

Similarly, we can write

$$D_2 = (a_2 - b_2)^2\alpha_4^2\beta_4^2 s_1 s_2 + g_2^2 \tag{3.138}$$

The two expressions can be cascaded to yield D_1D_2, as follows:

$$D_1D_2 = (a_1 - b_1)^2(a_2 - b_2)^2\alpha_2^2\beta_2^2\alpha_4^2\beta_4^2 s_1^2 s_2^2 + s_1s_2[(a_1 - b_1)^2\alpha_2^2\beta_2^2 g_2^2 + (a_2 - b_2)^2\alpha_4^2\beta_4^2 g_1^2] + g_1^2 g_2^2 \tag{3.139}$$

Therefore, the polynomial will be

$$D_{12} = D_1D_2 + k_1\frac{\partial(D_1D_2)}{\partial s_1} + k_2\frac{\partial(D_1D_2)}{\partial s_2} \tag{3.140}$$

which gives

$$\begin{aligned} D_{12} = {} & s_1^2 s_2^2\{(a_1 - b_1)^2(a_2 - b_2)^2\alpha_2^2\beta_2^2\alpha_4^2\beta_4^2\} \\ & + 2k_1 s_1 s_2^2\{(a_1 - b_1)^2(a_2 - b_2)^2\alpha_2^2\beta_2^2\alpha_4^2\beta_4^2\} \\ & + 2k_2 s_1^2 s_2\{(a_1 - b_1)^2(a_2 - b_2)^2\alpha_2^2\beta_2^2\alpha_4^2\beta_4^2\} \\ & + s_1 s_2\{(a_1 - b_1)^2\alpha_2^2\beta_2^2 g_2^2 + (a_2 - b_2)^2\alpha_4^2\beta_4^2 g_1^2\} \\ & + k_1 s_2\{(a_1 - b_1)^2\alpha_2^2\beta_2^2 g_2^2 + (a_2 - b_2)^2\alpha_4^2\beta_4^2 g_1^2\} \\ & + k_2 s_1\{(a_1 - b_1)^2\alpha_2^2\beta_2^2 g_2^2 + (a_2 - b_2)^2\alpha_4^2\beta_4^2 g_1^2\} + g_1^2 g_2^2 \end{aligned} \tag{3.141}$$

This is not a VSHP, so we must add another set of derivatives applied to equation (3.141), as follows:

$$D(s_1, s_2) = D_{12} + k_3\frac{\partial D_{12}}{\partial s_1} + k_4\frac{\partial D_{12}}{\partial s_2} \tag{3.142}$$

The polynomial then becomes

$$
\begin{aligned}
D(s_1, s_2) = {} & s_1^2 s_2^2\{(a_1 - b_1)^2(a_2 - b_2)^2\alpha_2^2\beta_2^2\alpha_4^2\beta_4^2\} \\
& + (2k_1 + 2k_3)s_1 s_2^2\{(a_1 - b_1)^2(a_2 - b_2)^2\alpha_2^2\beta_2^2\alpha_4^2\beta_4^2\} \\
& + (2k_2 + 2k_4)s_1^2 s_2\{(a_1 - b_1)^2(a_2 - b_2)^2\alpha_2^2\beta_2^2\alpha_4^2\beta_4^2\} \\
& + 2k_1k_3 s_2^2\{(a_1 - b_1)^2(a_2 - b_2)^2\alpha_2^2\beta_2^2\alpha_4^2\beta_4^2\} \\
& + 2k_2k_4 s_1^2\{(a_1 - b_1)^2(a_2 - b_2)^2\alpha_2^2\beta_2^2\alpha_4^2\beta_4^2\} \\
& + s_1 s_2\{(a_1 - b_1)^2\alpha_2^2\beta_2^2 g_2^2 + (a_2 - b_2)^2\alpha_4^2\beta_4^2 g_1^2 \\
& \qquad + [(a_1 - b_1)^2(a_2 - b_2)^2\alpha_2^2\beta_2^2\alpha_4^2\beta_4^2](4k_1k_4 + 4k_2k_3)\} \\
& + s_2(k_1 + k_3)\{(a_1 - b_1)^2\alpha_2^2\beta_2^2 g_2^2 + (a_2 - b_2)^2\alpha_4^2\beta_4^2 g_1^2\} \\
& + s_1\{(k_2 + k_4)[(a_1 - b_1)^2\alpha_2^2\beta_2^2 g_2^2 + (a_2 - b_2)^2\alpha_4^2\beta_4^2 g_1^2]\} \\
& + g_1^2 g_2^2 + (k_1 + k_2)[(a_1 - b_1)^2\alpha_2^2\beta_2^2 g_2^2 + (a_2 - b_2)^2\alpha_4^2\beta_4^2 g_1^2]
\end{aligned}
\tag{3.143}
$$

This is a VSHP polynomial with quantities k_1, k_2, k_3, and k_4 in equation (3.143) assigned to unity.

The concept of resistance matrices can be introduced into equation (3.132) (as suggested elsewhere[27]) to avoid the lengthy process of calculating the higher-order partial derivatives. Let us consider the matrix

$$D = A\Gamma A^t s_1 + B\Delta B^t s_2 + G + C \in C^t \tag{3.144}$$

where A, B, Γ, Δ, and G are matrices as defined by expressions (3.133a)-(3.133e) while matrices C and ε are given by

$$
C = \begin{bmatrix}
1 & c_{12} & c_{13} & c_{14} & \cdots & c_{1n} \\
0 & 1 & c_{23} & c_{24} & \cdots & c_{2n} \\
0 & 0 & 1 & c_{34} & \cdots & c_{3n} \\
\vdots & & & \ddots & & \vdots \\
 & & & & \ddots & \\
0 & 0 & 0 & & \cdots & 1
\end{bmatrix}
\tag{3.145}
$$

and

$$
\varepsilon = \begin{bmatrix}
\varepsilon_1^2 & 0 & 0 & 0 & \cdots & 0 \\
0 & \varepsilon_2^2 & 0 & 0 & \cdots & 0 \\
0 & 0 & \varepsilon_3^2 & 0 & \cdots & 0 \\
\vdots & & & \ddots & & \\
 & & & & \ddots & \\
0 & 0 & 0 & & \cdots & \varepsilon_n^2
\end{bmatrix}
\tag{3.146}
$$

Now under certain conditions det D can become a two-variable VSHP, as shown in the following example:

EXAMPLE 3.9. Consider

$$D = \begin{bmatrix} 1 & a_1 \\ 0 & 1 \end{bmatrix} \begin{bmatrix} \gamma_1^2 & 0 \\ 0 & \gamma_2^2 \end{bmatrix} \begin{bmatrix} 1 & 0 \\ a_1 & 1 \end{bmatrix} s_1 + \begin{bmatrix} 1 & b_1 \\ 0 & 1 \end{bmatrix} \begin{bmatrix} \delta_1^2 & 0 \\ 0 & \delta_2^2 \end{bmatrix} \begin{bmatrix} 1 & 0 \\ b_1 & 1 \end{bmatrix} s_2$$
$$+ \begin{bmatrix} 0 & g \\ -g & 0 \end{bmatrix} + \begin{bmatrix} 1 & c_1 \\ 0 & 1 \end{bmatrix} \begin{bmatrix} \varepsilon_1^2 & 0 \\ 0 & \varepsilon_2^2 \end{bmatrix} \begin{bmatrix} 1 & 0 \\ c_1 & 1 \end{bmatrix} \tag{3.147}$$

Taking the determinant of D and setting $\gamma_1 = \delta_1 = 0$ yields

$$\det D = s_1 s_2[\gamma_2^2\delta_2^2(b_1 - a_1)^2] + s_1[\varepsilon_1^2\gamma_2^2 + \gamma_2^2\varepsilon_2^2(c_1 - a_1)^2]$$
$$+ s_2[\varepsilon_1^2\delta_2^2 + \varepsilon_2^2\delta_2^2(c_1 - b_1)^2] + [g^2 + \varepsilon_1^2\varepsilon_2^2] \tag{3.148}$$

which is a two-variable VSHP.

An alternative approach for generating a two-variable VSHP is due to Abiri *et al.*[28] This too uses the properties of a symmetric positive-definite or positive-semidefinite matrix. We first consider the matrix C_n defined by

$$C_n = \sum_{i=1}^{n} \mu_i [A_i][A_i]^t + G \tag{3.149}$$

where μ_i, $i = 1, \ldots, n$ are complex variables, while A_i is a column matrix corresponding to variable μ_i and possessing the form

$$[A_i] = [a_{1i} \quad a_{2i} \quad a_{3i} \quad \cdots \quad a_{ni}]^t \tag{3.150}$$

G is a real skew-symmetric matrix given by

$$[G] = \begin{bmatrix} 0 & g_{12} & g_{13} & \cdots & g_{1n} \\ -g_{12} & 0 & g_{23} & \cdots & g_{2n} \\ -g_{13} & -g_{23} & 0 & \cdots & g_{3n} \\ \vdots & & & \ddots & \vdots \\ -g_{1n} & -g_{2n} & -g_{3n} & \cdots & 0 \end{bmatrix} \tag{3.151}$$

Matrix C_n in equation (3.149) is physically realizable and can be written in the alternative, compact form

$$C_n = A\mu A^t + G \tag{3.152}$$

where A is a square matrix of the form

$$A = [A_1 A_2 \cdots A_n] \tag{3.153}$$

or, in expanded form,

$$A = \begin{bmatrix} a_{11} & a_{12} & a_{13} & \cdots & a_{1n} \\ a_{21} & a_{22} & a_{23} & \cdots & a_{2n} \\ a_{31} & a_{32} & a_{33} & \cdots & a_{3n} \\ \vdots & & & & \vdots \\ a_{n1} & a_{n2} & a_{n3} & \cdots & a_{nn} \end{bmatrix} \tag{3.154}$$

Matrix μ is diagonal of order n and is given by

$$\mu = \mathrm{diag}[\mu_1 \mu_2 \mu_3 \cdots \mu_n] \tag{3.155}$$

We now examine the relationship

$$M_{\mathrm{e(o)}} = |C_n| = \det C_n = \det[A\mu A^{\mathrm{t}} + G] \tag{3.156}$$

where subscript e(o) stands for even (odd). If A is a nonsingular matrix (i.e., $|A| \neq 0$), then we can alternatively construct the following operation:

$$\det[A^{-1} C_n (A^{-1})^{\mathrm{t}}] = \det[A^{-1}(A\mu A^{\mathrm{t}} + G)(A^{-1})^{\mathrm{t}}] \tag{3.157}$$

or

$$|A^{-1}||C_n||(A^{-1})^{\mathrm{t}}| = \det[\mu + A^{-1} G (A^{-1})^{\mathrm{t}}] = \det(\mu + G') \tag{3.158}$$

where

$$G' = A^{-1} G (A^{-1})^{\mathrm{t}} \tag{3.159}$$

Since $\det A^{-1} = 1/\det A$ we have

$$|C_n| = |A|^2 \det(\mu + G') = |A|^2 \det(W) \tag{3.160}$$

where $W = \mu + G'$. But

$$(G')^{\mathrm{t}} = [A^{-1} G (A^{-1})^{\mathrm{t}}]^{\mathrm{t}} = A^{-1} G^{\mathrm{t}} (A^{-1})^{\mathrm{t}} = -[A^{-1} G (A^{-1})^{\mathrm{t}}] = -G' \tag{3.161}$$

as G is a skew-symmetric matrix (i.e., $G^{\mathrm{t}} = -G$). The equality in equation (3.161) implies that G' is also a skew-symmetric matrix.

Since W is already known to be a reactance matrix, it follows from equation (3.160) that C_n is also a reactance matrix (for a reactance matrix W, $W + W^t \equiv 0$, W^t being the transpose of W with μ_i changed to $-\mu_i$); thus C_n is always physically realizable. The determinant of C_n (det C_n) then constitutes either the even part or the odd part of an n-variable HP, depending on whether the order of the matrix A_i is even or odd, with the condition that $\det A \neq 0$. It can now be easily proved that $(\partial \det C/\partial \mu_i)/\det C$ is a reactance function and therefore

$$M = M(\mu_1, \mu_2, \mu_3, \ldots, \mu_n) = M_{e(o)} + \sum_{j=1}^{n} k_j \frac{\partial M_{e(o)}}{\partial \mu_j} \tag{3.162}$$

is an n-variable HP with k_j being positive constants.

Similarly, other HPs can be formed using higher-order derivatives. From equation (3.162), a two-variable HP is generated by putting some μ_i equal to s_1 and the rest equal to s_2. It should be noted that one of the properties of $M_{e(o)}$ (det C_n) is that the quantities μ_i for $i = 1, 2, \ldots, n$ are of degree unity. This means that the terms $\mu_1^k, \mu_2^k, \ldots, \mu_n^k$ for $k = 2, 3, \ldots, n$ are absent in det C_n. In fact, this is one of the conditions required for the generation of a VSHP from det C_n. The required VSHP is generated using higher-order partial derivatives as outlined elsewhere.[26]

The following steps would lead to the generation of a VSHP. First, we form

$$M_1 = M + k_{11} \frac{\partial M}{\partial s_1} + k_{21} \frac{\partial M}{\partial s_2} \tag{3.163}$$

where M is a two-variable HP obtained from equation (3.162). This is known to be a two-variable HP.

We now form

$$M_2 = M_1 + k_{12} \frac{\partial M_1}{\partial s_1} + k_{22} \frac{\partial M_1}{\partial s_2} \tag{3.164}$$

which involves higher-order partial derivatives of M. This will also be a one-variable HP.

By continuing this process we ultimately form

$$M_n = M_{n-1} + k_{1,n-1} \frac{\partial M_{n-1}}{\partial s_1} + k_{2,n-1} \frac{\partial M_{n-1}}{\partial s_2} \tag{3.165}$$

It should be noted that quantities k_{ij} in equations (3.163)-(3.165) are all positive constants. Equality (3.165) can easily be shown to be a VSHP.

EXAMPLE 3.10. Let us consider

$$C_3 = A\mu A^t + G \tag{3.166}$$

where

$$A = \begin{bmatrix} a_{111} & a_{121} & a_{131} \\ a_{211} & a_{221} & a_{231} \\ a_{311} & a_{321} & a_{331} \end{bmatrix} \tag{3.167}$$

$$G = \begin{bmatrix} 0 & g_{121} & g_{131} \\ -g_{121} & 0 & g_{231} \\ -g_{131} & -g_{231} & 0 \end{bmatrix} \tag{3.168}$$

and

$$\mu = \text{diag}[\mu_1\mu_2\mu_3] \tag{3.169}$$

By employing equations (3.166) and (3.162) and substituting $\mu_1 = \mu_2 = s_1$ and $\mu_3 = s_2$, we can write

$$D_1(s_1, s_2, \mathbf{a}, \mathbf{g}) = (d_{11} + d_{21})(s_1 + 1) + d_{31}(s_2 + 1) + d_{41}[s_1^2(s_2 + 1) + 2s_1s_2] \tag{3.170}$$

where

$$d_{11} = (a_{111}g_{231} - a_{121}g_{131} + a_{131}g_{121})^2$$

$$d_{21} = (a_{211}g_{231} - a_{221}g_{131} + a_{231}g_{121})^2$$

$$d_{31} = (a_{311}g_{231} - a_{321}g_{131} + a_{331}g_{121})^2$$

$$d_{41} = (a_{111}a_{221}a_{331} - a_{111}a_{231}a_{321} - a_{121}a_{211}a_{331} + a_{131}a_{211}a_{321} + a_{121}a_{231}a_{311} - a_{131}a_{221}a_{311})^2 \tag{3.171}$$

It is noteworthy that the degrees of s_1 and s_2 in $D(s_1, s_2, \mathbf{a}, \mathbf{g})$ are not equal. It is preferable to maintain the degrees of s_1 and s_2 the same, as this will influence the filter response, particularly with regard to the symmetry. They can be made equal when the first filter section is properly cascaded with another section, e.g., by substituting $\mu_1 = s_1$ and $\mu_2 = \mu_3 = s_2$ in equations (3.156) and (3.162). We can write

$$D_2(s_1, s_2, \mathbf{a}, \mathbf{g}) = d_{12}(s_1 + 1) + (d_{22} + d_{32})(s_2 + 1) + d_{42}[s_2^2(s_1 + 1) + 2s_1s_2] \tag{3.172}$$

where

$$d_{12} = (a_{112}g_{232} - a_{122}g_{131} + a_{132}g_{122})^2$$

$$d_{22} = (a_{212}g_{232} - a_{222}g_{132} + a_{232}g_{122})^2$$

$$d_{32} = (a_{312}g_{232} - a_{322}g_{132} + a_{332}g_{122})^2$$

$$d_{42} = (a_{112}a_{222}a_{332} - a_{112}a_{232}a_{322} - a_{122}a_{212}a_{332} + a_{132}a_{212}a_{322} + a_{122}a_{232}a_{312} - a_{132}a_{222}a_{312})^2 \quad (3.173)$$

By cascading $D_1(s_1, s_2, \mathbf{a}, \mathbf{g})$ and $D_2(s_1, s_2, \mathbf{a}, \mathbf{g})$ we obtain

$$D(s_1, s_2, \mathbf{a}, \mathbf{g}) = D_1(s_1, s_2, \mathbf{a}, \mathbf{g})D_2(s_1, s_2, \mathbf{a}, \mathbf{g}) \quad (3.174)$$

This equation constitutes a two-variable VSHP, since D_1 and D_2 are both two-variable VSHPs.

We note that a 1-D HP can be formed by using any of the above techniques of generating VSHPs by simply setting $s_1 = s_2 = s$.

REFERENCES

1. E. I. Jury, *Theory and Application of the z-Transform Method*, John Wiley and Sons, New York (1964).
2. H. W. Schussler, A stability theorem for discrete systems, *IEEE Trans. Acoust., Speech, Signal Process.* **ASSP-24,** 87–89 (1976).
3. V. Ramachandran and C. S. Gargour, Implementation of a stability test of 1-D discrete system based on Schussler's theorem and some consequent coefficient conditions, *J. Franklin Inst.* **317,** 341–358 (1984).
4. R. R. Read and S. Treitel, The stabilization of two-dimensional recursive filters via the discrete Hilbert transform, *IEEE Trans. Geosci. Electron.* **GE-11,** 153–160 (1973).
5. B. Gold and C. M. Rader, *Digital Processing of Signals*, McGraw-Hill, New York (1969).
6. V. Cizek, Discrete Hilbert transform, *IEEE Trans. Audio Electroacoust.* **AU-18,** 340–343 (1970).
7. C. Farmer and J. B. Bednar, Stability of spatial digital filters, *Math. Biosci.* **14,** 113–119 (1972).
8. J. L. Shanks, S. Treitel, and J. H. Justice, Stability and synthesis of two-dimensional recursive filters, *IEEE Trans. Audio Electroacoust.* **AU-20,** 115–128 (1972).
9. T. S. Huang, Stability of two-dimensional recursive filters, *IEEE Trans. Audio Electroacoust.* **AU-20,** 158–163 (1972).
10. H. G. Ansell, On certain two-variable generalizations of circuit theory, with applications to networks of transmission lines and lumped reactances, *IEEE Trans. Circuit Theory* **CT-11,** 214–223 (1964).
11. E. A. Guillemin, *Synthesis of Passive Networks*, John Wiley and Sons, New York (1957).

12. B. D. O. Anderson and E. I. Jury, Stability test for two-dimensional recursive filters, *IEEE Trans. Audio Electroacoust.* **AU-21,** 366–372 (1973).
13. E. I. Jury, *Theory and Application of the z-Transform Method,* John Wiley and Sons, New York (1964).
14. E. I. Jury, A note on the reciprocal zeros of a real polynomial with respect to the unit circle, *IEEE Trans. Circuit Theory* **CT-11,** 292–294 (1964).
15. G. A. Maria and M. M. Fahmy, On the stability of two-dimensional digital filters, *IEEE Trans. Audio Electroacoust.* **AU-21,** 470–472 (1973).
16. D. Goodman, Some stability properties of two-dimensional linear shift-invariant digital filters, *IEEE Trans. Circuits Syst.* **CAS-24,** 201–208 (1977).
17. E. Robinson, *Statistical Communication and Detection,* Hafner, New York (1967).
18. R. A. Wiggins, On Factoring the Correlations of Discrete Multivariable Stochastic Processes, Mass. Inst. Tech., Cambridge, Sci. Report 9 of Contract AF, 19(604) 7378, 127–152 (1965).
19. Y. Genin and Y. Kamp, Comments on, "On the stability of the least mean-square inverse process in two-dimensional digital filters," *IEEE Trans. Acoust., Speech, Signal Process* **ASSP-25,** 92–93 (1977).
20. J. L. Shanks and R. R. Read, Techniques for Stabilizing Two-Dimensional Finite Difference Operator, Proc. of Colloque Int. Point Transformations and Their Application, Toulouse, France (September 1973).
21. P. Pistor, Stability criterion for recursive filters, *IBM J. Res. Dev.* **18,** 59–71 (1974).
22. S. Chakrabarti and S. K. Mitra, Design of two-dimensional filters via spectral transformations, *Proc. IEEE* **65,** 905–914 (1977).
23. D. Goodman, Some difficulties with the double bilinear transformation in 2-D recursive filter design, *Proc. IEEE* **66,** 796–797 (1978).
24. P. K. Rajan, H. C. Reddy, M. N. S. Swamy, and V. Ramachandran, Generation of two-dimensional digital functions without nonessential singularities of the second kind, *IEEE Trans. Acoust., Speech, Signal Process* **ASSP-28,** 216–223 (1980).
25. V. Ramachandran and M. Ahmadi, Design of 2-D stable analog and recursive digital filters using properties of the derivative of even or odd parts of Hurwitz polynomials, *J. Franklin Inst.* **315,** 259–267 (1983).
26. M. Ahmadi and V. Ramachandran, New method for generating two-variable VSHPs and its application in the design of two-dimensional recursive digital filters with prescribed magnitude and constant group delay responses, *IEE Proc.* **131,** Part G, 151–155 (1984).
27. M. Ahmadi, M. T. Boraie, and V. Ramachandran, Generation of 2-Variable VSHPs using Properties of the Resistance Matrices and its Application in Recursive 2-D Filter Design, Invited Paper, Proc. 1985 IEEE Int. Conf. on Decision and Control, Fort Lauderdale, Florida, USA, 1590–1593 (1985).
28. M. A. Abiri, M. Ahmadi, and V. Ramachandran, An alternative approach in generating a 2-variable very strictly Hurwitz polynomial (VSHP) and its application, *J. Franklin Inst.* **324,** 187–203 (1987).

4

Recursive Filters

PART ONE. ONE-DIMENSIONAL FILTERS

4.1. INTRODUCTION

In this chapter we shall discuss techniques for designing recursive digital filters, with the restriction that the designed filter be realizable and stable. Part One of this chapter deals with different approaches for the design of one-dimensional (1-D) recursive filters.

A 1-D recursive filter is characterized, as discussed in Section 1.6.1, by its z-transfer function, as follows:

$$H_D(z) = \frac{N(z)}{D(z)} = \frac{\sum_{i=0}^{M} a_i z^{-i}}{\sum_{i=0}^{N} b_i z^{-i}} \tag{4.1}$$

where **a** and **b** are vectors of real numbers and are known as the coefficients of the filter; also in most cases $M \leq N$. Without loss of generality, we can assume that $M = N$. By the term "designing a 1-D filter" we mean the calculation of a filter's coefficients **a** and **b** in such a way that the magnitude and/or phase (or group delay) of the designed filter approximates a desired one while maintaining the stability of the designed filter. The latter requires

$$D(z) \neq 0 \qquad \text{for } |z| \geq 1 \tag{4.2}$$

The magnitude response of the 1-D recursive filter is expressed as

$$M(\omega) = |H_D[e^{j\omega T}]| = |H_D(z)H_D(z^{-1})|^{1/2}_{z=e^{j\omega T}} \tag{4.3}$$

and the phase response as

$$\phi_{\mathrm{D}}(\omega) = \phi_{\mathrm{D}}(e^{j\omega T}) = \arctan\left\{\frac{\mathrm{Im}[H_{\mathrm{D}}(z)]}{\mathrm{Re}[H_{\mathrm{D}}(z)]}\right\}_{z=e^{j\omega T}} \tag{4.4}$$

which can be expressed in the alternative form

$$\phi_{\mathrm{D}}(e^{j\omega T}) = \frac{1}{2j}\ln[H_{\mathrm{D}}(z)/H_{\mathrm{D}}(z^{-1})]_{z} = e^{j\omega T} \tag{4.5}$$

Finally, the group delay of a filter is defined as

$$\tau_{\mathrm{D}}(e^{j\omega T}) = -\frac{d\phi_{\mathrm{D}}(e^{j\omega T})}{d\omega} = -jz\frac{d\phi_{\mathrm{D}}}{dz}\bigg|_{z=e^{j\omega T}} \tag{4.6}$$

Replacing ϕ_{D} in equation (4.6) by expression (4.5) gives

$$\begin{aligned}\tau_{\mathrm{D}}(e^{j\omega T}) &= -\mathrm{Re}\left[z\frac{dH_{\mathrm{D}}(z)/dz}{H_{\mathrm{D}}(z)}\right]_{z=e^{j\omega T}} \\ &= -\mathrm{Re}\left\{z\frac{d}{dz}[\ln H_{\mathrm{D}}(z)]\right\}_{z=e^{j\omega T}}\end{aligned} \tag{4.7}$$

It can be seen from equation (4.6) that if the phase response of the filter is linear over some frequency band(s), the group delay response will be constant over the same frequency band(s).

There are two approaches to the design of recursive digital filters: the direct and indirect methods. In direct techniques, the desired discrete-time transfer function is obtained directly from the given magnitude, with or without a specified phase specification, generally through the use of an iterative method based on linear or nonlinear programming.[1,2,3,6] In indirect methods, on the other hand, one of the classical analogue-filter approximations (such as Butterworth, Chebyshev, elliptic) is used to generate an analogue transfer function, which is subsequently discretized by the use of some method, such as the mapping of differentials, the impulse-invariant technique, the bilinear transformation, or the matched z-transform method.

The four different discretization procedures will now be discussed in detail.

4.2. TECHNIQUES FOR DISCRETIZATION OF ANALOGUE TRANSFER FUNCTIONS

4.2.1. Mapping of Differentials

We assume that the analogue linear time-invariant filter is specified by the differential equation

$$\sum_{i=0}^{M} b_i \frac{\mathrm{d}^i y(t)}{\mathrm{d}t^i} = \sum_{i=0}^{M} a_i \frac{\mathrm{d}^i u(t)}{\mathrm{d}t^i} \tag{4.8}$$

This filter has input $u(t)$ and output $y(t)$, and can be characterized by its transfer function $H_a(s)$, which is derived by taking the Laplace transform of equation (4.8) assuming zero initial conditions:

$$H_a(s) = \sum_{i=0}^{M} a_i s^i \Big/ \sum_{i=0}^{M} b_i s^i \tag{4.9}$$

We use backward differences to approximate the derivatives, since forward differences have been shown to impair the stability of the filter.[2] The first backward difference $\nabla_1[\cdot]$ is defined by

$$\nabla_1[y(n)] = [y(n) - y(n-1)]/T \tag{4.10}$$

Now, higher-order backward differences are found by applying the first backward difference repeatedly:

$$\nabla_i[y(n)] = \nabla_1[\nabla_{i-1}[y(n)]] \tag{4.11}$$

Now replacing the ith-order differences as approximations to the derivatives in equation (4.8), we obtain

$$\sum_{i=0}^{M} b_i \nabla_i[y(n)] = \sum_{i=0}^{M} a_i \nabla_i[u(n)] \tag{4.12}$$

The latter equation is a numerical approximation to equation (4.8). The z transform of equations (4.10) and (4.11) yields

$$Z\{\nabla_1[y(n)]\} = Z\{[y(n) - y(n-1)]/T\} = Y(z)\frac{(1 - z^{-1})}{T} \tag{4.13}$$

$$Z\{\nabla_i[y(n)]\} = Y(z)[(1 - z^{-1})/T]^i \tag{4.14}$$

On taking the z transform of equation (4.12) and replacing $\nabla_i[y(n)]$ by its equivalent in equation (4.14), and similarly for $\nabla_i[u(n)]$, we get

$$\sum_{i=0}^{M} b_i \left(\frac{1-z^{-1}}{T}\right)^i Y(z) = \sum_{i=0}^{M} a_i \left(\frac{1-z^{-1}}{T}\right)^i U(z) \tag{4.15}$$

and the transfer function can be deduced from equation (4.15) as follows:

$$H(z) = \frac{Y(z)}{U(z)} = \frac{\sum_{i=0}^{M} a_i \left(\frac{1-z^{-1}}{T}\right)^i}{\sum_{i=0}^{M} b_i \left(\frac{1-z^{-1}}{T}\right)^i} \tag{4.16}$$

A comparison of equations (4.16) and (4.9) leads to

$$H(z) = H(s)|_{s=(1-z^{-1})/T} \tag{4.17}$$

This corresponds to mapping from the s plane to the z plane and from the z plane to the s plane using backward differences:

$$s = \frac{1-z^{-1}}{T} \tag{4.18}$$

and

$$z = \frac{1}{1-sT} \tag{4.19}$$

Now, we have to check whether two desirable properties of any mapping from the analogue to the digital domain are met under this transformation. These properties are:

1. The $j\Omega$ axis in the s plane should be mapped to the circumference of the unit circle in the z plane.
2. Points in the left half of the s plane $\{\mathrm{Re}(s) < 0\}$ should be mapped inside the unit circle ($|z| < 1$).

Property (1) preserves the frequency characteristics of the analogue filter in the digital domain, while property (2) preserves stability of the system under this transformation.

To check for property (1), we put $s = j\Omega$ in equation (4.19) and obtain

$$z = \frac{1}{1-j\Omega T} = \frac{1}{1+\Omega^2 T^2} + j\frac{\Omega T}{1+\Omega^2 T^2} = a + jb \tag{4.20}$$

It can easily be shown that the real and imaginary parts of z are related by

$$(a - \tfrac{1}{2})^2 + b^2 = \tfrac{1}{4} \tag{4.21}$$

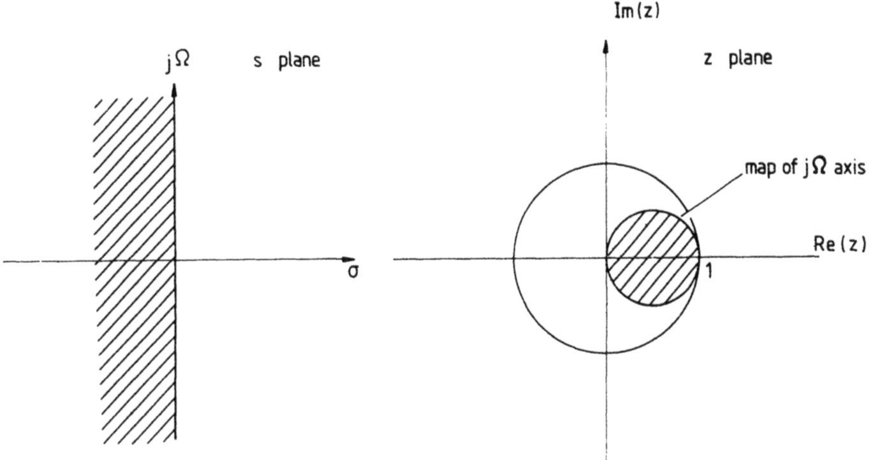

Figure 4.1. Mapping from the s plane to the z plane using backward differences.

Thus, the map of the jΩ axis of the s plane in the z plane is the circumference of a circle of radius $\frac{1}{2}$ as shown in Figure 4.1. Except for very small values of Ω, the image of the jΩ axis in the s plane is off the unit circle in the z plane. Therefore, property (1) is not satisfied.

To check whether property (2) of the mapping is satisfied, we let

$$sT = c + \mathrm{j}d \tag{4.22}$$

where c and d are real, and for stability $c < 0$. Substitution of equation (4.22) in relation (4.19) gives

$$z = \frac{1}{1 - c - \mathrm{j}d} \tag{4.23}$$

or

$$|z| = \frac{1}{[(1-c)^2 + d^2]^{1/2}} < 1 \tag{4.24}$$

which means that the second property is preserved under this transformation.

4.2.2. Impulse Invariant Transformation

In order to discretize an analogue filter using the impulse invariant transformation, the analogue transfer function (4.9) should be rewritten as

$$H(s) = \sum_{i=1}^{M} \frac{c_i}{s + d_i} \tag{4.25}$$

where the coefficients c_i are calculated using

$$c_i = H(s)(s + d_i)|_{s=-d_i} \tag{4.26}$$

quantity d_i being the ith pole of the transfer function. It is noteworthy that the order of the numerator of the transfer function is always less than or equal to the order of the denominator, and that multiple poles in the system are not permitted.

The corresponding impulse response $h(t)$ of the analogue filter of equation (4.25) can be expressed as

$$h(t) = \sum_{i=1}^{M} c_i \, e^{-d_i t} u_{-1}(t) \tag{4.27}$$

where $u_{-1}(t)$ is the unit step function. By sampling equation (4.27), the digital impulse response $h(nT)$ can be obtained in the form

$$h(nT) = \sum_{i=1}^{M} c_i \, e^{-d_i nT} u_{-1}(nT) \tag{4.28}$$

If the z transform of both sides of equation (4.28) is taken, we obtain

$$H(z) \sum_{i=1}^{M} c_i \sum_{n=0}^{\infty} (e^{-d_i T} z^{-1})^n \tag{4.29}$$

or

$$H(z) = \sum_{i=1}^{M} \frac{c_i}{1 - e^{-d_i T} z^{-1}} \tag{4.30}$$

A quick comparison of equations (4.30) and (4.25) gives the following mapping relation:

$$\frac{1}{s + d_i} \rightarrow \frac{1}{1 - z^{-1} e^{-d_i T}} \tag{4.31}$$

The frequency response of a digital system, derived from an impulse invariant transformation, is an aliased version of the frequency response of the analogue filter from which it was derived. Thus we can write

$$H(e^{j\Omega T}) = \frac{1}{T} \sum_{n=-\infty}^{\infty} H(j\Omega + jn\omega_s) \tag{4.32}$$

where $\omega_s = 2\pi/T$ is the sampling frequency of the digital system in radians.

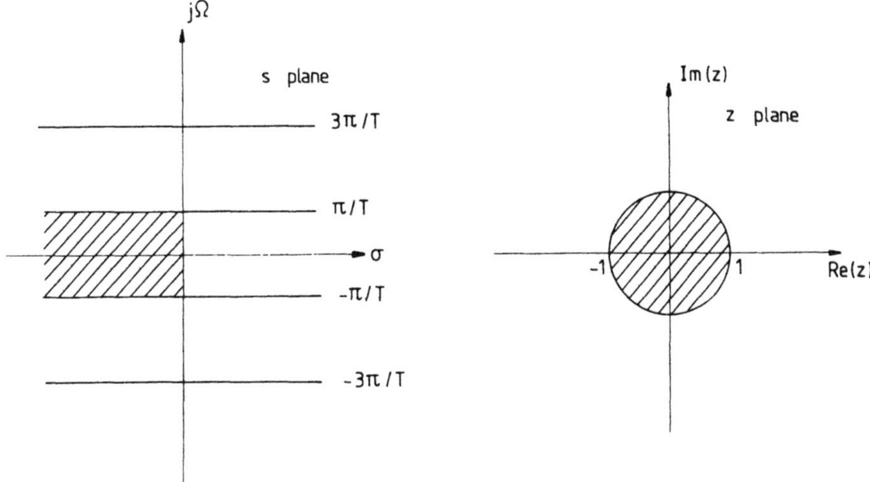

Figure 4.2. Mapping from the s plane to the z plane using impulse invariant transformation.

Figure 4.2 shows the mapping from the s plane to the z plane via the impulse invariant transformation. Each horizontal strip of the s plane of width $2\pi/T$ is mapped into the entire z plane, with the left-hand side of the strip being mapped to the interior of the unit circle in the z plane; this implies preservation of stability. A thorough treatment of this can be found elsewhere.[2]

4.2.3. Bilinear Transformation

A simple conformal mapping from the s plane to the z plane and vice versa which preserves the desired algebraic form of the analogue system is the bilinear transformation defined by

$$s \rightarrow \frac{2}{T} \frac{1 - z^{-1}}{1 + z^{-1}} \tag{4.33}$$

and

$$z \rightarrow \frac{2/T + s}{2/T - s} \tag{4.34}$$

Using this transformation, the entire $j\Omega$ axis in the s plane is mapped onto the circumference of the unit circle in the z plane; the left-half s plane is mapped inside the unit circle and the right-half s plane is mapped outside the unit circle in the z plane, as shown in Figure 4.3. The preservation of

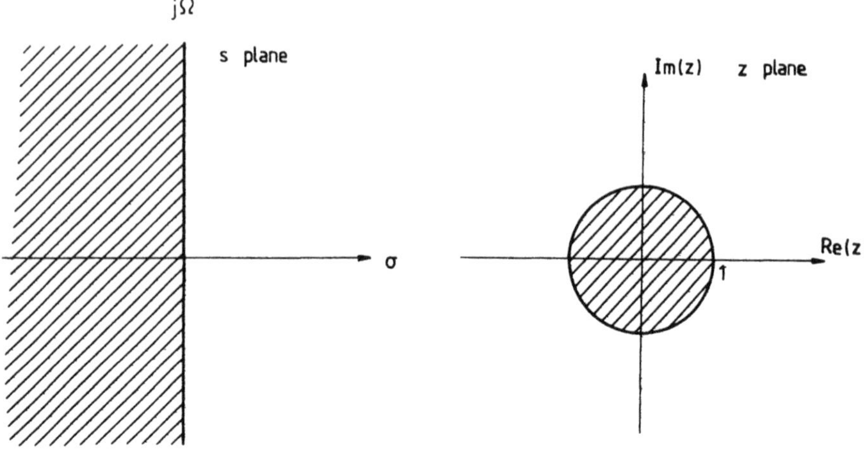

Figure 4.3. Mapping from the s plane to the z plane using the bilinear transformation.

stability through this transformation can be verified by replacing $s = \mathrm{j}\Omega$ in equation (4.34), which gives

$$z = \frac{2/T + \mathrm{j}\Omega}{2/T - \mathrm{j}\Omega} \tag{4.35}$$

It is seen from this equation that when $\Omega = 0$ and $\Omega = \infty$, $z = 1$ and $z = -1$ respectively; between these limits the angle of z varies monotonically from 0 to π. Now letting $s = \sigma + \mathrm{j}\Omega$ in equation (4.34) gives

$$z = \frac{2/T + \sigma + \mathrm{j}\Omega}{2/T - \sigma - \mathrm{j}\Omega} \tag{4.36}$$

It is clear from the above that for $\sigma < 0$ (left-half plane), $|z| < 1$, and for $\sigma > 0$ (right-half plane), $|z| > 1$.

It should be noted that the digital transfer function obtained from an analogue counterpart through bilinear transformation will have the same order denominator, but the order of the numerator may differ from that of the analogue transfer function if the order of the numerator of this is less than the order of the denominator. As an example, we consider

$$H(s) = \frac{s + a}{s^2 + b_1 s + b_2}$$

which has a numerator of the first order and a second-order denominator.

Assuming $T = 2$ in equation (4.33), $H(z)$ is obtained from the bilinear transformation as

$$H(z) = \frac{(1 + z^{-1})[(1 + a) + (a - 1)z^{-1}]}{(1 + b_1 + b_2) + 2(b_2 - 1)z^{-1} + (1 - b_1 + b_2)z^{-2}}$$

which has a denominator and numerator of second order. In fact, the numerator has a zero at $z = -1$. A close look at $H(s)$ would indicate that it has a zero at $s = \infty$; the bilinear transformation maps this zero to $z = -1$, as expected.

This transformation is free of aliasing errors, which are inherent with the impulse invariant method, since the entire $j\Omega$ axis of the s plane is mapped onto the circumference of the unit circle in the z plane; also, the unit circle in the z plane is mapped from the entire left-half s plane, rather than from a single strip of the s plane, as in the impulse invariant transformation.

There are, however, two problems associated with the bilinear transformation.

1. Although the magnitude response of the analogue filter will somehow be preserved under the bilinear transformation, the phase response will be greatly distorted under this transformation, i.e., if the analogue filter has a linear-phase characteristic, the digital transfer function derived from this analogue filter using the bilinear transformation will not have linear-phase characteristics.
2. There is a highly nonlinear relationship between analogue frequency Ω and digital frequency ω. This nonlinearity is evident by evaluating transformation (4.33) for $z = e^{j\omega T}$ and $s = j\Omega$, giving

$$j\Omega \rightarrow \frac{2}{T}\frac{1 - e^{-j\omega T}}{1 + e^{-j\omega T}} \tag{4.37}$$

or

$$\Omega \rightarrow \frac{2}{T}\tan(\omega T/2) \tag{4.38}$$

The form of this nonlinearity is shown in Figure 4.4 for $T = 2$. It is seen that, for small values of ω, the mapping is nearly linear, but for large values of ω the mapping is highly nonlinear. This creates a restriction on the use of the bilinear transformation. It implies that only special types of analogue filters with piecewise-constant amplitude response should be used for discretization through the bilinear transformation. This effect of nonlinearity is named the "warping effect." To compensate for the warping

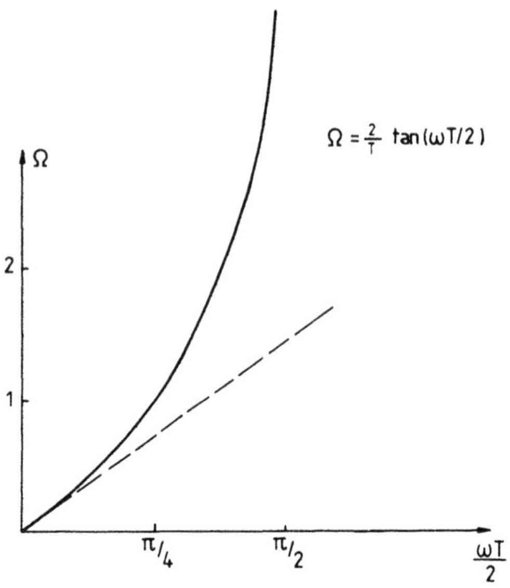

Figure 4.4. The relation between analogue and digital frequency scales for the bilinear transformation.

effect due to the bilinear transformation, the desired digital frequencies should be prewarped using equation (4.38), so that the analogue filter after warping will produce the desired frequency response. A pictorial representation of prewarping is shown in Figure 4.5.

4.2.4. Matched z Transform

The last technique for discretizing an analogue filter is named the matched z transformation. In this method, poles as well as zeros in the s plane are mapped to poles and zeros in the z plane using the relationship

$$s + a \rightarrow 1 - z^{-1}\,\mathrm{e}^{-aT} \tag{4.39}$$

where T is the sampling period. It should be noted that the poles of the impulse invariant method and those of the matched z transform are identical. The matched z transform is simple to apply to an analogue system, but its suitability for certain analogue systems remains in question. It is therefore preferable to use the impulse invariant method for discretization of an all-pole analogue system, and also for analogue filters with zeros having frequencies greater than the half sampling frequency.

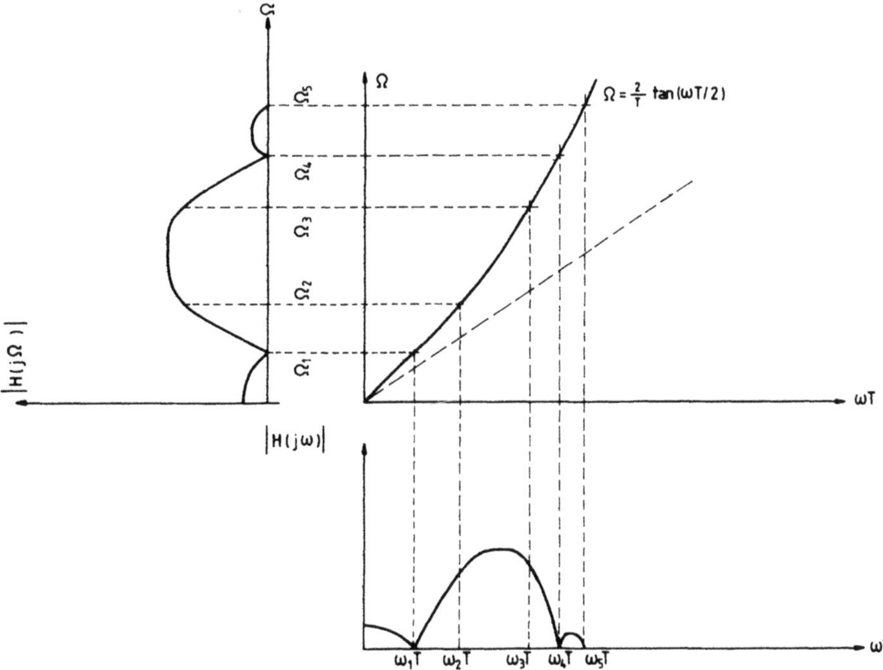

Figure 4.5. Method of prewarping a frequency specification for an analogue filter undergoing a bilinear transformation.

4.3. DIRECT DESIGN OF RECURSIVE DIGITAL FILTERS

4.3.1. Linear Programming Approach

Linear programming techniques can be utilized for the approximation of a recursive digital filter satisfying a prescribed magnitude response with or without a prespecified phase characteristic. First, we assume that magnitude-only specification is required; then we modify the technique to include simultaneous approximation of magnitude and phase (group-delay) response.

If the filter's transfer function is given in the form of equation (4.1), then the magnitude-squared response of the filter can be obtained through the following steps:

$$M_{\mathrm{D}}^{2}(\omega) = H_{\mathrm{D}}(z)H_{\mathrm{D}}(z^{-1})\big|_{z=\mathrm{e}^{\mathrm{j}\omega T}} = \left|\frac{N(z)N(z^{-1})}{D(z)D(z^{-1})}\right|_{z=\mathrm{e}^{\mathrm{j}\omega T}} \tag{4.40a}$$

where

$$H_D(z)H_D(z^{-1}) = \frac{\left(\sum_{i=0}^{M} a_i z^{-i}\right)\left(\sum_{i=0}^{M} a_i z^{i}\right)}{\left(\sum_{i=0}^{M} b_i z^{-i}\right)\left(\sum_{i=0}^{M} b_i z^{i}\right)}$$

$$= \frac{\sum_{i=-M}^{M} c_i z^{-i}}{\sum_{i=-M}^{M} d_i z^{-i}} \tag{4.40b}$$

in which the new coefficients c_i and d_i are functions of a_i and b_i respectively and are centrally symmetric; that is

$$c_i = c_{-i} \tag{4.41}$$

and

$$d_i = d_{-i} \tag{4.42}$$

The magnitude-squared function is obtained by evaluating $H_D(z)\,H_D(z^{-1})$ on the unit circle and is of the form

$$M_D^2(\omega) = |H_D(e^{j\omega T})|^2 = \frac{N_1(j\omega)}{D_1(j\omega)} = \frac{c_0 + \sum_{i=1}^{M} 2c_i \cos i\omega}{d_0 + \sum_{i=1}^{M} 2d_i \cos i\omega} \tag{4.43}$$

It is seen from this equation that $N_1(j\omega)$ and $D_1(j\omega)$ are linear combinations of c_i and d_i respectively. Now, given a magnitude-squared characteristic $M_I(\omega)$, then the approximation problem is that of finding the filter coefficients such that

$$-\varepsilon(\omega) \leq \frac{N_1(j\omega)}{D_1(j\omega)} - M_I(\omega) \leq \varepsilon(\omega) \tag{4.44}$$

where $\varepsilon(\omega)$ is the tolerance function.

Equation (4.44) can be expressed as a set of linear inequalities in the coefficients c_i and d_i as follows:

$$N_1(j\omega) - D_1(j\omega)M_I(\omega) \leq \varepsilon(\omega)D_1(\omega)$$

$$-N_1(j\omega) + D_1(j\omega)M_I(\omega) \leq \varepsilon(\omega)D_1(j\omega) \tag{4.45}$$

or

$$N_1(j\omega) - D_1(j\omega)[M_I(\omega) + \varepsilon(\omega)] \le 0$$

$$-N_1(j\omega) + D_1(j\omega)[M_I(\omega) - \varepsilon(\omega)] \le 0 \quad (4.46)$$

with the additional inequalities

$$-N_1(j\omega) \le 0 \quad (4.47)$$

and

$$-D_1(j\omega) \le 0 \quad (4.48)$$

The equiripple approximation problem is thus defined. To solve the system of linear inequalities (4.45)–(4.48), a dummy variable δ is subtracted from the left side of each constraint in the system, and this variable is then minimized. If the resulting value of δ is 0, then a solution exists to the approximation problem, and the filter coefficients may be obtained directly as the output of the linear programming routine. If $\delta < 0$, then no solution to the approximation problem exists, and either $M_I(\omega)$ or $\varepsilon(\omega)$, or both, must be modified in order to obtain a solution. The problem associated with this design formulation is that the derived digital transfer function is not guaranteed to be stable. Therefore, at the end of the design procedure, a stability test as well as stabilization procedure should, if needed, be carried out. This technique also fails in its present format to address the design of a linear-phase 1-D filter satisfying a prescribed magnitude specification.

4.3.2. Modified Linear Programming Approach

Chottera and Jullien[3] have modified the linear programming approach of Rabiner *et al.*[1] by including the stability constraint in the process of optimization, as well as including the constant group-delay (linear-phase) response in the design process.

If $M(\omega_i)$ and $\phi(\omega_i)$ are the given magnitude and phase specifications respectively, where $\phi(\omega_i)$ is defined as

$$\phi(\omega_i) = -\tau_D\omega_i \qquad \text{for } i = 1, 2, \ldots, k \quad (4.49)$$

then the real and imaginary components of the frequency domain specifications, $Y_R(\omega_i)$ and $Y_I(\omega_i)$ respectively, can be written as

$$Y_R(\omega_i) = M(\omega_i)\cos[\phi(\omega_i)] = M(\omega_i)\cos(-\tau_D\omega_i) \quad (4.50)$$

and

$$Y_I(\omega_i) = M(\omega_i)\sin[\phi(\omega_i)] = M(\omega_i)\sin(-\tau_D\omega_i) \tag{4.51}$$

We now define a complex error $r(\omega_i)$, between the specification and the designed filter function given by equation (4.40a), such that

$$r(\omega_i) = Y_R(\omega_i) + jY_I(\omega_i) - N(e^{-j\omega_i})/D(e^{-j\omega_i}) \tag{4.52}$$

Multiplication of this equation by $D(e^{-j\omega_i})$ yields a weighted complex error

$$r(\omega_i)D(e^{-j\omega_i}) = [Y_R(\omega_i) + jY_I(\omega_i)]D(e^{-j\omega_i}) - N(e^{-j\omega_i}) \tag{4.53}$$

We set

$$r(\omega_i)D(e^{-j\omega_i}) = e_R(\omega_i) + je_I(\omega_i) \tag{4.54}$$

$$N(e^{-j\omega_i}) = \sum_{n=0}^{M} a_n\cos(n\omega_i) - j\sum_{n=0}^{M} a_n\sin(n\omega_i) \tag{4.55}$$

and

$$D(e^{-j\omega_i}) = \sum_{m=0}^{M} b_m\cos(m\omega_i) - j\sum_{m=0}^{M} b_m\sin(m\omega_i) \tag{4.56}$$

On substituting relations (4.53), (4.55), and (4.56) into equation (4.54), we obtain the real and imaginary components of the complex weighted error as

$$e_R(\omega_i) = \sum_{m=0}^{M} b_m[Y_R(\omega_i)\cos(m\omega_i) + Y_I(\omega_i)\sin(m\omega_i)] - \sum_{n=0}^{M} a_n\cos(n\omega_i) \tag{4.57}$$

and

$$e_I(\omega_i) = \sum_{m=0}^{M} b_m[Y_I(\omega_i)\cos(m\omega_i) - Y_R(\omega_i)\sin(m\omega_i)] + \sum_{n=0}^{M} a_n\sin(n\omega_i) \tag{4.58}$$

for $i = 1, 2, \ldots, k$.

A good approximation to the desired specifications can now be obtained by minimizing a positive variable ε such that

$$|e_R(\omega_i)| \le \varepsilon \tag{4.59}$$

and

$$|e_I(\omega_i)| \leq \varepsilon \tag{4.60}$$

from which the following inequalities can be obtained:

$$e_R(\omega_i) - \varepsilon \leq 0 \tag{4.61}$$

$$-e_R(\omega_i) - \varepsilon \leq 0 \tag{4.62}$$

$$e_I(\omega_i) - \varepsilon \leq 0 \tag{4.63}$$

$$-e_I(\omega_i) - \varepsilon \leq 0 \tag{4.64}$$

for $i = 1, 2, \ldots, k$.

The above approximation problem is amenable to linear programming, since $e_R(\omega_i)$ and $e_I(\omega_i)$ are linear in terms of the coefficients a_i and b_i. Therefore, by letting $\xi = -\varepsilon$, the linear programming approximation problem can be stated as follows:

Maximize

$$g = [b_1, b_2, b_3, \ldots, b_M, a_0, a_1, \ldots, a_N, \xi]\begin{bmatrix} 0 \\ 0 \\ \vdots \\ 0 \\ 1 \end{bmatrix} \tag{4.65}$$

subject to

$$\sum_{m=0}^{M} b_m\{M(\omega_i)\cos[\omega_i(m+\tau_D)]\} - \sum_{n=0}^{M} a_n \cos(n\omega_i) + \xi \leq 0 \tag{4.66}$$

$$\sum_{m=0}^{M} b_m\{-M(\omega_i)\cos[\omega_i(m+\tau_D)]\} + \sum_{n=0}^{M} a_n \cos(n\omega_i) + \xi \leq 0 \tag{4.67}$$

$$-\sum_{m=0}^{M} b_m\{M(\omega_i)\cos[\omega_i(m+\tau_D)]\} + \sum_{n=0}^{M} a_n \cos(n\omega_i) + \xi \leq 0 \tag{4.68}$$

$$\sum_{m=0}^{M} b_m\{-M(\omega_i)\sin[\omega_i(m+\tau_D)]\} - \sum_{n=0}^{M} a_n \sin(n\omega_i) + \xi \leq 0 \tag{4.69}$$

where $b_0 = 1$ and $i = 1, 2, \ldots, k$.

Clearly, the upper bound on g is zero and maximizing g minimizes the real and imaginary error components e_I and e_R. If g is greater than zero, then the solution is meaningless. The constraints given above are sufficient for the approximation, but they do not ensure filter stability. Therefore additional constraints are necessary which enable stable filter design.

Stability constraints. Since the approximation procedure involves linear programming, the stability constraints that need to be incorporated for stable design must be linear in form. There are two types of constraint that satisfy the requirement of linearity:

1. Monotonicity of denominator coefficients,[4] i.e.,

$$b_0 > b_1 > b_2 > \cdots > b_{m-1} > b_m > 0 \tag{4.70}$$

2. The real part of the denominator polynomial must be greater than zero for all $|z| = 1$,[5] i.e.,

$$\text{Re}[D(z)] > 0 \qquad \text{for } |z| = 1 \tag{4.71}$$

The latter constraint has been used successfully in the design of a digital phase network.

It should be noted that constraints (4.70) and (4.71) are sufficient conditions for filter stability, and therefore filters designed subject to these constraints would belong to a subclass of possible stable filters. Although constraint (4.70) is linear in form, it designs filters that belong to a smaller subclass of stable filters, in comparison to constraint (4.71). Therefore, constraint (4.71) has been used in the design algorithm for stable filter design.

The actual constraint that is incorporated in the design procedure is a slightly modified form of (4.71) and is given by

$$\text{Re}[D(z)] \geq \delta \qquad \text{for } |z| = 1 \tag{4.72}$$

where δ is a small positive quantity. With $b_0 = 1$, the above constraint can be written as

$$-\sum_{m=1}^{M} b_m \cos(m\omega_i) \leq 1 - \delta \qquad \text{for } 0 \leq \omega_i \leq \pi \tag{4.73}$$

Thus constraint (4.73), together with relations (4.65)-(4.69), completely define the linear programming design problem for magnitude and linear-phase (constant group-delay) approximation for recursive digital filters.

Design procedure. The foregoing discussion can be summarized in the following steps:

1. Specify the desired magnitude characteristics.
2. Specify the order of the filter, say N, and set $\tau_D = N - 1$.
3. Solve the linear programming problem given by system (4.65), (4.69), and (4.73) to obtain the maximizing function $g = g_a$ and the coefficient set c_1.
4. Set $\tau_D = \tau_D - 1$ and solve the linear programming problem to obtain $g = g_b$ and the coefficient set c_2.
5. If $g_b > g_a$, store c_2 in c_1 and go to step (4); if not, go to step (6).
6. $g^* = g_a$ and denotes the optimum g, and the corresponding optimum group delay $\tau_D^* = \tau_D + 1$. The coefficient set c_1 is the desired coefficient set. Finally, compute errors in magnitude and group delay.

EXAMPLE 4.1. *Low-Pass Filter.* The specifications for the design are as follows:

$$M(\omega) = 1.0, \qquad 0 \leq \omega \leq 0.5\pi$$

$$M(\omega) = e^{-k(0.5\pi - \omega)}, \qquad 0.5\pi < \omega \leq \pi$$

The above desired specification was chosen in this and other examples because it provides magnitude values in the transition band and gives control over the cutoff characteristics by varying the value for k. Also, the approximation has been found to improve with some specifications in the transition region.

In the above specification, the value of k is set equal to 186.6 to give a sharp cutoff characteristic, and δ in equation (4.73) is set equal to 1.0×10^{-6}. The transition bandwidth is approximately equal to 0.05. The desired gain at and beyond 0.58π is to be less than -40 dB. A filter of order 18 was chosen. The design algorithm terminated for a group delay of $\tau_D^* = 15$. The corresponding (maximizing function) $g^* = -1.4963 \times 10^{-2}$. Figures 4.6a and 4.6b show the designed magnitude and group-delay characteristics. Tables 4.1a and 4.1b give the coefficient values and pole/zero positions.

EXAMPLE 4.2. The specifications are for a five-band filter, with three stop bands and two pass bands. The desired specifications are as follows:

$$M(\omega) = e^{-k(0.2\pi - \omega)}, \qquad 0 \leq \omega < 0.2\pi$$

$$M(\omega) = 1, \qquad 0.2\pi \leq \omega \leq 0.3\pi$$

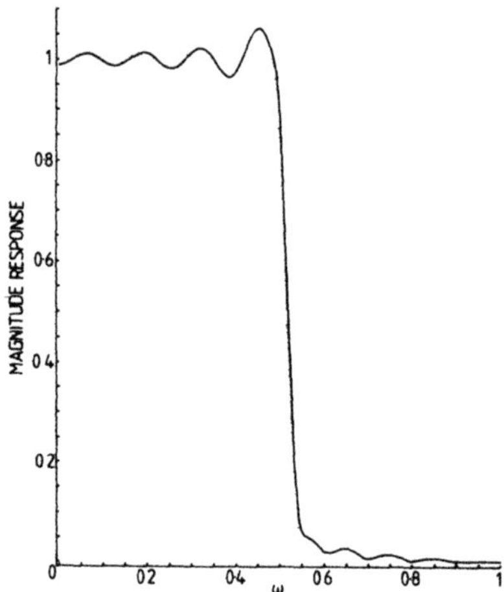

Figure 4.6a. Magnitude response characteristics for low-pass filter example.

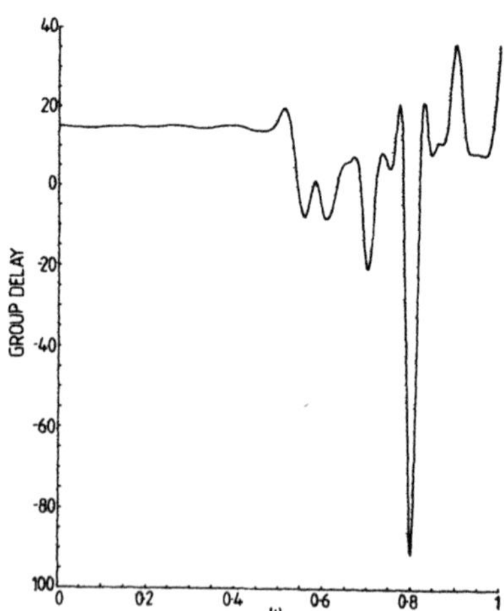

Figure 4.6b. Group-delay response (in the pass band) of the designed low-pass filter.

Table 4.1a. Coefficients of the Designed Low-Pass Filter

$a_0 =$ 0.7485878 E − 01	$b_0 =$ 0.2864426 E − 02
$a_1 =$ 0.2242485 E00	$b_1 =$ −0.1508292 E − 01
$a_2 =$ 0.3734605 E00	$b_2 =$ 0.3978608 E − 01
$a_3 =$ 0.3753272 E00	$b_3 =$ −0.7412958 E − 01
$a_4 =$ 0.2102609 E00	$b_4 =$ 0.1174653 E00
$a_5 =$ 0.2152715 E − 01	$b_5 =$ −0.1764883 E00
$a_6 =$ −0.4565689 E − 01	$b_6 =$ 0.2544858 E00
$a_7 =$ −0.7424634 E − 02	$b_7 =$ −0.3424158 E00
$a_8 =$ 0.2178770 E − 01	$b_8 =$ 0.4302391 E00
$a_9 =$ 0.3254443 E − 02	$b_9 =$ −0.5249142 E00
$a_{10} =$ −0.1397011 E − 01	$b_{10} =$ 0.6380558 E00
$a_{11} =$ −0.7479091 E − 03	$b_{11} =$ −0.7573233 E00
$a_{12} =$ 0.9584092 E − 02	$b_{12} =$ 0.8552690 E00
$a_{13} =$ 0.9662641 E − 04	$b_{13} =$ −0.9376680 E00
$a_{14} =$ −0.7434126 E − 02	$b_{14} =$ 0.1045841 E01
$a_{15} =$ 0.1640602 E − 03	$b_{15} =$ −0.1099716 E01
$a_{16} =$ 0.7836282 E − 02	$b_{16} =$ 0.1435216 E01
$a_{17} =$ −0.5112606 E − 02	$b_{17} =$ −0.6491873 E00
$a_{18} =$ −0.1432080 E − 01	$b_{18} =$ 0.1000000 E01

Table 4.1b. Poles and Zeros of the Designed Filter

Zeros	Poles
Z(1) = −1.0327946 E00 + j0.0000000	P(1) = 0.9784455 E00 − j0.2068859 E00
Z(2) = 1.4053055 E00 + j0.0000000	P(2) = 0.9784455 E00 + j0.2068859 E00
Z(3) = −0.9529882 E00 − j0.3030076 E00	P(3) = 0.9865953 E00 − j0.1502099 E00
Z(4) = −0.9529882 E00 + j0.3030076 E00	P(4) = 0.9865953 E00 + j0.1502099 E00
Z(5) = 0.6849739 E00 − j0.7285676 E00	P(5) = −0.8863566 E00 − j0.4630035 E00
Z(6) = 0.6849739 E00 + j0.7285676 E00	P(6) = −0.8863566 E00 + j0.4630035 E00
Z(7) = −0.5925453 E00 − j0.8055365 E00	P(7) = −0.4099355 E00 − j0.9121145 E00
Z(8) = −0.5925453 E00 + j0.8055365 E00	P(8) = −0.4099355 E00 + j0.9121145 E00
Z(9) = 0.9185387 E00 − j0.3953306 E00	P(9) = 0.7052577 E00 − 0.7089528 E00
Z(10) = 0.9185387 E00 + j0.3953306 E00	P(10) = 0.7052577 E00 + 0.7089528 E00
Z(11) = −0.1630025 E00 − j0.0986625 E01	P(11) = 0.1448659 E00 − j0.9894553 E00
Z(12) = −0.1630025 E00 + j0.0986625 E01	P(12) = 0.1448659 E00 + j0.9894553 E00
Z(13) = −0.3200112 E00 − j0.9474138 E00	P(13) = 0.8920994 E00 − j0.4518983 E00
Z(14) = −0.3200112 E00 + j0.9474138 E00	P(14) = 0.8920994 E00 + j0.4518983 E00
Z(15) = −0.8114263 E00 − j0.5844922 E00	P(15) = −0.6829389 E00 − j0.7304458 E00
Z(16) = −0.8114263 E00 + j0.5844922 E00	P(16) = −0.6829389 E00 + j0.7304458 E00
Z(17) = 0.3163798 E00 − j0.9443573 E00	P(17) = 0.5398856 E00 − j0.8417519 E00
Z(18) = 0.3163798 E00 + j0.9443573 E00	P(18) = 0.5398856 E00 + j0.8417519 E00

$$M(\omega) = e^{-k(\omega - 0.3\pi)}, \qquad 0.3\pi < \omega \leq 0.45\pi$$

$$M(\omega) = e^{-k(0.6\pi - \omega)}, \qquad 0.45\pi < \omega < 0.6\pi$$

$$M(\omega) = 1.0, \qquad 0.6\pi \leq \omega \leq 0.8\pi$$

$$M(\omega) = e^{-k(\omega - 0.8\pi)}, \qquad 0.8\pi < \omega \leq \pi$$

The value of k was set equal to 46.6 to give a transition bandwidth of 0.1π and gain -40 dB at the edge of the transition band. A filter order of 20 was chosen. The design algorithm terminated for a group-delay value of 16 (i.e., $\tau_D^* = 16$) and $g^* = -2.205432 \times 10^{-2}$. Figures 4.7a and 4.7b show the designed magnitude and group-delay characteristics. Tables 4.2a and 4.2b show the coefficient values and pole/zero positions.

4.3.3. Nonlinear Programming Approach

Assume that the IIR filter's transfer function is of the form

$$H(z) = \sum_{i=0}^{M} a_i z^{-i} \Big/ \sum_{i=0}^{M} b_i z^{-i} \tag{4.74}$$

We let the ideal magnitude response of the filter be denoted as

$$|H_I(e^{j\omega T})| = M_I(\omega) \tag{4.75}$$

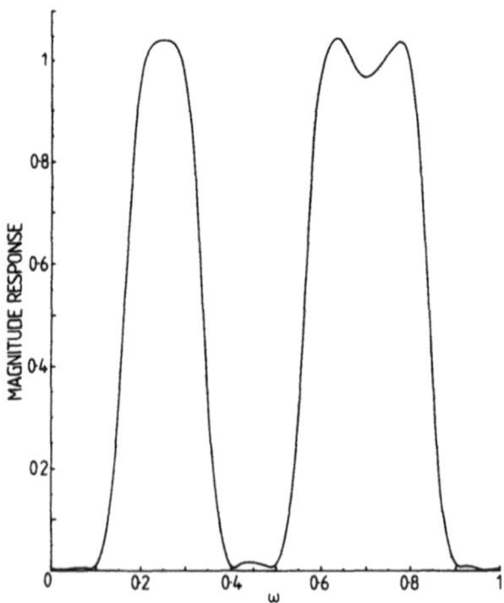

Figure 4.7a. Magnitude response characteristics for the five-band filter example.

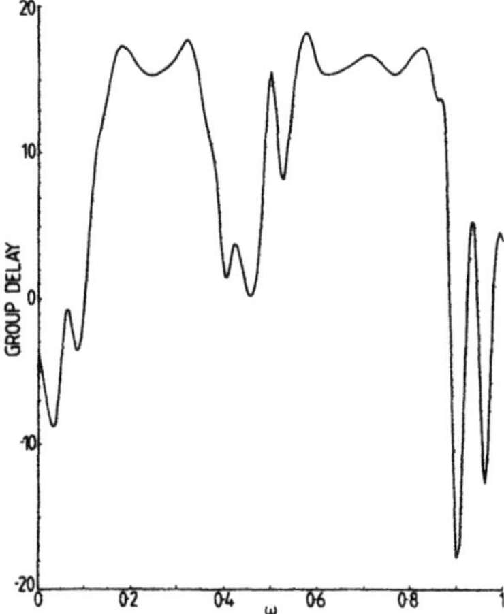

Figure 4.7b. Group-delay response (in the pass band) of the designed five-band filter.

Table 4.2a. Coefficients of the Five-Band Filter

$a_0 = -0.7212126$ E − 01	$b_0 = 0.4822142$ E − 02
$a_1 = 0.3733674$ E − 01	$b_1 = -0.9949651$ E − 02
$a_2 = 0.2149973$ E − 01	$b_2 = 0.4310815$ E − 01
$a_3 = -0.5386061$ E − 01	$b_3 = 0.2391319$ E − 02
$a_4 = 0.2032603$ E00	$b_4 = 0.6244975$ E − 01
$a_5 = -0.1104029$ E − 01	$b_5 = -0.5886914$ E − 01
$a_6 = -0.1195799$ E00	$b_6 = 0.1989163$ E00
$a_7 = 0.7865602$ E − 01	$b_7 = 0.4310714$ E − 01
$a_8 = -0.1585760$ E00	$b_8 = 0.2938179$ E00
$a_9 = -0.6233907$ E − 01	$b_9 = -0.1730217$ E00
$a_{10} = 0.7893807$ E − 01	$b_{10} = 0.4896968$ E00
$a_{11} = 0.1777164$ E − 01	$b_{11} = 0.1188005$ E00
$a_{12} = 0.5812104$ E − 01	$b_{12} = 0.7834710$ E00
$a_{13} = 0.4462991$ E − 02	$b_{13} = -0.2660012$ E00
$a_{14} = 0.2373366$ E − 01	$b_{14} = 0.7038747$ E00
$a_{15} = -0.2725624$ E − 01	$b_{15} = 0.1067267$ E00
$a_{16} = -0.2721384$ E − 01	$b_{16} = 0.1058714$ E01
$a_{17} = 0.2454230$ E − 01	$b_{17} = -0.1410559$ E00
$a_{18} = -0.1691916$ E − 01	$b_{18} = 0.5570049$ E00
$a_{19} = -0.8274231$ E − 02	$b_{19} = -0.7348579$ E − 01
$a_{20} = -0.1319616$ E − 01	$b_{20} = 0.1000000$ E01

Table 4.2b. Poles and Zeros of the Designed Five-Band Filter

Zeros	Poles
Z(1) = −0.9564665 E00 − j0.2922242 E00	P(1) = 0.8687688 E00 − j0.4952051 E00
Z(2) = −0.9564665 E00 + j0.2922242 E00	P(2) = 0.8687688 E00 + j0.4952051 E00
Z(3) = 0.6949541 E00 − j0.7192390 E00	P(3) = −0.6360668 E00 − j0.7716560 E00
Z(4) = 0.6949541 E00 + j0.7192390 E00	P(4) = −0.6360668 E00 + j0.7716560 E00
Z(5) = −0.5057666 E00 − j0.8623853 E00	P(5) = −0.8863637 E00 − j0.4635920 E00
Z(6) = −0.5057666 E00 + j0.8623853 E00	P(6) = −0.8863637 E00 + j0.4635920 E00
Z(7) = 0.9952496 E00 − j0.1025428 E00	P(7) = 0.5107741 E00 − j0.8598031 E00
Z(8) = 0.9952496 E00 + j0.1025428 E00	P(8) = 0.5107741 E00 + j0.8598031 E00
Z(9) = −0.9944223 E00 − j0.1061929 E00	P(9) = 0.8337398 E00 − j0.5558211 E00
Z(10) = −0.9944223 E00 + j0.1061929 E00	P(10) = 0.8337398 E00 + j0.5558211 E00
Z(11) = −0.6013383 E00 − j0.7991816 E00	P(11) = −0.4000965 E00 − j0.9132533 E00
Z(12) = −0.6013383 E00 + j0.7991816 E00	P(12) = −0.4000965 E00 + j0.9132533 E00
Z(13) = 0.9605041 E00 − j0.2785551 E00	P(13) = −0.8340336 E00 − j0.5530252 E00
Z(14) = 0.9605041 E00 + j0.2785551 E00	P(14) = −0.8340336 E00 + j0.5530252 E00
Z(15) = 0.0249226 E00 − j0.9997278 E00	P(15) = 0.6359656 E00 − j0.7717720 E00
Z(16) = 0.0249226 E00 + j0.9997278 E00	P(16) = 0.6359656 E00 + j0.7717720 E00
Z(17) = 0.1192637 E00 − j0.9929243 E00	P(17) = 0.3996778 E00 − j0.9166750 E00
Z(18) = 0.1192637 E00 + j0.9929243 E00	P(18) = 0.3996778 E00 + j0.9166750 E00
Z(19) = 0.2943162 E00 − j0.9558086 E00	P(19) = −0.2153330 E00 − j0.9769188 E00
Z(20) = 0.2943162 E00 + j0.9558086 E00	P(20) = −0.2153330 E00 + j0.9769188 E00

and the designed magnitude response of the filter as

$$|H_D(e^{j\omega T})| = M_D(\omega) \tag{4.76}$$

while the ideal and designed group-delay response are $\tau_I(\omega)$ and $\tau_D(\omega)$, respectively.

Further, $\{\omega_i,\ i = 1, 2, \ldots, N\}$ is the discrete set of frequencies, equally or nonequally spaced, at which $E_M(j\omega_i)$, the error in magnitude response, and $E_\tau(j\omega_i)$, the error in group-delay response, between the designed and ideal values, are evaluated:

$$E_M(j\omega_i) = |M_I(\omega_i) - M_D(\omega_i)| \tag{4.77}$$

and

$$E_\tau(j\omega_i) = \tau_I T - \tau_D(j\omega_i) \tag{4.78}$$

In order to formulate the design problem, two cases will be considered.

1. *Formulation of the design problem for approximation of the magnitude response only.* In this case, we use the least mean-square error criterion (l_2

norm) for computing the cost function, using equation (4.77) in the following relationship:

$$E_{l2}(j\omega_i, \mathbf{a}, \mathbf{b}) = \sum_{i \in I_{ps}} E_m^2(j\omega_i) \tag{4.79}$$

where I_{ps} is the set of all discrete frequency points along the ω axis in the pass band and the stop band of the 1-D filter. Now the problem is to calculate the filter coefficients **a** and **b** in such a way so as to minimize E_{l2} in equation (4.79). This is a simple nonlinear optimization problem and can be implemented using any of the existing nonlinear optimization routines (e.g., Fletcher and Powell). The problem with this formulation is that no constraints are imposed on the coefficients of the filter transfer function to ensure stability of the designed filter.

One may argue that, if the designed filter is found to be unstable, it may be stabilized by replacing the poles located outside the unit circle by their mirror images with respect to the unit circle in the z plane.

2. *Formulation of the design problem for approximation of the magnitude and group-delay response of the filter.* In this case, the general mean-square error E_G is calculated using equations (4.77) and (4.78) in the following manner:

$$E_G(j\omega_i, \mathbf{a}, \mathbf{b}) = \sum_{i \in I_{ps}} E_m^2(j\omega_i) + \sum_{i \in I_p} E_\tau^2(j\omega_i) \tag{4.80}$$

where I_{ps} is as previously defined, and I_p is the set of all discrete frequency points along the ω axis in the pass band of the 1-D filter. Again, in the design of a 1-D filter satisfying prescribed magnitude and constant group-delay response, **a** and **b** should be calculated in such a way that E_G in equation (4.80) is minimized. This is also a simple nonlinear optimization technique, but for this particular formulation it has a major drawback, namely, if the designed filter is found to be unstable, the stabilization method through pole replacement will distort the phase response completely, and therefore there is a need to impose a constraint on the coefficients of the 1-D filter to ensure the stability of the designed filter.

This can be done simply by using the stability test due to Ramachandran and Gargour[7] in the following manner, as suggested elsewhere.[8] The denominator polynomial will take one of the two forms below depending upon whether the order of the filter is even or odd:

$$D(z) = K_e \prod_{i=1}^{n} (z^2 - 2\alpha_i z + 1) + (z^2 - 1) \prod_{i=1}^{n-1} (z^2 - 2\beta_i z + 1), \qquad n = \frac{m}{2}\,(m \text{ even}) \tag{4.81}$$

with $K_e > 1$ and the stability constraint

$$1 > \alpha_1 > \beta_1 > \alpha_2 > \beta_2 > \cdots > \beta_{n-1} > \alpha_n > -1 \qquad (4.82)$$

or

$$D(z) = K_o(z+1) \prod_{i=1}^{n} (z^2 - 2\alpha_i z + 1) + (z-1) \prod_{i=1}^{n} (z^2 - 2\beta_i z + 1) \qquad n = \frac{m-1}{2}\ (m \text{ odd}) \qquad (4.83)$$

with $K_o > 1$ and the stability constraint

$$1 > \alpha_1 > \beta_1 > \alpha_2 > \beta_2 > \cdots > \alpha_n > \beta_n > -1 \qquad (4.84)$$

and also in both cases $|b_0/b_m| > 1$.

The design problem now becomes that of calculating $(\mathbf{a}, \boldsymbol{\alpha}, \boldsymbol{\beta})$, the numerator and the new denominator coefficients, in such a way that E_G in equation (4.80) is minimized, subject to linear constraint (4.82) or (4.84), depending upon whether the filter has even or odd order. This new formulation is also a simple nonlinear constrained optimization problem, which can be solved either by utilization of any suitable optimization procedure with a linear constraint, or by transforming the problem to an unconstrained optimization problem by using the following N-variable substitution method. For example, if m is even we may write

$$\begin{aligned} \alpha_n &= \cos \pi \exp(-\theta_1^2) \\ \beta_{n-1} &= \cos \pi \exp[-(\theta_1^2 + \theta_2^2)] \\ \alpha_{n-1} &= \cos \pi \exp[-(\theta_1^2 + \theta_2^2 + \theta_3^2)] \\ &\vdots \\ \alpha_1 &= \cos \pi \exp\left[-\sum_{i=1}^{2n-1} \theta_i^2\right] \end{aligned} \qquad (4.85)$$

Similar conditions can be derived if m is odd. Now any unconstrained optimization technique can be employed to compute the new filter coefficients $\mathbf{a}$ and $\boldsymbol{\theta}$ to minimize E_G in equation (4.80).

EXAMPLE 4.3. Design a low-pass filter with the magnitude specification

$$|H(j\omega)| = \begin{cases} 1 & \text{for } 0 \le |\omega| < 1 \text{ rad/s} \\ 0 & \text{for } 2.5 \le |\omega| \le \omega_s/2 \text{ rad/s} \end{cases}$$

and constant group-delay response. The order of the filter is chosen to be 4. Figures 4.8a and 4.8b show the plot of the designed amplitude and group-delay response of the filter, while Table 4.3 gives the coefficients of the filter respectively.

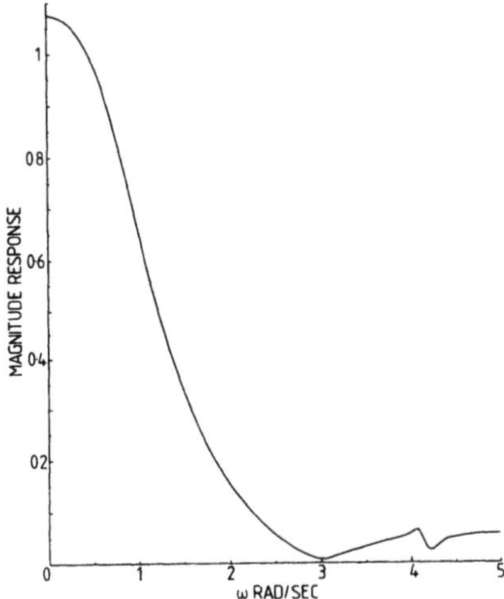

Figure 4.8a. Amplitude response of the low-pass filter with pass-band and stop-band edges of 1.0 rad/s and 2.5 rad/s.

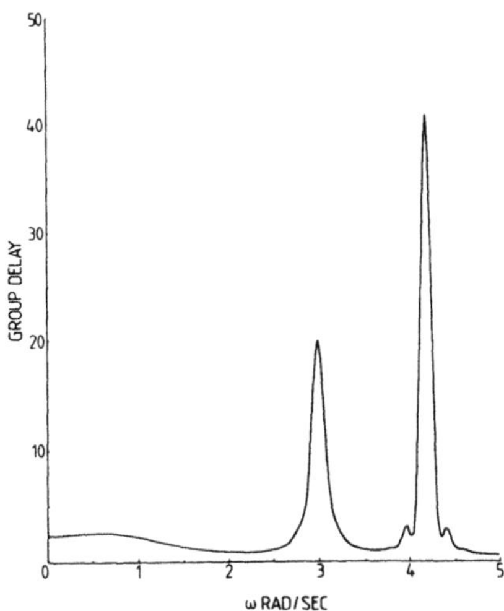

Figure 4.8b. Group-delay response of the low-pass filter.

Table 4.3. Coefficients of the Designed Low-Pass Filter with Constant Group-Delay Characteristics

Numerator coefficient	Denominator coefficient	Typical values of θ
$a_0 = 0.10940$		
$a_1 = -0.09563$	$\alpha_2 = -0.92450$	$\theta_1 = 0.36461$
$a_2 = 0.07519$	$\beta_2 = -0.76160$	$\theta_2 = 0.38417$
$a_3 = 0.25029$	$\alpha_1 = -0.58240$	$\theta_3 = 0.32485$
$a_4 = -0.07875$		

4.4. ANALOGUE FILTER APPROXIMATIONS

A very straightforward approach to the design of recursive digital filters is to apply one of the discretization methods discussed in Section 4.2 to an analogue counterpart. For this reason, in this section, procedures for the design of analogue Butterworth and Chebyshev filters will be presented, while references for the design of elliptic filters are provided for interested readers. A table for transformation from analogue low-pass to low-pass, as well as to band-pass, band-stop, and high-pass filters is also given. Readers are encouraged to familiarize themselves with different types of analogue filters, namely, low-pass, band-stop, band-pass, and high-pass filters, as depicted in Figure 4.9. As can be seen from Figure 4.9, the actual analogue filters are approximations to the ideal response within a prespecified tolerance, which is determined by the design and the application for which this filter has been designed.

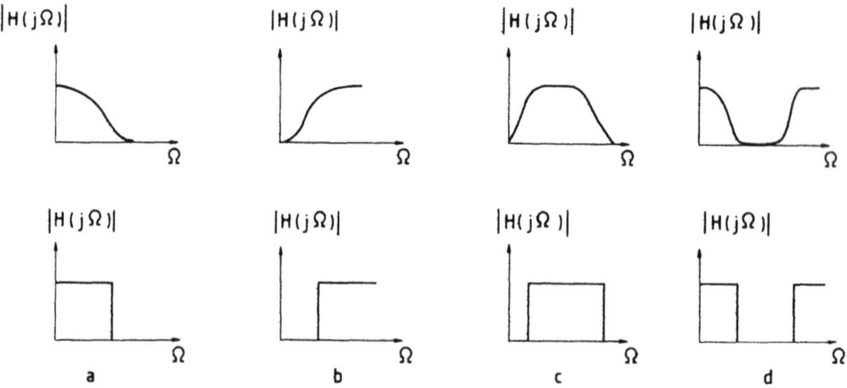

Figure 4.9. Basic types of ideal and realizable analogue filter magnitude responses for (a) low pass, (b) high pass, (c) band pass, (d) band stop.

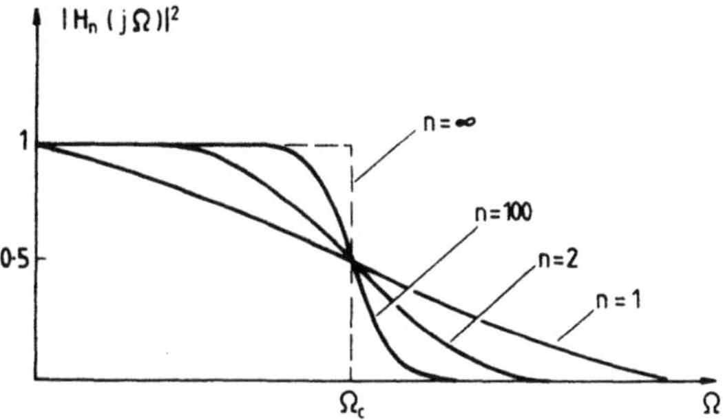

Figure 4.10. Plot of the magnitude-squared function for a Butterworth filter.

4.4.1. The Butterworth Approximation

The magnitude response of the Butterworth filter of order n is characterized by

$$|H_n(j\Omega)| = \left[\frac{1}{1 + (\Omega/\Omega_c)^{2n}}\right]^{1/2} \tag{4.86}$$

where Ω_c is the cutoff frequency of the filter. Figure 4.10 shows the magnitude-squared frequency response of the Butterworth filter for several different values of n.

The following are the properties of the Butterworth filter.

1. Zeros of the Butterworth filter are located at infinity.
2. Binomial series expansion of $|H(j\Omega)|$ from equation (4.86) yields

$$|H(j\Omega)| = 1 - \frac{1}{2}\left(\frac{\Omega}{\Omega_c}\right)^{2n} + \frac{3}{8}\left(\frac{\Omega}{\Omega_c}\right)^{4n} - \frac{5}{16}\left(\frac{\Omega}{\Omega_c}\right)^{6n} + \cdots \tag{4.87}$$

   It is easily seen that the first $2n - 1$ derivatives of equation (4.87) are zero, and therefore $|H_n(j\Omega)|$ is maximally flat at the origin.
3. The magnitude-squared function $|H_n(j\Omega)|^2$ is always equal to unity at the origin ($\Omega = 0$) for all n.
4. The magnitude-squared function $|H_n(j\Omega)|^2_{\Omega=\Omega_c} = \frac{1}{2}$ for all finite n. This implies that $|H_n(j\Omega)|_{\Omega=\Omega_c} = 0.707$ and $20 \log|H_n(j\Omega)|_{\Omega=\Omega_c} = -3.0103$ dB.
5. $|H_n(j\Omega)|^2$ is a monotonically decreasing function of Ω.
6. The magnitude-squared frequency response of a Butterworth filter approaches an ideal low-pass filter when $n \to \infty$.

In this section, we concentrate on the design of the normalized low-pass Butterworth filter for which $\Omega_c = 1$. Other types of Butterworth filter (such as low-pass, high-pass, band-pass, and band-stop) with different cutoff frequencies can be obtained, as will be discussed later, through transformation of the normalized low-pass filter.

Given the magnitude-squared frequency response of equation (4.86) for $\Omega_c = 1$, we would like to find $H(s)$, the filter transfer function, that gives this Butterworth response. We may obtain the frequency response of the required filter by setting $s = j\Omega$ in the transfer function $H(s)$ of the given analogue filter. Thus, if Ω is replaced by s/j, the filter transfer function is obtained. Knowing that for $\Omega_c = 1$

$$|H_n(j\Omega)|^2 = H_n(s)H_n(-s)|_{s=j\Omega} = H_n(j\Omega)H_n(-j\Omega) = 1/(1+\Omega^{2n})$$

then

$$H_n(s)H_n(-s) = \frac{1}{1+(s/j)^{2n}} \tag{4.88}$$

The poles of $H_n(s)H_n(-s)$ can be obtained from the roots of the denominator as follows. Setting

$$1+(s/j)^{2n} = 0$$

we obtain

$$s^{2n} = -1(j)^{2n} = (-1)^{n+1} \tag{4.89}$$

The roots of the above equation can be given as follows:

$$\text{for } n \text{ odd:} \qquad s_l = 1\underline{/l\pi/n}, \qquad l = 0, 1, 2, \ldots, 2n-1$$

$$\text{for } n \text{ even:} \qquad s_l = 1\underline{/\pi/2n + l\pi/n}, \qquad l = 0, 1, 2, \ldots, 2n-1$$

These poles for n odd and even are illustrated in Figure 4.11. Note that, for n odd, we have a pole of $H_n(s)H_n(-s)$ at $s = 1$, and then poles equally spaced on the unit circle π/n in angle, while for n even, the first pole is at $1\underline{/\pi/2n}$ and the remaining poles equally spaced around the unit circle by π/n.

To ensure stability, all the poles in the left-half plane (LHP) should be allocated to $H_n(s)$, and the remaining poles associated with $H_n(-s)$. Therefore $H_n(s)$ can be written in the form

$$H_n(s) = \frac{1}{\prod_{\text{LHP}} (s - s_l)} \tag{4.90}$$

where s_l are all the left-half-plane poles of $H_n(s)H_n(-s)$.

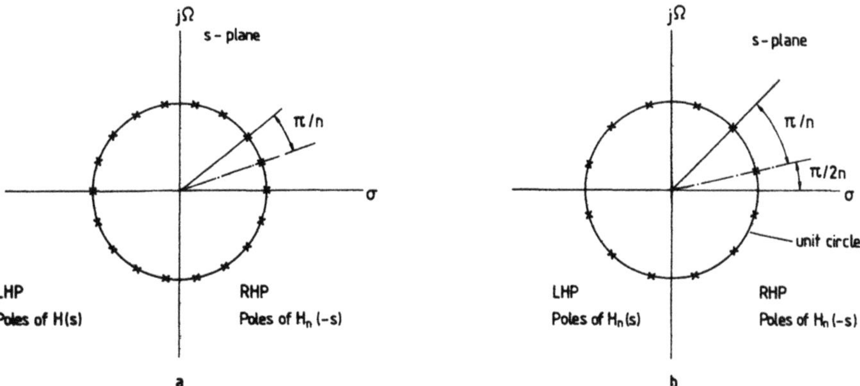

Figure 4.11. Poles of $H_n(s)H_n(-s)$ for a low-pass Butterworth filter with a cutoff frequency of 1.0 rad/s: (a) n odd, (b) n even.

EXAMPLE 4.4. Find the transfer function $H_3(s)$ for the normalized Butterworth filter of order 3.

The poles of $H_3(s)H_3(-s)$ are derived by

$$s_l = 1\underline{/l\pi/n}, \qquad l = 0, 1, 2, 3, 4, 5$$

These poles are shown in Figure 4.12 and, using the left-half-plane poles, we can express the transfer function as follows:

$$H_3(s) = \frac{1}{(s+1)(s+\frac{1}{2}-j\sqrt{3}/2)(s+\frac{1}{2}+j\sqrt{3}/2)}$$

$$= \frac{1}{s^3 + 2s^2 + 2s + 1}$$

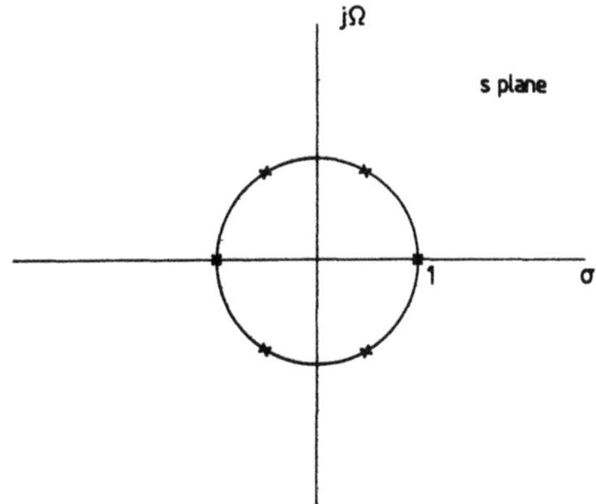

Figure 4.12. Poles for a third-order Butterworth filter.

Butterworth filters with cutoff frequencies other than 1 rad/s can be obtained by applying analogue-to-analogue transformation to the normalized low-pass prototype. This will be discussed in the next section.

4.4.2. Analogue-to-Analogue Transformation

In the previous section, we showed how a normalized low-pass Butterworth filter of any order with cutoff frequency of 1 rad/s can be obtained. In this section, we show that, by analogue-to-analogue transformation, we may also achieve nonnormalized Butterworth filters with different cutoff frequencies, be they low-pass, high-pass, or band-stop filters. It should be noted that this transformation is not limited in its application to Butterworth filters or to a normalized filter. In the following presentation, for the purpose of illustration, a normalized low-pass filter will be used as the prototype filter. If in $H(s)$, the transfer function for a normalized low-pass filter, s is replaced by s/Ω_p, we get a new transfer function $H'(s)$ given by

$$H'(s) = H(s)|_{s \to s/\Omega_p} = H(s/\Omega_p) \tag{4.91}$$

If the magnitude of the transfer function $H'(s)$ at $s = j\Omega$ is evaluated, we can write

$$|H'(j\Omega)| = |H(j\Omega/\Omega_p)| \tag{4.92}$$

We note that, at the value of $\Omega = \Omega_p$, we have

$$|H'(j\Omega_p)| = |H(j\Omega_p/\Omega_p)| = |H(j1)|$$

In other words, the cutoff frequency of the new transfer function has moved from $\Omega = 1$ to $\Omega = \Omega_p$, which represents a scaling of the frequency axis. Similar transformations can be defined for taking low-pass transfer functions to high-pass, band-pass, and band-stop transfer functions. Table 4.4 gives

Table 4.4. Analogue-to-Analogue Transformations

Prototype response	Transformed filter response	Design equations
$20 \log\lvert H(j\Omega)\rvert$ 0, K_1, K_2; 1, Ω_a, Ω Low-pass $H(s)$, $s \to s/\Omega_p$	$20 \log\lvert H'(j\Omega)\rvert$ 0, K_1, K_2; Ω_p, Ω_a', Ω Low-pass $H'(s)$	Forward: $\Omega_a' = \Omega_a\Omega_p'$ Backward: $\Omega_a = \Omega_a'/\Omega_p$

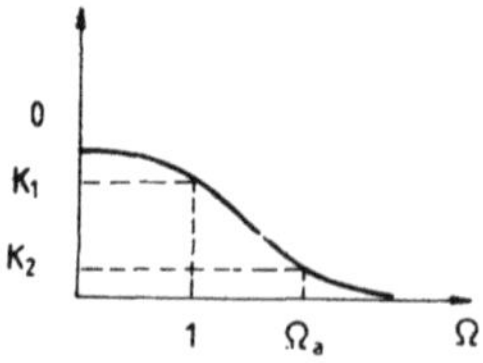

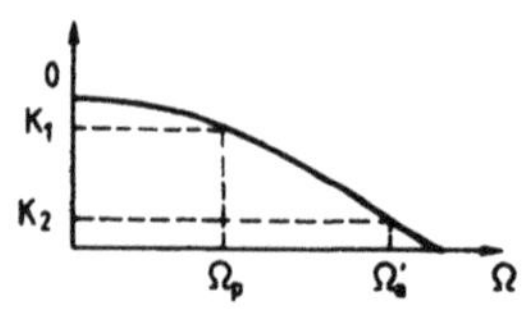

Table 4.4—continued

Prototype response	Transformed filter response	Design equations
$20\log\lvert H(j\Omega)\rvert$ 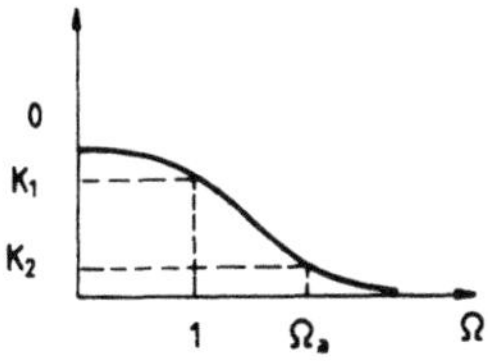 0, K_1, K_2, 1, Ω_a, Ω Low-pass $H(s)$, $s \to \Omega_p/s$	$20\log\lvert H'(j\Omega)\rvert$ 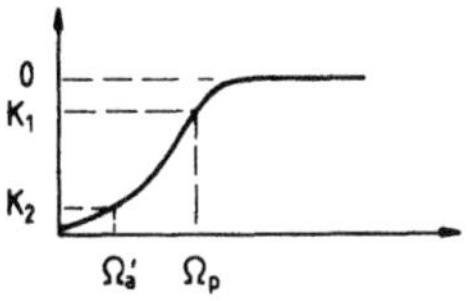 0, K_1, K_2, Ω'_a, Ω_p High-pass $H'(s)$	Forward: $\Omega'_a = \Omega_p/\Omega_a$ Backward: $\Omega_a = \Omega_p/\Omega'_a$
$20\log\lvert H(j\Omega)\rvert$ 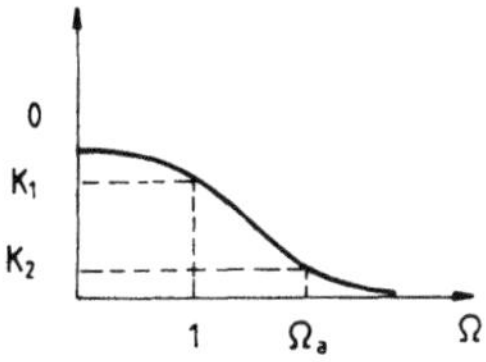 0, K_1, K_2, 1, Ω_a, Ω Low-pass $H(s)$,	$20\log\lvert H'(j\Omega)\rvert$ 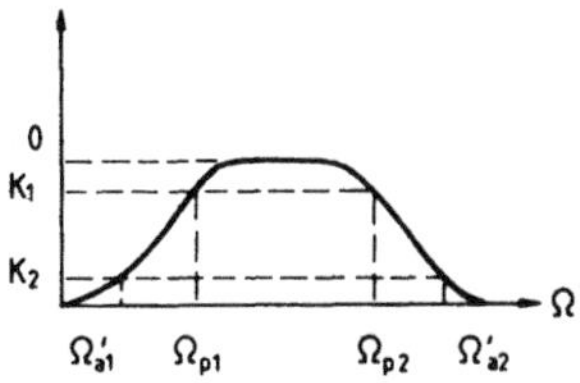 0, K_1, K_2, Ω'_{a1}, Ω_{p1}, Ω_{p2}, Ω'_{a2}, Ω Band-pass $H'(s)$ $s \to \dfrac{s^2 + \Omega_{p1}\Omega_{p2}}{s(\Omega_{p2} - \Omega_{p1})}$	Forward: $\Omega_{av} = (\Omega_{p2} - \Omega_{p1})/2$ $\Omega'_{a1} = (\Omega_a^2\Omega_{av}^2 + \Omega_{p2}\Omega_{p1})^{1/2} - \Omega_{av}\Omega_a$ $\Omega'_{a2} = (\Omega_a^2\Omega_{av}^2 + \Omega_{p2}\Omega_{p1})^{1/2} + \Omega_{av}\Omega_a$ Backward: $\Omega_a = \min\{\lvert A\rvert, \lvert B\rvert\}$ $A = \dfrac{-\Omega'^2_{a1} + \Omega_{p1}\Omega_{p2}}{\Omega'_{a1}(\Omega_{p2} - \Omega_{p1})}$ $B = \dfrac{\Omega'^2_{a2} - \Omega_{p1}\Omega_{p2}}{\Omega'_{a2}(\Omega_{p2} - \Omega_{p1})}$
$20\log\lvert H(j\Omega)\rvert$ 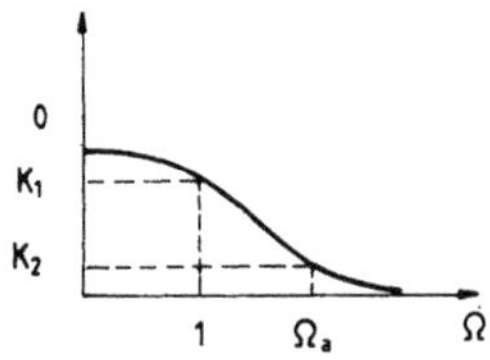 0, K_1, K_2, 1, Ω_a, Ω Low-pass $H(s)$,	$20\log\lvert H'(j\Omega)\rvert$ 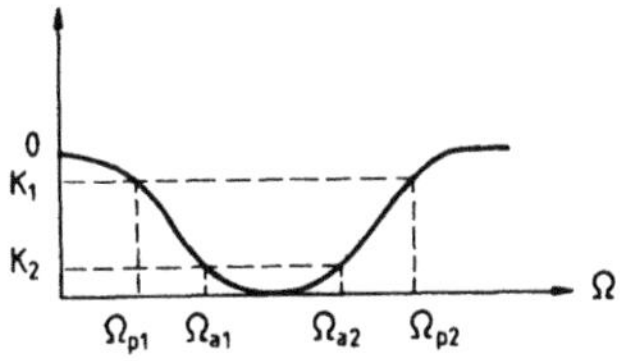 0, K_1, K_2, Ω_{p1}, Ω_{a1}, Ω_{a2}, Ω_{p2}, Ω Band-stop $H'(s)$ $s \to \dfrac{s(\Omega_{p2} - \Omega_{p1})}{s^2 + \Omega_{p2}\Omega_{p1}}$	Forward: $\Omega_{av} = (\Omega_{p2} - \Omega_{p1})/2$ $\Omega_1 = \left(\dfrac{\Omega_{av}}{\Omega_a} + \Omega_{p1}\Omega_{p2}\right)^{1/2} - \dfrac{\Omega_{av}}{\Omega_a}$ $\Omega_2 = \left(\dfrac{\Omega_{av}}{\Omega_a} + \Omega_{p1}\Omega_{p2}\right)^{1/2} + \dfrac{\Omega_{av}}{\Omega_a}$ Backward: $\Omega_a = \min\{\lvert A\rvert, \lvert B\rvert\}$ $A = \Omega'_{a1}\dfrac{\Omega_{p2} - \Omega_{p1}}{-\Omega'^2_{a1} + \Omega_{p1}\Omega_{p2}}$ $B = \Omega'_{a2}\dfrac{\Omega_{p2} - \Omega_{p1}}{\Omega'^2_{a2} - \Omega_{p1}\Omega_{p2}}$

these transformations, along with design equations for both forward and backward development. For example, if the transformation $s \to s/\Omega_p$ is applied to the low-pass structure, as shown at the top of Table 4.4, the stop-band frequency Ω_a will be transformed forward into the new stop-band frequency Ω_a', which is Ω_a times Ω_p.

In a similar manner, if Ω_a' is the desired stop-band frequency of the transformed filter, the backward equation gives the value of Ω_a that must be used, such that going through the transformation $s \to s/\Omega_p$ results in the required Ω_a'. We have Ω_a equals Ω_a'/Ω_p. The following examples demonstrate the design of low-pass and band-pass filters with arbitrary cutoff frequencies. Similar treatment can be adopted by using Table 4.4 for the high-pass and band-stop filters.

4.4.3. Design of Low-Pass Butterworth Filters

The filter specifications are usually given in terms of gains K_1 and K_2 at pass-band (Ω_p) and stop-band (Ω_a) frequencies, respectively. Figure 4.13 shows the desired filter characteristics in dB for a low-pass filter which can be described as

$$0 \geq 20\log|H(j\Omega)| \geq K_1 \qquad \text{for all } \Omega \leq \Omega_p \tag{4.93}$$

$$20\log|H(j\Omega)| \leq K_2 \qquad \text{for all } \Omega \geq \Omega_a \tag{4.94}$$

It is seen from equation (4.86) that the Butterworth low-pass frequency response is characterized by only two parameters: n, the order, and Ω_c, the cutoff frequency of the filter.

If we replace $H(j\Omega)$ in equations (4.93) and (4.94) by the Butterworth magnitude-squared function (4.86) and assume that the equalities hold,

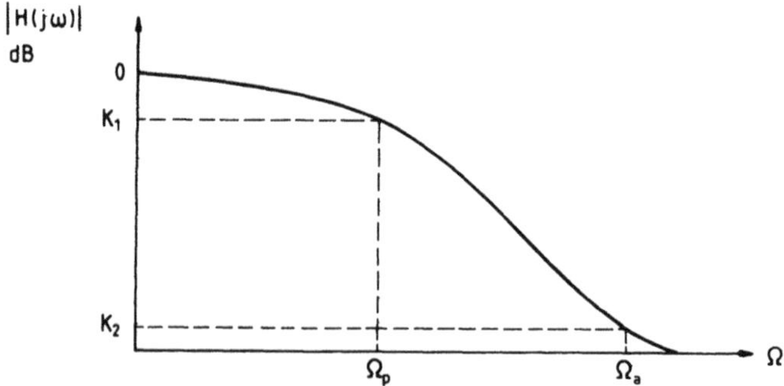

Figure 4.13. A low-pass filter magnitude specification.

then n and Ω_c must satisfy

$$10 \log\left[\frac{1}{1+(\Omega_p/\Omega_c)^{2n}}\right] = K_1 \tag{4.95}$$

and

$$10 \log\left[\frac{1}{1+(\Omega_a/\Omega_c)^{2n}}\right] = K_2 \tag{4.96}$$

Dividing both sides of the above equations by 10, taking the antilog, and simplifying yields

$$(\Omega_p/\Omega_c)^{2n} = 10^{-K_1/10} - 1 \qquad \text{and} \qquad (\Omega_a/\Omega_c)^{2n} = 10^{-K_2/10} - 1 \tag{4.97}$$

On dividing the above two equations to eliminate Ω_c, we obtain the following relationship linking Ω_p, Ω_a, K_1, K_2, and n:

$$(\Omega_p/\Omega_a)^{2n} = \frac{10^{-K_1/10} - 1}{10^{-K_2/10} - 1} \tag{4.98}$$

where n can be calculated from the above using

$$n = \frac{\log_{10}[(10^{-K_1/10} - 1)/(10^{-K_2/10} - 1)]}{2 \log_{10}(\Omega_p/\Omega_a)} \tag{4.99}$$

If n is not an integer, we use the next larger integer. Defining $\lceil \cdot \rceil$ to indicate this operation we get

$$n = \left\lceil \frac{\log_{10}[(10^{-K_1/10} - 1)/(10^{-K_2/10} - 1)]}{2 \log_{10}(\Omega_p/\Omega_a)} \right\rceil \tag{4.100}$$

This value for n results in two different selections for Ω_c, as seen in equation (4.97). If we wish to satisfy our requirement at Ω_p exactly and do better than our requirement at Ω_a, we use

$$\Omega_c = \Omega_p/(10^{-K_1/10} - 1)^{1/2n} \tag{4.101}$$

while if we wish to satisfy our requirement at Ω_a and exceed our requirement at Ω_p, we use

$$\Omega_c = \Omega_a/(10^{-K_2/10} - 1)^{1/2n} \tag{4.102}$$

Choosing a value of Ω_c in between these gives a function $H(j\Omega)$ that exceeds both requirements. An example of the design of a low-pass Butterworth filter using the procedure just described is now given.

EXAMPLE 4.5. Design an analogue low-pass Butterworth filter that has a −2 dB or better cutoff frequency of 40 rad/s and at least 10 dB attenuation at 60 rad/s.

SOLUTION. The specifications are as follows: $\Omega_p = 40$, $K_1 = -2$, $\Omega_a = 60$, and $K_2 = -10$. Substitution of these requirements into equation (4.100) gives

$$n = \left\lceil \frac{\log_{10}[(10^{0.2} - 1)/(10^{+1} - 1)]}{2\log_{10}(40/60)} \right\rceil = \lceil 3.3709 \rceil = 4$$

Quantity Ω_c can now be calculated using equation (4.101) to yield

$$\Omega_c = 40/(10^{0.2} - 1)^{1/8} = 42.7736$$

The transfer function for the normalized low-pass Butterworth filter ($\Omega_c = 1$) for $n = 4$ can be formed using the procedure described earlier, namely

$$H_4(s) = \frac{1}{(s^2 + 0.76536s + 1)(s^2 + 1.84776s + 1)}$$

Applying a low-pass to low-pass transformation $s \to s/\Omega_c$, with $\Omega_c = 42.7736$, to the above equation would yield the desired transfer function:

$$\begin{aligned} H(s) &= H_4(s)\big|_{s \to s/42.7736} \\ &= \frac{1}{\left(\dfrac{s}{42.7736}\right)^2 + 0.76536\left(\dfrac{s}{42.7736}\right) + 1} \\ &\quad \times \frac{1}{\left(\dfrac{s}{42.7736}\right)^2 + 1.84776\left(\dfrac{s}{42.7736}\right) + 1} \\ &= \frac{3.347366 \times 10^6}{(s^2 + 32.7372s + 1829.576)(s^2 + 79.0352s + 1829.576)} \end{aligned}$$

4.4.4. Design of Band-Pass Butterworth Filters

The design of a band-pass filter is also based on applying the band-pass transformation to a low-pass normalized Butterworth filter of the proper order.

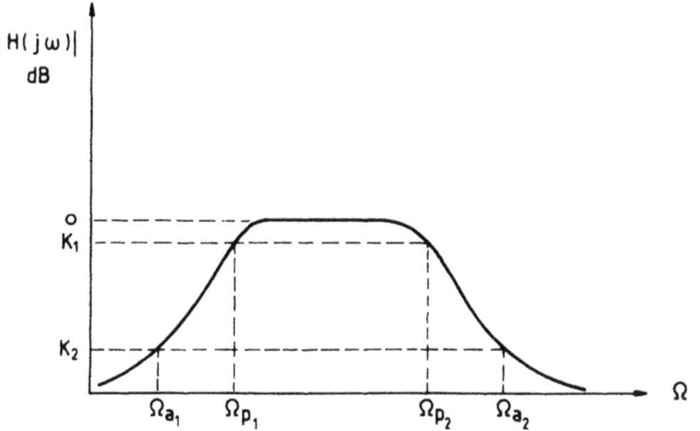

Figure 4.14. A typical specification for a band-pass filter.

The typical band-pass filter specifications are depicted in Figure 4.14, and can be described as

$$20\log|H(j\Omega)| \leq K_2, \qquad \Omega \leq \Omega_{a_1}$$

$$0 \leq 20\log|H(j\Omega)| \leq K_1, \qquad \Omega_{p_1} \leq \Omega \leq \Omega_{p_2}$$

$$20\log|H(j\Omega)| \leq K_2, \qquad \Omega \geq \Omega_{a_2}$$

If $H_{Lp}(s)$ represents a unit-bandwidth low-pass filter with stop-band radian frequency Ω_a, then, from Table 4.1, the transfer function of the required band-pass filter can be calculated using

$$H_{Bp}(s) = H_{Lp}(s)|_{s\to(s^2+\Omega_{p_1}\Omega_{p_2})/[s(\Omega_{p_2}-\Omega_{p_1})]}$$

For the band-pass filter to satisfy the K_2 requirement at Ω_{a_1}, we must have equality within the transformation, i.e.,

$$j\Omega_a = [(j\Omega_{a_1})^2 + \Omega_{p_1}\Omega_{p_2}]/[j\Omega_{a_1}(\Omega_{p_2} - \Omega_{p_1})]$$

Solving the above equation for Ω_a, and a similar equation to satisfy the K_2 requirement at Ω_{a_2}, gives

$$\Omega_a = (\Omega_{a_1}^2 - \Omega_{p_1}\Omega_{p_2})/[\Omega_{a_1}(\Omega_{p_2} - \Omega_{p_1})]$$

and

$$\Omega_a = (\Omega_{a_2}^2 - \Omega_{p_1}\Omega_{p_2})/[\Omega_{a_2}(\Omega_{p_2} - \Omega_{p_1})]$$

Depending upon the size of Ω_{a_1} and Ω_{a_2} with respect to the product $\Omega_{p_1}\Omega_{p_2}$, the quantities Ω_a above could correspond to either positive or negative frequency values. Also, in most cases, the Ω_a values resulting from these equations will not be equal, and we must choose the most restrictive value, thus giving a filter that exceeds our less restrictive requirement. Therefore, the selection of Ω_a becomes that given in the backward design equations for the low-pass to band-pass transformation part of Table 4.1:

$$\Omega_a = \min\{|A|, |B|\} \tag{4.103a}$$

where

$$A = (-\Omega_{a_1}^2 + \Omega_{p_1}\Omega_{p_2})/[\Omega_{a_1}(\Omega_{p_2} - \Omega_{p_1})] \tag{4.103b}$$

and

$$B = (\Omega_{a_2}^2 - \Omega_{p_1}\Omega_{p_2})/[\Omega_{a_2}(\Omega_{p_2} - \Omega_{p_1})] \tag{4.103c}$$

The procedure for the design of a band-pass filter, $H_{BP}(s)$, to satisfy the given set of specifications comprises two steps. First, design a low-pass filter $H_{LP}(s)$ with Ω_a determined above, and second, apply the low-pass to band-pass transformation using the desired Ω_{p_2} and Ω_{p_1}.

4.5. CHEBYSHEV FILTERS

There are two types of Chebyshev filters: those which contain ripples in their pass band (type I) and those which contain ripples in their stop band (type II). A normalized, type I low-pass Chebyshev filter is characterized by its magnitude-squared frequency response given by

$$|H_n(j\Omega)|^2 = \frac{1}{1 + \varepsilon^2 T_n^2(\Omega)} \tag{4.104}$$

where $T_n(\Omega)$ is the nth-order Chebyshev polynomial defined by the recursive equation

$$T_n(x) = 2xT_{n-1}(x) - T_{n-2}(x), \qquad n > 2 \tag{4.105}$$

with

$$T_0(x) = 1 \qquad \text{and} \qquad T_1(x) = x$$

Other Chebyshev polynomials of higher order can be generated using equation (4.105). A list of the first twelve Chebyshev polynomials is given in Table 4.5.

Table 4.5. The First Twelve Chebyshev Polynomials

n	$T_n(x)$
0	1
1	x
2	$2x^2 - 1$
3	$4x^3 - 3x$
4	$8x^4 - 8x^2 + 1$
5	$16x^5 - 20x^3 + 5x$
6	$32x^6 - 48x^4 + 18x^2 - 1$
7	$64x^7 - 112x^5 + 56x^3 - 7x$
8	$128x^8 - 256x^6 + 160x^4 - 32x^2 + 1$
9	$256x^9 - 576x^7 + 432x^5 - 120x^3 + 9x$
10	$512x^{10} - 1280x^8 + 1120x^6 - 400x^4 + 50x^2 - 1$
11	$1024x^{11} - 2816x^9 + 2816x^7 - 1232x^5 + 220x^3 - 11x$
12	$2048x^{12} - 6144x^{10} + 6912x^8 - 3584x^6 + 840x^4 - 72x^2 + 1$

In Figure 4.15, polynomial $T_5(\Omega)$ and its corresponding magnitude-squared function for the Chebyshev filter of type I is plotted. It can be seen from Figure 4.15a that $T_5(\Omega)$ oscillates between plus and minus one within the interval $-1 \leq \Omega \leq 1$, while outside that interval it grows toward plus and minus infinity. The same type of oscillation takes place for every Chebyshev polynomial and causes the equiripple magnitude in $|H_n(\mathrm{j}\Omega)|^2$ as shown in Figure 4.15b. The amplitude of $|H_n(\mathrm{j}\Omega)|^2$ oscillates between 1 and $1/(1 + \varepsilon^2)$ as Ω varies between -1 and $+1$, but the periods of the oscillations are not equal. Function $|H_n(\mathrm{j}\Omega)|^2$ approaches zero outside the intervals -1 or $+1$, since $T_n(\Omega) \to \infty$ outside the same interval.

It can easily be seen from Table 4.5 that $T_n(x) = 1$ at $x = 0$ for n even and $T_n(x) = 0$ at $x = 0$ for n odd, causing $|H_n(\mathrm{j}\Omega)|^2$ to be $1/(1 + \varepsilon^2)$ at $\Omega = 0$ for n even and 1 at $\Omega = 0$ for n odd.

The above discussions can be summarized in the following properties for the Chebyshev filters of type I.

1. The magnitude-squared frequency response oscillates between 1 and $1/(1 + \varepsilon^2)$ within the pass band, the so-called equiripple, and has a value of $1/(1 + \varepsilon^2)$ at $\Omega = 1$, the so-called cutoff frequency.
2. The magnitude-squared frequency response $|H_n(\mathrm{j}\Omega)|^2$ is monotonic outside the pass band, including both the transition band and the stop band. The stop band begins at Ω_a with magnitude-squared frequency response at value $1/A^2$. (These are depicted in Figure 4.16.)

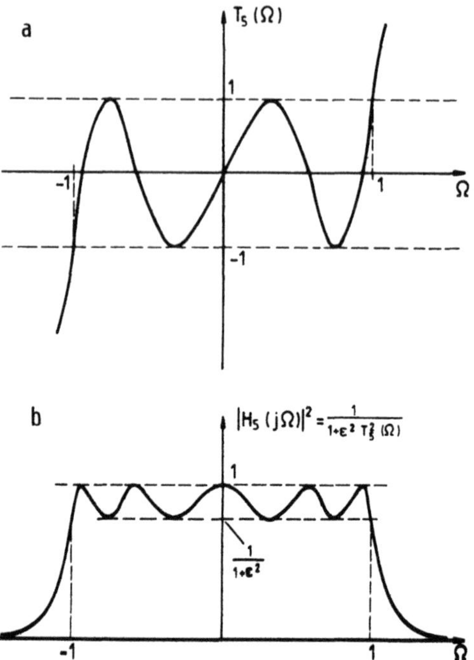

Figure 4.15. (a) Plot of the fifth-order Chebyshev polynomial T_5; (b) plot of the corresponding function $|H_5(j\Omega)|^2$ for the Chebyshev filter.

To obtain the causal stable transfer function $H_n(s)$ that corresponds to the Chebyshev magnitude-squared function in equation (4.104), we must find the poles of $H_n(s)H_n(-s)$ and select the left-half-plane poles for $H_n(s)$. The poles of $H_n(s)H_n(-s)$ are obtained by finding the roots of

$$1 + \varepsilon^2 T_n^2(s/j) = 0 \tag{4.106}$$

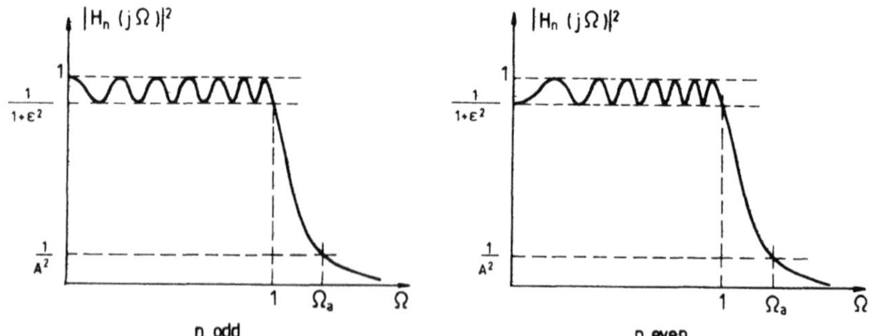

Figure 4.16. Magnitude-squared frequency responses for the normalized Chebyshev filter of odd and even n.

given by $s_k = \sigma_k + j\Omega_k$ for $k = 1, n$. These lie on an ellipse satisfying the equation

$$\sigma_k^2/a^2 + \Omega_k^2/b^2 = 1 \tag{4.107}$$

where a, b and σ_k, Ω_k are given by

$$a = \tfrac{1}{2}\{[1 + (1 + \varepsilon^2)^{1/2}]/\varepsilon\}^{1/n} - \tfrac{1}{2}\{[1 + (1 + \varepsilon^2)^{1/2}]/\varepsilon\}^{-1/n}$$

$$b = \tfrac{1}{2}\{[1 + (1 + \varepsilon^2)^{1/2}]/\varepsilon\}^{1/n} + \tfrac{1}{2}\{[1 + (1 + \varepsilon^2)^{1/2}]/\varepsilon\}^{-1/n} \tag{4.108}$$

and

$$\begin{aligned} \sigma_k &= -a \sin[(2k-1)\pi/2n] \\ \Omega_k &= b \cos[(2k-1)\pi/2n] \end{aligned} \qquad k = 1, 2, \ldots, 2n \tag{4.109}$$

If the left-half-plane poles only are employed, then function $H_n(s)$ can be written in the form

$$H_n(s) = \frac{k}{\prod_{\substack{\text{LHP} \\ \text{poles}}} (s - s_k)} = \frac{k}{F_n(s)} \tag{4.110}$$

where

$$F_n(s) = s^n + f_{n-1}s^{n-1} + \cdots + f_1 s + f_0$$

The normalizing constant k has a value which makes $H(0)$ equal to 1 for n odd and $1/(1 + \varepsilon^2)^{1/2}$ for n even; thus

$$k = F_n(0) = f_0, \qquad n \text{ odd}$$

$$k = F_n(0)/(1 + \varepsilon^2)^{1/2}, \qquad n \text{ even}$$

We see that the normalized Chebyshev low-pass filter has only two parameters, ε and n. In the following, we show how these two parameters can be selected so that the filter meets a prespecified set of specifications.

Selection of ε. This quantity may be determined by noting that the maximum pass-band ripple is equal to $1/(1 + \varepsilon^2)$.

Selection of n. For low-pass filter design, we are usually interested not only in the cutoff frequency, but also in the stop-band attenuation. Usual specifications might be that the magnitude-squared frequency response is less than a certain value $1/A$ at a frequency Ω_a in the stop band. Given the ripple factor ε and the stop-band gain $1/A$ at Ω_a, n can be calculated using the relationship

$$n = \left\lceil \frac{\log_{10}[g + (g^2 - 1)^{1/2}]}{\log_{10}[\Omega_a + (\Omega_a^2 - 1)^{1/2}]} \right\rceil \tag{4.111}$$

where

$$A = 1/|H_n(j\Omega_a)| \qquad \text{and} \qquad g = [(A^2 - 1)/\varepsilon^2]^{1/2} \tag{4.112}$$

Example 4.6. To illustrate how the above procedure can be used to design a normalized Chebyshev filter, we design a 1-D low-pass filter with the following specifications:

1. $\Omega_p = 1$ rad/s.
2. Acceptable pass-band ripple of 1 dB.
3. $\Omega_a = 2$ rad/s.
4. Attenuation of 30 dB or greater beyond the stop-band frequency of 2 rad/s.

Solution. To satisfy the ripple and stop-band attenuation requirements, we have

$$20 \log|H_n(j1)| = 20 \log\left[\frac{1}{1+\varepsilon^2}\right]^{1/2} = 10 \log\frac{1}{1+\varepsilon^2} = -1$$

and

$$20 \log|H_n(j\Omega_a)| = 20 \log(1/A) = -30$$

the solution of which gives $\varepsilon = 0.5088$ and $A = 31.62$. Equations (4.111) and (4.112) enable the corresponding g and n to be determined as follows:

$$g = \left\{\frac{[(31.62)^2 - 1]}{(0.5088)^2}\right\}^{1/2} = 62.14$$

$$n = \left\lceil \frac{\log\{62.14 + [(62.14)^2 - 1]^{1/2}\}}{\log[2 + (2^2 - 1)^{1/2}]} \right\rceil = 4$$

The procedure described earlier can now be followed to arrive at $H_4(s)$, the transfer function of the filter. The detailed procedure can be found elsewhere.[2,9,10]

4.6. DIGITAL-TO-DIGITAL TRANSFORMATION

As in the continuous case, there exists a set of simple transformations suitable for converting a digital low-pass filter (of cutoff frequency ω_c) to another low-pass, high-pass, band-pass, or band-stop filter with desired frequency specifications. The transformations are summarized below.

1. For a low-pass to low-pass filter

$$z^{-1} \to \frac{z^{-1} - \alpha}{1 - \alpha z^{-1}} \tag{4.113}$$

where

$$\alpha = \frac{\sin\{[(\omega_c - \omega_u)/2]T\}}{\sin\{[(\omega_c + \omega_u)/2]T\}} \tag{4.114}$$

and ω_u is the new low-pass cutoff frequency.

2. For a low-pass to high-pass filter

$$z^{-1} \to \left(\frac{z^{-1} + \alpha}{1 + \alpha z^{-1}}\right) \tag{4.115}$$

where

$$\alpha = -\frac{\cos\{[(\omega_c - \omega_u)/2]T\}}{\cos\{[(\omega_c + \omega_u)/2]T\}} \tag{4.116}$$

and ω_u is the desired high-pass cutoff frequency.

3. For a low-pass to band-pass filter

$$z^{-1} \to -\frac{\{z^{-2} - [2\alpha k/(k+1)]z^{-1} + (k-1)/(k+1)\}}{\{[(k-1)/(k+1)]z^{-2} - [2\alpha k/(k+1)]z^{-1} + 1\}} \tag{4.117}$$

where

$$\alpha = \cos(\omega_0 T) = \frac{\cos\{[(\omega_u + \omega_l)/2]T\}}{\cos\{[(\omega_u - \omega_l)/2]T\}} \tag{4.118}$$

$$k = \cos\left[\left(\frac{\omega_u - \omega_l}{2}\right)T\right]\tan\left(\frac{\omega_c T}{2}\right) \tag{4.119}$$

ω_0 is the center frequency of the band-pass filter, while ω_u and ω_l are the defined upper and lower cutoff frequencies, respectively.

4. For a low-pass to band-stop filter

$$z^{-1} \to \frac{\{z^{-2} - [2\alpha/(1+k)]z^{-1} + (1-k)/(1+k)\}}{\{[(1-k)/(1+k)]z^{-2} - [2\alpha/(1+k)]z^{-1} + 1\}} \tag{4.120}$$

where

$$\alpha = \cos(\omega_0 T) = \frac{\cos\{[(\omega_u - \omega_l)/2]T\}}{\cos\{[(\omega_u + \omega_l)/2]T\}} \tag{4.121}$$

and

$$k = \tan\left[\left(\frac{\omega_u - \omega_l}{2}\right)T\right]\tan\left(\frac{\omega_c T}{2}\right) \tag{4.122}$$

The transformations in equations (4.113)–(4.122) are easily seen to be of the all-pass type, i.e., the unit circle is mapped into itself one or more times. Thus although the frequency scale is warped by the mapping, the magnitude response of the original low-pass filter is preserved. Thus, for example, the Chebyshev low-pass filter is transformed into new Chebyshev filters.

PART TWO. TWO-DIMENSIONAL FILTERS

4.7. INTRODUCTION

A linear two-dimensional (2-D) recursive digital filter can be characterized by its transfer function

$$H_D(z_1, z_2) = \frac{N(z_1, z_2)}{D(z_1, z_2)} = \frac{\sum_{i=0}^{M_1} \sum_{j=0}^{N_1} a(i,j) z_1^{-i} z_2^{-j}}{\sum_{i=0}^{M_2} \sum_{j=0}^{N_2} b(i,j) z_1^{-i} z_2^{-j}} \tag{4.123}$$

If $z_i = e^{j\omega_i T_i}$, where the sampling period $T_i = 2\pi/\omega_{s_i}$ for $i = 1, 2$ with ω_{s_i} the sampling frequencies, **a** and **b** are vectors of real numbers and known as the coefficients of the filter. Without loss of generality we can assume $M_1 = M_2 = N_1 = N_2 = M$ and $T_1 = T_2 = T$. Designing a 2-D filter involves calculations of the filter's coefficients **a** and **b** in such a way that the magnitude response and/or the phase response (group delay) of the designed filter approximates an ideal one while maintaining the stability of the designed filter. The latter requires

$$D(z_1, z_2) \neq 0 \qquad \text{for } \bigcap_{i=1}^{2} |z_i| \geq 1 \tag{4.124}$$

The magnitude response of the 2-D filter is expressed as

$$M(\omega_1, \omega_2) = |H_D(e^{j\omega_1 T}, e^{j\omega_2 T})| \tag{4.125}$$

the phase response as

$$\phi_D(\omega_1, \omega_2) = \arg H(e^{j\omega_1 T}, e^{j\omega_2 T}) \tag{4.126}$$

and the two group-delay functions as

$$\tau_i(\omega_1, \omega_2) = -\frac{d\phi_D(\omega_1, \omega_2)}{d\omega_i}, \qquad i = 1, 2 \tag{4.127}$$

The transfer function of the 2-D recursive digital filter can involve two subclasses of the general form shown in equation (4.123), namely, the separable product transfer function

$$\begin{aligned} H(z_1, z_2) &= H_1(z_1)H_2(z_2) \\ &= \left(\frac{\sum_{i=0}^{M} a_{1i}z_1^{-i}}{\sum_{i=0}^{M} b_{1i}z_1^{-i}}\right)\left(\frac{\sum_{j=0}^{N} a_{2j}z_2^{-j}}{\sum_{j=0}^{N} b_{2j}z_2^{-j}}\right) \end{aligned} \tag{4.128}$$

and the separable denominator, nonseparable numerator transfer function given by

$$H(z_1, z_2) = \frac{\sum_{i=0}^{M} \sum_{j=0}^{M} a(i, j) z_1^{-i} z_2^{-j}}{\left(\sum_{i=0}^{M} b_{1i}z_1^{-i}\right)\left(\sum_{j=0}^{M} b_{2j}z_2^{-j}\right)} \tag{4.129}$$

The stability constraints for the above two transfer functions are those of two 1-D ones. These are easy to check and easy to stabilize if the designed filter is found to be unstable. The design of 2-D filters is therefore divided into three classes: that of the separable product, the nonseparable numerator and separable denominator type, and the general case of nonseparable numerator and denominator filters.

4.8. SEPARABLE PRODUCT FILTERS

The design of 2-D digital filters can be simplified considerably by choosing constants $a(i, j)$ and $b(i, j)$ in such a way so as to permit the factorization of $H(z_1, z_2)$ into a product of two 1-D transfer functions, as in equation (4.128)[11]:

$$H(z_1, z_2) = H_1(z_1)H_2(z_2) \tag{4.130}$$

In these circumstances, the desired 2-D filter can be obtained by cascading two 1-D filters. The two 1-D filters can easily be designed by using known techniques discussed in Section 4.4. It should be noted that filters designed with a separable product transfer function will have their pass-band and stop-band contours rectangular in shape.

It is important to know how the specification of the 2-D filter in terms of the pass-band ripple and stop-band loss affects the constraints which must be imposed on the two 1-D filters that are cascaded to form the 2-D transfer function. Let us assume that we are to design a 2-D low-pass filter with the following specification:

$$|H(\omega_1, \omega_2)| = \begin{cases} 1 \pm \Delta p & \text{for } 0 \leq |\omega_i| \leq \omega_{p_i} \\ \Delta a & \text{for } \omega_{a_i} \leq |\omega_i| \leq \omega_{s_i}/2 \end{cases} \qquad i = 1, 2 \tag{4.131}$$

where ω_{p_i} and ω_{a_i}, $i = 1, 2$ are pass-band and stop-band edges along the ω_1, ω_2 axis, respectively. The two 1-D filters which are cascaded to form a 2-D filter are specified by ω_{p_i}, ω_{a_i}, δ_{p_i} and δ_{a_i} $(i = 1, 2)$ as the pass-band edge, stop-band edge, pass-band ripple, and stop-band loss, respectively. From equation (4.130) we have

$$\max\{M(\omega_1, \omega_2)\} = \max\{M_1(\omega_1)\} \max\{M_2(\omega_2)\}$$

and

$$\min\{M(\omega_1, \omega_2)\} = \min\{M_1(\omega_1)\} \min\{M_2(\omega_2)\}$$

hence the derived 2-D filter will satisfy the specifications of equation (4.131) if the following constraints are satisfied:

$$(1 + \delta_{p_1})(1 + \delta_{p_2}) \leq 1 + \Delta p \tag{4.132}$$

$$(1 - \delta_{p_1})(1 - \delta_{p_2}) \geq 1 - \Delta p \tag{4.133}$$

$$(1 + \delta_{p_1})\delta_{a_2} \leq \Delta a \tag{4.134}$$

$$(1 + \delta_{p_2})\delta_{a_1} \leq \Delta a \tag{4.135}$$

$$\delta_{a_1}\delta_{a_2} \leq \Delta a \tag{4.136}$$

Constraints (4.132) and (4.133) can be expressed respectively in the alternative form

$$\delta_{p_1} + \delta_{p_2} + \delta_{p_1}\delta_{p_2} \leq \Delta p \tag{4.137}$$

and

$$\delta_{p_1} + \delta_{p_2} - \delta_{p_1}\delta_{p_2} \leq \Delta p \tag{4.138}$$

Hence if relation (4.137) is satisfied, relation (4.138) is also satisfied. Similarly, constants (4.134)-(4.136) will be satisfied if

$$\max\{(1 + \delta_{p_1})\delta_{a_2}, (1 + \delta_{p_2})\delta_{a_1}\} \leq \Delta a \tag{4.139}$$

since

$$(1 + \delta_{p_1}) \gg \delta_{a_1} \qquad \text{and} \qquad (1 + \delta_{p_2}) \gg \delta_{a_2} \tag{4.140}$$

Now if we assume that

$$\delta_{p_1} = \delta_{p_2} = \delta_p \tag{4.141}$$

and

$$\delta_{a_1} = \delta_{a_2} = \delta_a \tag{4.142}$$

then we can assign

$$\delta_p = (1 + \Delta p)^{1/2} - 1 \tag{4.143}$$

and

$$\delta_a = \Delta a/(1 + \Delta p)^{1/2} \tag{4.144}$$

so as to satisfy constraints (4.132)-(4.136). With the specifications of the two 1-D filters, however, the two filters can now be designed using any of the approaches of the previous sections. Similar treatments for band-pass, band-stop, and high-pass filters can be carried out. It should be noted that the only drawback with this technique is that the circular-symmetric specification is not possible using separable product transfer function.

4.9. SEPARABLE DENOMINATOR, NONSEPARABLE NUMERATOR FILTERS

It has been shown recently[12-14] that circular-symmetric 2-D filters can be designed using a separable denominator, nonseparable numerator transfer function of the form

$$H(z_1, z_2) = \frac{\sum_{i=0}^{M} \sum_{j=0}^{M} a(i,j) z_1^{-i} z_2^{-j}}{\sum_{i=0}^{M} b_{1i} z_1^{-i} \sum_{j=0}^{M} b_{2j} z_2^{-j}} \tag{4.145}$$

This would also reduce the stability problem associated with 2-D filters to that of the 1-D case which can be tested at the end of the design process. The stabilization procedure will require replacement of the poles located outside the unit circles in the z_1 and z_2 planes by their mirror images with respect to the unit circles in the z_1 and z_2 planes. This stabilization procedure unfortunately changes the group-delay (or phase) characteristics of the designed filter, i.e., if the designed filter has constant group-delay responses (linear phase), at the end of this process the group-delay responses will no longer remain constant; this is undesirable in image-processing applications.[15] With this transfer function, also, a reduction of $(M^2 - 1)$ multiplier coefficients can be achieved; this leads to a considerable saving of computation time, and also a reduction in hardware costs.

In the technique to be presented here, the two denominator factors are given by equation (3.15) or (3.17). These are, for even values of M and $j = 1, 2$,

$$D_j(z_j) = K_{ej} \prod_{i=1}^{n} (z_j^2 - 2\alpha_{ji} z_j + 1) + (z_j^2 - 1) \prod_{i=1}^{n-1} (z_j^2 - 2\beta_{ji} z_j + 1) \tag{4.146}$$

where $n = M/2$, with the stability constraint

$$1 > \alpha_1 > \beta_1 > \alpha_2 > \beta_2 > \cdots > \beta_{n-1} > \alpha_n > -1 \tag{4.147}$$

or, for odd values,

$$D_j(z_j) = K_{oj}(z_j + 1) \prod_{i=1}^{n} (z_j^2 - 2\alpha_{ji} z_j + 1)$$

$$+ (z_j - 1) \prod_{i=1}^{n} (z_j^2 - 2\beta_{ji} z_j + 1) \tag{4.148}$$

where $n = (M - 1)/2$, with the stability constraint

$$1 > \alpha_1 > \beta_1 > \alpha_2 > \beta_2 > \cdots > \alpha_N > \beta_N > -1 \tag{4.149}$$

In the above equations K_{ej} and K_{oj} are positive numbers and $|d_0/d_n| < 1$. The numerator is left unchanged. As discussed in Chapter 3, all poles of equation (4.146) or (4.148) are located inside the unit circle in the z plane if condition (4.147) or (4.149) is satisfied.

Now the error between the ideal and designed magnitude response of the 2-D filter is calculated using the relationship

$$E_{\text{Mag}}(j\omega_{1m}, j\omega_{2n}) = M_{\text{I}}(\omega_{1m}, \omega_{2n}) - M_{\text{D}}(\omega_{1m}, \omega_{2n}) \tag{4.150}$$

where E_{Mag} is the error of the magnitude response while M_{I} and M_{D} are the magnitude responses of the ideal and designed filter, respectively. Also E_{τ_i} $(i = 1, 2)$ is the error between the ideal and designed group-delay response of the filter and is defined as

$$E_{\tau_i}(j\omega_{1m}, j\omega_{2n}) = \tau_{\text{I}} T - \tau_i(\omega_{1m}, \omega_{2n}) \tag{4.151}$$

where τ_{I} is a constant representing the ideal group-delay response of the filter and its value is usually chosen to be equal to the order of the filter; τ_i $(i = 1, 2)$ is the group-delay response of the designed filter. To formulate the design problem for simultaneous approximation of the magnitude and group-delay response of the filter, we choose the general mean-square error, which is calculated using equations (4.150) and (4.151) in the following relationship:

$$\begin{aligned} E_{\text{G}}(j\omega_{1m}, j\omega_{2n}, \mathbf{a}, \boldsymbol{\alpha}, \boldsymbol{\beta}) = &\sum_{m,n \in I_{\text{ps}}}\sum E^2_{\text{Mag}}(j\omega_{1m}, j\omega_{2n}) + \sum_{m,n \in I_{\text{p}}}\sum E^2_{\tau_1}(j\omega_{1m}, j\omega_{2n}) \\ &+ \sum_{m,n \in I_{\text{p}}}\sum E^2_{\tau_2}(j\omega_{1m}, j\omega_{2n}) \end{aligned} \tag{4.152}$$

where I_{ps} is a set of all discrete frequency pairs along the ω_1 and ω_2 axes covering the pass band and stop band of the filter, while I_{p} is a set of all discrete frequency pairs along the ω_1 and ω_2 axes covering only the pass band of the filter. (Obviously I_{p} is a subset of I_{ps}.) To design a 2-D filter satisfying a prescribed magnitude and constant group-delay responses, quantities $\boldsymbol{\alpha}$, $\boldsymbol{\beta}$, and $\mathbf{a}$ should be calculated in such a way that E_{G} in equation (4.152) is minimized subject to constraint (4.147) or (4.149). This constrained optimization problem can be transformed to an unconstrained optimization problem by employing the variable substitution, described in equation (4.85), as follows:

$$\begin{aligned}
\alpha_{n,j} &= \cos \pi \exp(-\theta_{1j}^2) \\
\beta_{n-1,j} &= \cos \pi \exp[-(\theta_{1j}^2 + \theta_{2j}^2)] \\
\alpha_{n-1,j} &= \cos \pi \exp[-(\theta_{1j}^2 + \theta_{2j}^2 + \theta_{3j}^2)] \\
&\vdots \\
\alpha_{1,j} &= \cos\left[\pi \exp\left(-\sum_{i=1}^{2n-1} \theta_{i,j}^2\right)\right] \qquad \text{for } j = 1, 2
\end{aligned} \tag{4.153}$$

Now any unconstrained optimization method can be utilized to compute the new filter's coefficients $\mathbf{a}$ and $\boldsymbol{\theta}$ by minimizing E_G in equation (4.152). It should be noted that, if magnitude-only specification is required, calculation of the sum of the squares of group-delay errors in the last two terms of equation (4.152) will be omitted.

EXAMPLE 4.7. We design a 2-D low-pass filter with the following magnitude specification:

$$M_I(\omega_1, \omega_2) = \begin{cases} 1 & \text{for } 0 \le (\omega_1^2 + \omega_2^2)^{1/2} \le 1.0 \text{ rad/s} \\ 0 & \text{for } 2.5 \le (\omega_1^2 + \omega_2^2)^{1/2} \le \omega_s/2 = 5.0 \text{ rad/s} \end{cases}$$

and constant group-delay characteristics. The order of the filter is chosen to be 4. Consequently τ_I in equation (4.151) is set equal to 4. The parameters of the designed filter using the technique discussed earlier are shown in Table 4.6, and the 3-D plot of the magnitude response, as well as the group-delay responses of the designed filter, are given in Figures 4.17a, 4.17b, and 4.17c, respectively.

4.10. NONSEPARABLE NUMERATOR AND DENOMINATOR FILTERS

In the earlier two classes, in order to reduce the stability problem to that of the 1-D case, the denominator of the 2-D transfer function is chosen to have two 1-D polynomials in z_1 and z_2 variables in cascade. However,

Table 4.6. Values of the Coefficients of the Designed Two-Dimensional Low-Pass Filter with Constant Group Delays

Numerator coefficients		Denominator coefficients
$n_{00} = -0.49500$	$n_{23} = 3.06599$	
$n_{01} = 0.13000$	$n_{24} = 1.30000$	
$n_{02} = 0.96500$	$n_{30} = 0.12000$	
$n_{03} = 0.69500$	$n_{31} = 0.60000$	$\alpha_{12} = -0.29546$
$n_{04} = 0.24500$	$n_{32} = 1.60999$	$\beta_{11} = 0.15016$
$n_{10} = 0.41500$	$n_{33} = 1.25498$	$\alpha_{11} = 0.37127$
$n_{11} = 0.12499$	$n_{34} = 0.74000$	$\alpha_{22} = -0.28257$
$n_{12} = 2.10499$	$n_{40} = -0.05000$	$\beta_{21} = 0.08156$
$n_{13} = 1.33499$	$n_{41} = -0.30000$	$\alpha_{21} = 0.32189$
$n_{14} = 0.71500$	$n_{42} = 0.14500$	
$n_{20} = 0.85500$	$n_{43} = 0.32500$	
$n_{21} = 1.56499$	$n_{44} = -0.36000$	
$n_{22} = 1.98500$		

in the general formulation, this oversimplification is removed. The filters of this class are generally designed either through transformation of 1-D filters, or through a linear programming or nonlinear programming approach.

4.11. TRANSFORMATION TECHNIQUES

4.11.1. Rotated Filters

The design technique proposed by Shanks *et al.*[16] consists of mapping 1-D into 2-D filters with arbitrary directivity in a 2-D frequency response plane. These filters are called rotated filters because they are obtained by rotating 1-D filters.

Suppose a 1-D continuous filter, whose impulse response is real, is given in its factored form

$$H_1(s) = H_0 \frac{\prod_{i=1}^{m} (s - q_i)}{\prod_{i=1}^{n} (s - p_i)} \tag{4.154}$$

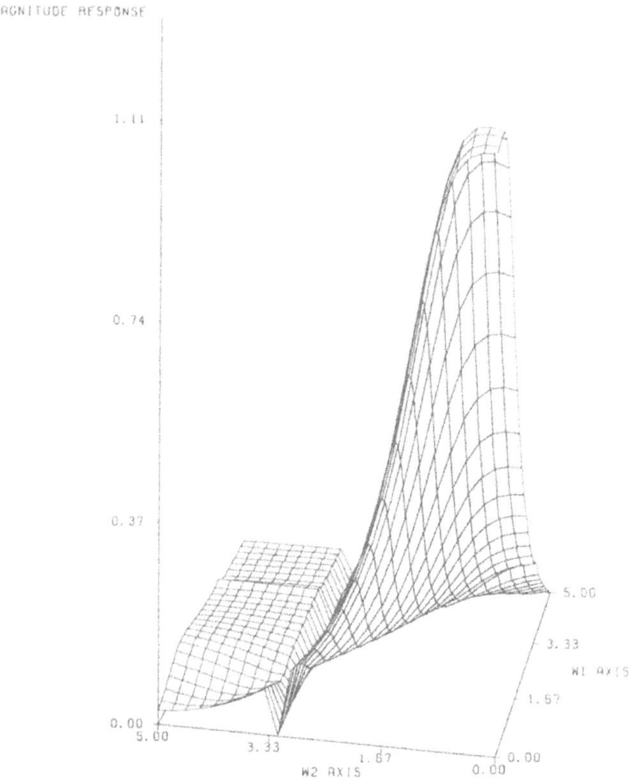

Figure 4.17a. Three-dimensional plot of the magnitude response of a two-dimensional low-pass filter.

where H_0 is a scalar gain constant. The zero locations q_i and pole locations p_i may be complex, in which case their conjugates are also present in the corresponding product. The cutoff frequency for this filter is assumed to be unity.

The filter given in equation (4.154) can also be viewed as a 2-D filter that varies in one dimension only and could be written as follows:

$$H_2(s_1, s_2) = H_1(s_2) = H_0 \frac{\prod_{i=1}^{m} (s_2 - q_i)}{\prod_{i=1}^{n} (s_2 - p_i)} \tag{4.155}$$

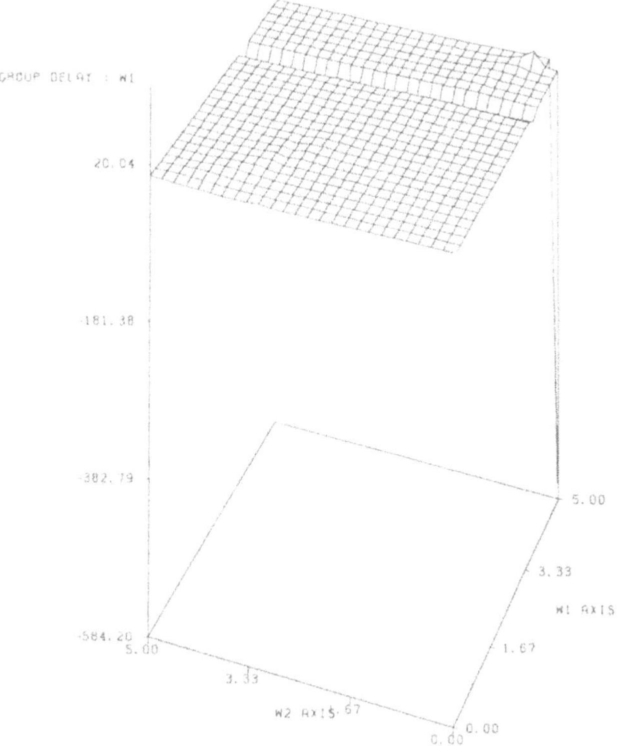

Figure 4.17b. Three-dimensional plot of the group-delay response with respect to ω_1 of the two-dimensional filter.

Clockwise rotation of the s_1, s_2 axis through an angle β by means of the transformation

$$s_1 = s_1' \cos\beta + s_2' \sin\beta \tag{4.156}$$

$$s_2 = -s_1' \sin\beta + s_2' \cos\beta \tag{4.157}$$

will result in a filter whose frequency response is rotated by an angle $-\beta$ with respect to the frequency response of filter (4.155). Thus

$$H_2(s_1', s_2') = H_0 \frac{\prod_{i=1}^{m} [(s_2' \cos\beta - s_1' \sin\beta) - q_i]}{\prod_{i=1}^{n} [(s_2' \cos\beta - s_1' \sin\beta) - p_i]} \tag{4.158}$$

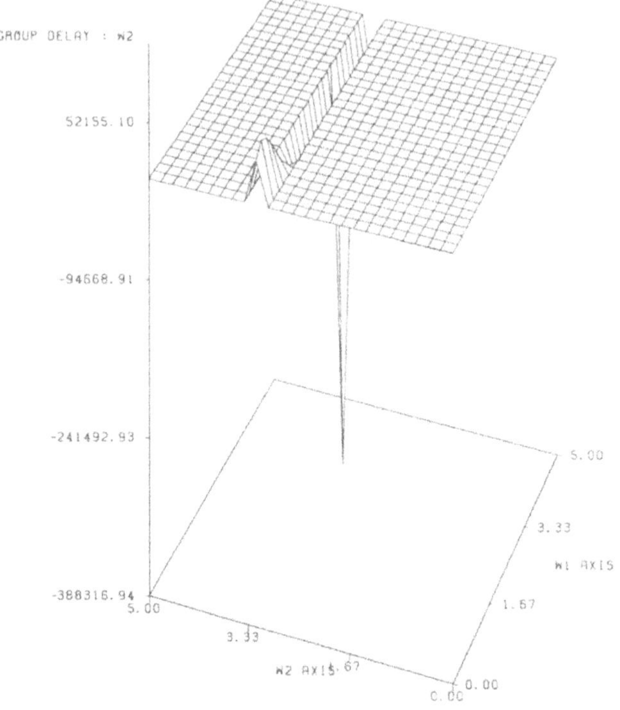

Figure 4.17c. Three-dimensional plot of the group-delay response with respect to ω_2 of the two-dimensional filter.

This function describes a continuous 2-D filter in the new coordinate system of s_1' and s_2'. The corresponding digital version of the above filter is obtained through application of the 2-D bilinear transformation

$$s_1' = \frac{2}{T_1}\frac{1 - z_1^{-1}}{1 + z_1^{-1}} \tag{4.159}$$

$$s_2' = \frac{2}{T_2}\frac{1 - z_2^{-1}}{1 + z_2^{-1}} \tag{4.160}$$

It is also frequently assumed that $T_1 = T_2 = T$, namely, the sample interval is the same in both directions. Substituting equations (4.159) and (4.160) into filter (4.158) will result in

$$H(z_1, z_2) = A \prod_{i=1}^{M} \frac{a_{00}^{(i)} - a_{10}^{(i)} z_1^{-1} + a_{01}^{(i)} z_2^{-1} + a_{11}^{(i)} z_1^{-1} z_2^{-1}}{b_{00}^{(i)} + b_{10}^{(i)} z_1^{-1} + b_{01}^{(i)} z_2^{-1} + b_{11}^{(i)} z_1^{-1} z_2^{-1}} \tag{4.161}$$

where

$$A = H_0(T/2)^{(n-m)} \quad \text{and} \quad M = \max(m, n)$$

$$\left.\begin{aligned} a_{00}^{(i)} &= \cos\beta - \sin\beta - Tq_i/2 \\ a_{10}^{(i)} &= \cos\beta + \sin\beta - Tq_i/2 \\ a_{01}^{(i)} &= -\cos\beta - \sin\beta - Tq_i/2 \\ a_{11}^{(i)} &= -\cos\beta + \sin\beta - Tq_i/2 \end{aligned}\right\} \quad \text{for } 1 \le i < m$$

$$a_{00}^{(i)} = a_{10}^{(i)} = a_{01}^{(i)} = a_{11}^{(i)} = 1 \qquad \text{for } m < i \le M$$

$$\left.\begin{aligned} b_{00}^{(i)} &= \cos\beta - \sin\beta - Tp_i/2 \\ b_{10}^{(i)} &= \cos\beta + \sin\beta - Tp_i/2 \\ b_{01}^{(i)} &= -\cos\beta - \sin\beta - Tp_i/2 \\ b_{11}^{(i)} &= -\cos\beta + \sin\beta - Tp_i/2 \end{aligned}\right\} \quad \text{for } 1 \le i \le n$$

$$b_{00}^{(i)} = b_{10}^{(i)} = b_{01}^{(i)} = b_{11}^{(i)} = 1 \qquad \text{for } n < i \le M \tag{4.162}$$

Unfortunately, despite its simplicity, this technique as it stands neither guarantees the stability of the designed filter nor achieves a circular-symmetric cutoff boundary. Costa and Venetsanopoulos[17] have modified the rotated filter approach by cascading a number of rotated filters whose angles of rotation are uniformly distributed over 180°, to achieve a magnitude response which approximates a circular-symmetric cutoff boundary by a polygon. This polygon has an even number of sides, because each rotated filter contributes two sides of the polygon. It has been proved that stability of the designed filter is ensured if the following two conditions hold:

1. $270° \le \beta \le 360°$, where β is the angle of rotation.
2. $C_i < 0$ for $i = 1, 2, \ldots, M$, where $C_i = \text{Re}[(T/2)p_i]$ and p_i represents the location of a pole.

4.11.2. Transformation Technique using a Two-Variable Reactance Function

The design of 2-D recursive zero-phase low-pass digital filters can be accomplished using a method described by Ahmadi *et al.*[18] in this method, a 1-D low-pass analogue transfer function

$$H(s) = \sum_{i=0}^{M} c_i s^i \Big/ \sum_{j=0}^{N} d_i s^i \tag{4.163}$$

is transformed into a 2-D first-quadrant filter by using the second-order two-variable reactance function

$$s \to g(s_1, s_2) = \frac{a_1 s_1 + a_2 s_2}{b_1 + b_2 s_1 s_2} \tag{4.164}$$

where coefficients a_1, a_2, b_1, and b_2 are all taken to be nonzero positive real constants to ensure a low-pass to low-pass property for equation (4.164). It can easily be shown that, without loss of generality, b_1 can be set equal to unity so as to reduce the number of transformation coefficients to a minimum of three. Thus, the transformation to be considered below is

$$s \to g(s_1, s_2) = \frac{a_1 s_1 + a_2 s_2}{1 + b s_1 s_2} \tag{4.165}$$

A zero-phase 2-D low-pass analogue filter can be obtained by forming the transfer function

$$\hat{H}(s_1, s_2) = H(s_1, s_2)H(-s_1, s_2)H(-s_1, -s_2)H(s_1, -s_2) \tag{4.166}$$

The discrete version of the latter equation can now be obtained as

$$H_d(z_1, z_2) = \hat{H}(s_1, s_2)|_{s_i=(2/T_i)(1-z_i^{-1})/(1+z_i^{-1})} \quad \text{for } i = 1, 2 \tag{4.167}$$

Let us consider a 2-D filter designed by using a 1-D analogue low-pass filter with cutoff frequency Ω_c. Equation (4.165) then gives

$$\Omega_2 = \frac{\Omega_c - a_1\Omega_1}{a_2 + b\Omega_c\Omega_1} \tag{4.168}$$

The mapping of $\Omega = \Omega_c$ onto the (Ω_1, Ω_2) plane for various values of b is depicted in Figure 4.18. The cutoff frequencies along the Ω_1 and Ω_2 axes

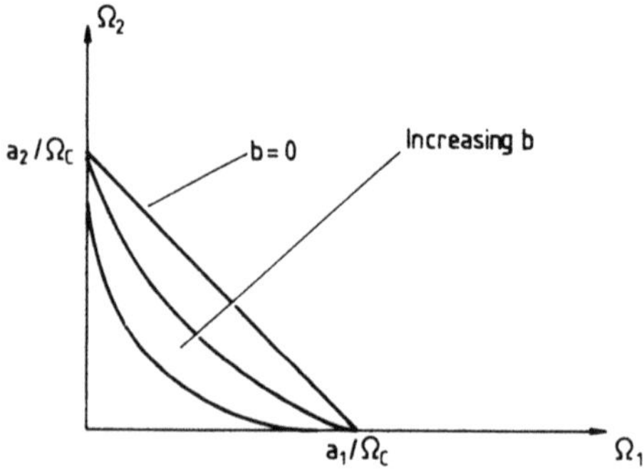

Figure 4.18. Plot of $\Omega_2 = (\Omega_c - a_1\Omega_1)/(a_2 + b\Omega_c\Omega_1)$ for different values of b.

can be adjusted by simply varying a_1 and a_2. On the other hand, the convexity of the boundary can be adjusted by varying b. We note that b must be greater than zero to preserve stability. Also, it should be noted that transformation (4.165) becomes a low-pass to band-stop transformation if $s_2 = s_1$, and therefore the designed filter will behave like a band-stop filter along $\Omega_2 = \Omega_1$. This problem can be overcome by using a guard filter of any order.

EXAMPLE 4.8. A 2-D low-pass filter is designed by transforming a 1-D third-order Butterworth filter possessing transfer function

$$H(s) = \frac{1}{s^3 + 2s^2 + 2s + 1}$$

into a 2-D first-quadrant analogue filter using the transformation

$$s = \frac{s_1 + s_2}{1 + bs_1}$$

Figures 4.19–4.24 show the amplitude response and contour plot of the designed 2-D filter for the values $b = 0.1$, 0.2, and 0.6.

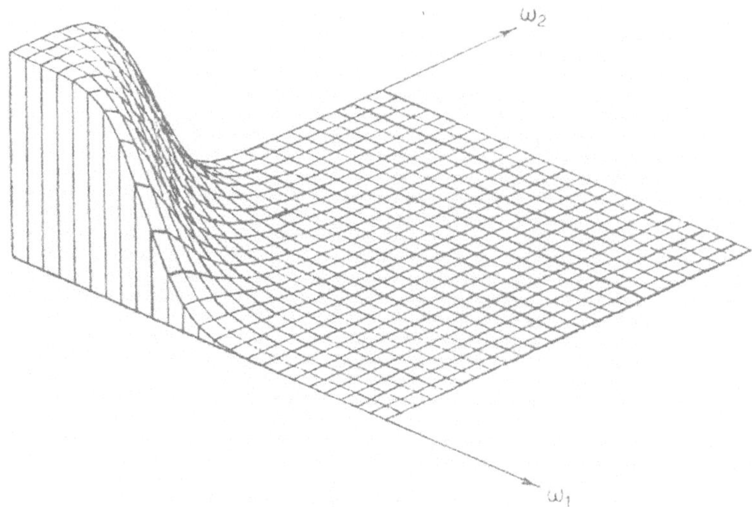

Figure 4.19. Amplitude plot of a two-dimensional recursive filter obtained from a third-order Butterworth filter with $a_1 = a_2 = 1$ and $b = 0.1$ via s transforming to $(a_1s_1 + a_2s_2)/(1 + bs_1s_2)$.

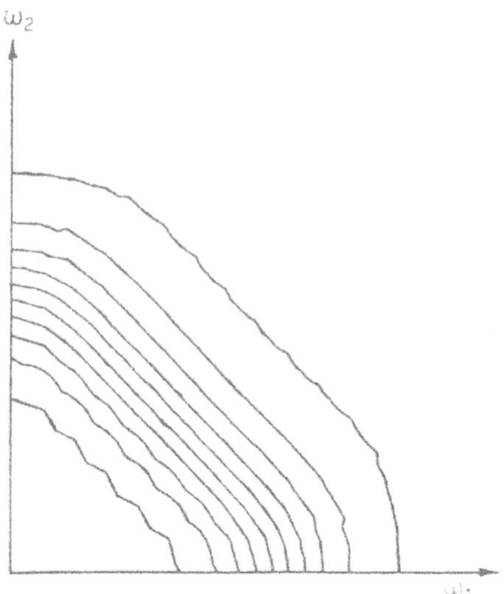

Figure 4.20. Contour plot of a two-dimensional recursive filter obtained from a third-order Butterworth filter with $a_1 = a_2 = 1$ and $b = 0.1$.

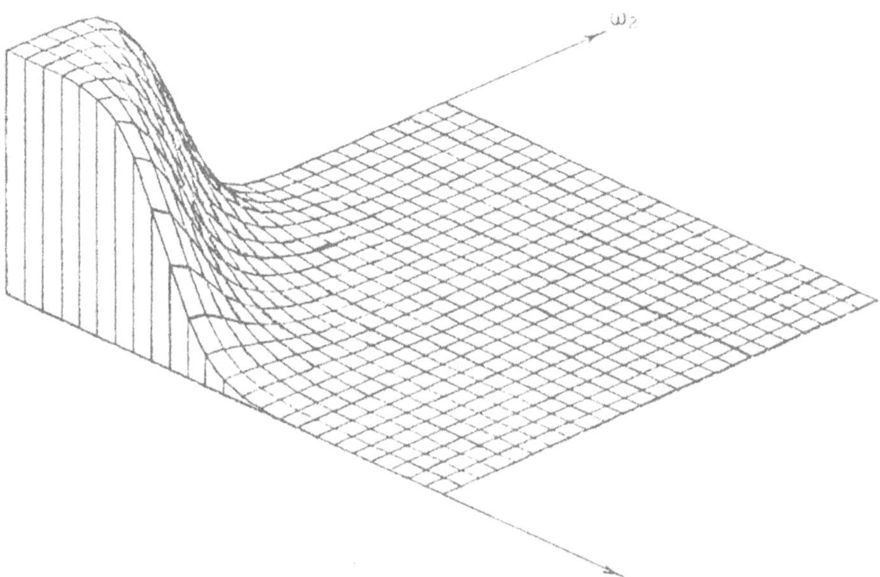

Figure 4.21. Amplitude plot of a two-dimensional recursive filter obtained from a third-order Butterworth filter with $a_1 = a_2 = 1$ and $b = 0.2$.

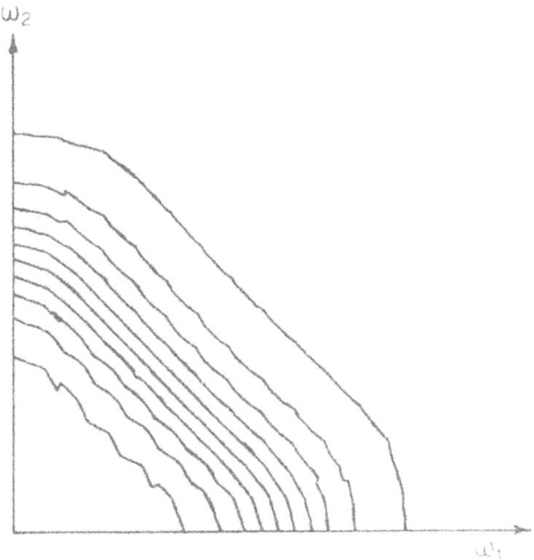

Figure 4.22. Contour plot of a two-dimensional recursive filter obtained from a third-order Butterworth filter with $a_1 = a_2 = 1$ and $b = 0.2$.

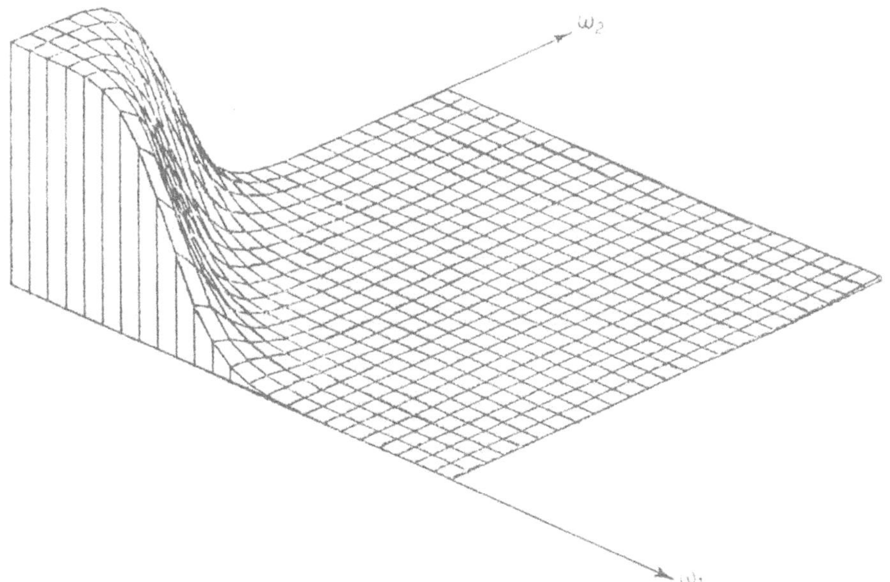

Figure 4.23. Amplitude plot of a two-dimensional recursive filter obtained from a third-order Butterworth filter with $a_1 = a_2 = 1$ and $b = 0.6$.

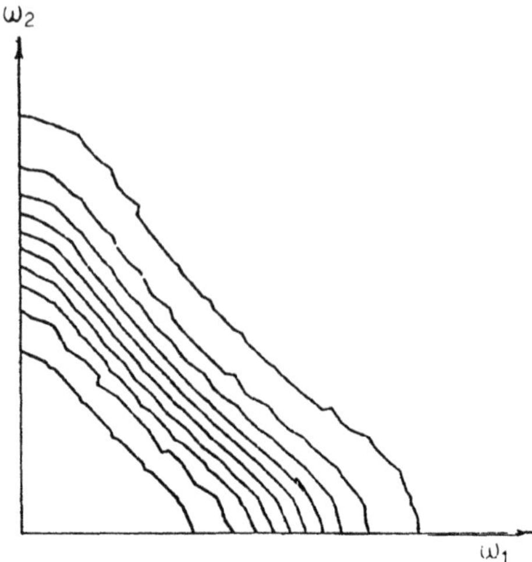

Figure 4.24. Contour plot of a two-dimensional recursive filter obtained from a third-order Butterworth filter with $a_1 = a_2 = 1$ and $b = 0.6$.

King and Kayran[19] have modified the above technique by using a higher-order reactance function of the form

$$s = g(s_1, s_2) = \frac{a_1 s_1 + a_2 s_2}{1 + b_1(s_1^2 + s_2^2) + b_2 s_1 s_2} \tag{4.169}$$

and they proved that the stability of $g(s_1, s_2)$ is ensured if

$$b_1 > 0 \tag{4.170}$$

and

$$b_1 > b_2^2/4 - b_1^2 > 0 \tag{4.171}$$

However, it is necessary to include a guard filter, which may have the simple form

$$G(z_1, z_2) = \frac{(1 + z_1)(1 + z_2)}{(d_1 + z_1)(d_2 + z_2)}$$

in order to remove the high-pass regions along all radii except the coordinate axes.

Then, through an optimization procedure, the coefficients of $g(s_1, s_2)$ and $G(z_1, z_2)$ are calculated subject to constraints (4.170) and (4.171), so that the cutoff frequency of the 1-D filter is mapped into a desired cutoff boundary in the (Ω_1, Ω_2) plane.

Table 4.7. Values of the Parameters of the Two-Variable Reactance Transformation Function $T(s_1, s_2) = [a_1 s_1 + a_2 s_2]/[1 + b_1(s_1^2 + s_2^2) + b_2 s_1 s_2]$

$a_1 = 0.96572$	$T_1/2 = 0.80112$
$a_2 = 1.00363$	$T_2/2 = 0.77385$
$b_1 = 0.10890$	$d_1 = 11.27362$
$b_2 = 0.29020$	$d_2 = 11.25402$

Example 4.9. A 2-D circular-symmetric low-pass filter is to be designed. The 1-D analogue prototype filter is a third-order Butterworth filter

$$H(s) = \frac{1}{s^3 + 2s^2 + 2s + 1}$$

and the parameters of $g(s_1, s_2)$ after optimization are as shown in Table 4.7. Figure 4.25 shows the amplitude response of the designed 2-D filter while Figure 4.26 shows the contour plot of the same filter.

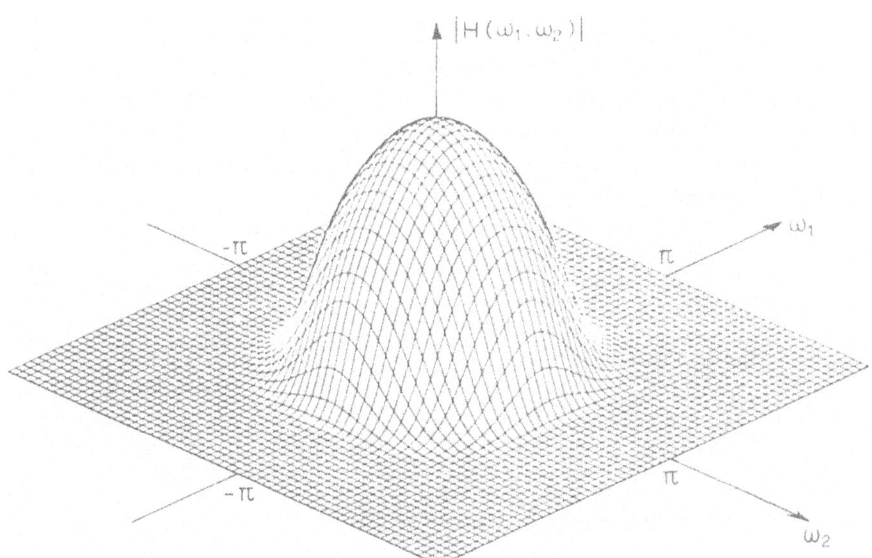

Figure 4.25. Amplitude response of a two-dimensional recursive filter using the second-order reactance transformation.

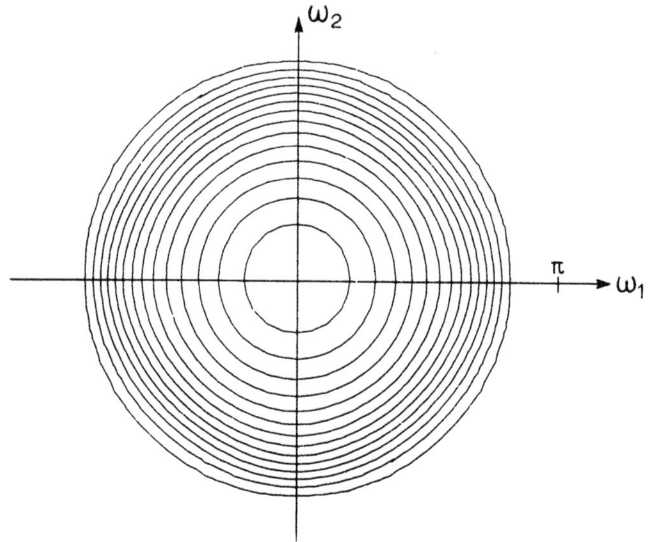

Figure 4.26. Contour plot of the amplitude response of a two-dimensional recursive filter using the second-order reactance transformation.

4.11.3. Complex Transformations and Their Applications in Two-Dimensional Fan-Filter Design

A technique for designing recursive 2-D fan filters based on complex transformation applied to a 1-D low-pass filter has been presented elsewhere.[20] A set of transformed filters with their appropriate combination to form a zero-phase fan filter is also discussed.

We consider a causal and stable 2-D filter, described as in equation (4.123) to be generated according to

$$H(z_1, z_2) = H_1(z) \tag{4.172}$$

via the transformation

$$z = e^{\phi} z_1^{\alpha_1/\beta_1} z_2^{\alpha_2/\beta_2} \tag{4.173}$$

The corresponding frequency transformation is

$$\exp(j\omega) \rightarrow \exp\left[j\left(\phi + \frac{\alpha_1}{\beta_1}\omega_1 + \frac{\alpha_2}{\beta_2}\omega_2\right)\right] \tag{4.174}$$

or

$$\omega \to \phi + \frac{\alpha_1}{\beta_1}\omega_1 + \frac{\alpha_2}{\beta_2}\omega_2 \tag{4.175}$$

For example, if the prototype filter is a low-pass filter with cutoff frequency $\omega_c = \pi/2$, the amplitude plot of the frequency response after transformations $\alpha_1/\beta_1 = \frac{1}{2}$, $\alpha_2/\beta_2 = \frac{1}{2}$, $\phi = 90°$ is shown in Figure 4.27.

There are three effects of transformation (4.173) on the resulting filter:

1. *Frequency shifting along the ω_1 axis.* The frequency response of the resulting filter will be shifted by ϕ along the ω_1 axis.
2. *Rotation of the frequency response.* The angle of rotation is

$$\theta = \arctan(\alpha_2/\beta_2) \tag{4.176}$$

   Since the original filter (4.172) is 1-D and a function of z_1, the angle of rotation will be defined by the fractional power of z_2.
3. *Scaling the frequency response along the ω_1 axis.* The fractional power of z_1 will scale the frequency response by a factor β_1/α_1. However, the periodicity of the frequency response will be $(\alpha_1/\beta_1)2\pi$ instead of 2π.

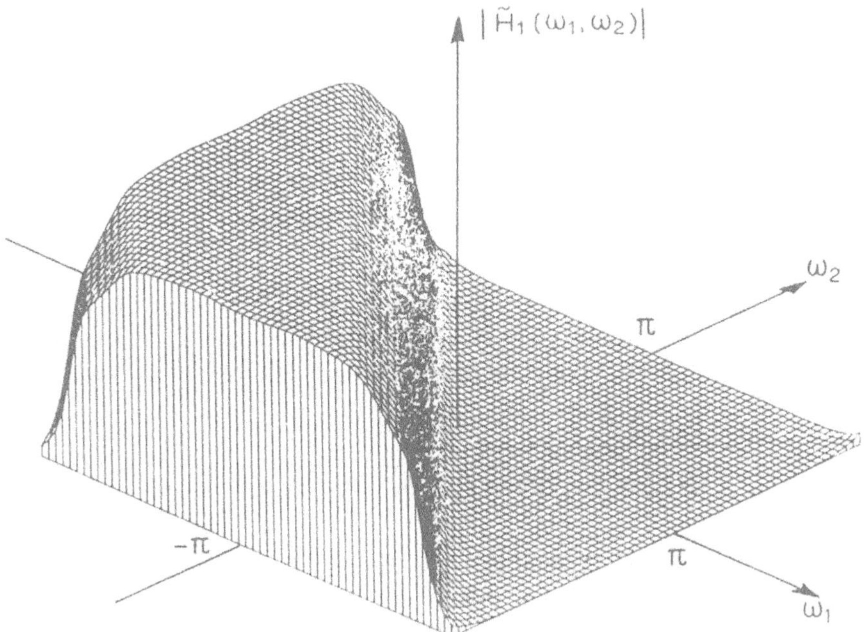

Figure 4.27. Frequency response of a low-pass filter after the complex transformation $z = j(z_1 z_2)^{1/2}$.

The other effects of the transformation on the resulting filter may be specified as follows:

4. When $\alpha_1/\beta_1 > 0$ and $\alpha_2/\beta_2 > 0$, the transformation is causal; otherwise it is noncausal. However, the transformation $\alpha_1/\beta_1 < 0$ or (and) $\alpha_2/\beta_2 < 0$ may be implemented in the spatial time domain, for a finite array, by reorienting the input signal array.
5. When $\phi = 0$, the resulting filter alone cannot be implemented on its own in the spatial time domain. On the other hand, such transformed filters may be combined in appropriate ways so that complex values do not exist in the final transfer function.
6. Both rotation and frequency scaling are equivalent to the rotation of the recursion direction with a new sampling interval.
7. The stability of the resulting filter is unaffected by $\alpha_1/\beta_1 > 0$ and $\alpha_2/\beta_2 > 0$. For $\alpha_1/\beta_1 < 0$ or $\alpha_2/\beta_2 < 0$, the transformed filter will be unstable for the causal recursion direction. However, if the orientation of the finite area array (input) is changed, there is always 4 noncausal recursion direction where the filter function is stable.

4.11.3.1. Symmetric Fan-Filter Design

A symmetric fan filter will be designed employing the above-proposed complex transformation (4.173). For this design, the ideal fan-filter specification is

$$H_f(e^{j\omega_1 T}, e^{j\omega_2 T}) = \begin{cases} 1 & \text{for } |\omega_1| \geq |\omega_2| \\ 0 & \text{otherwise} \end{cases} \tag{4.177}$$

By using as prototype a low-pass filter with a cutoff frequency at $\omega_c = \pi/2$, and transformation (4.173), one obtains the shifted, scaled, and rotated characteristics in the frequency domain. We denote the transformed filter by

$$\tilde{H}(z_1, z_2; \alpha_1/\beta_1, \alpha_2/\beta_2, \phi) = H_1(z)\big|_{z = e^{j\phi} z_1^{\alpha_1/\beta_1} z_2^{\alpha_2/\beta_2}} \tag{4.178}$$

In general, the filter coefficients in function $\tilde{H}$ will be complex and the variables z_1 and z_2 will have rational noninteger powers. However, appropriate combinations of transformed filters will remove both these difficulties.

We introduce transfer functions $\tilde{H}_1(z_1, z_2)$, $\tilde{H}_2(z_1, z_2)$, $\tilde{H}_3(z_1, z_2)$, and $\tilde{H}_4(z_1, z_2)$ of four filters generated by equation (4.178) with $(\alpha_1/\beta_1, \alpha_2/\beta_2) =$

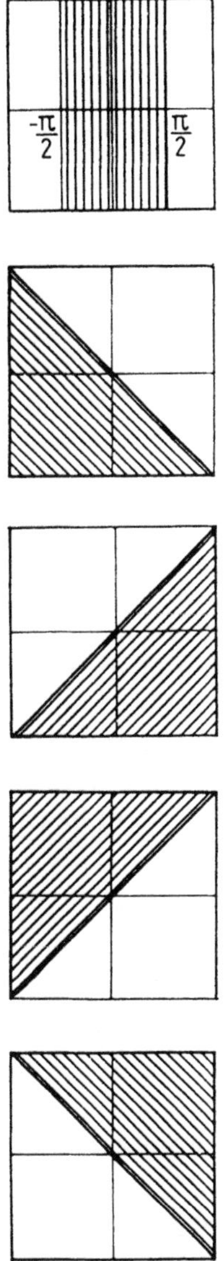

Figure 4.28. Basic building blocks for fan-filter design.

$(\frac{1}{2}, \frac{1}{2})$, $(-\frac{1}{2}, \frac{1}{2})$, $(\frac{1}{2}, -\frac{1}{2})$, and $(-\frac{1}{2}, -\frac{1}{2})$, respectively, and $\phi = \pi/2$. The frequency responses of the transformed filters, $\tilde{H}_i(z_1, z_2)$, $i = 1, 2, 3, 4$, can be found in Figure 4.28.

In this design procedure, the filters in Figure 4.28 will be used as the main building blocks for a fan filter with specification (4.177). One can construct the following filter characteristics:

$$H_{11}(z_1, z_2) = \tilde{H}_1(z_1, z_2)\tilde{H}_2(z_1, z_2)\tilde{H}_3^*(z_1, z_2)\tilde{H}_4^*(z_1, z_2) + \tilde{H}_1^*(z_1, z_2)\tilde{H}_2^*(z_1, z_2)\tilde{H}_3(z_1, z_2)\tilde{H}_4(z_1, z_2) \quad (4.179)$$

and

$$H_{22}(z_1, z_2) = \tilde{H}_1(z_1, z_2)\tilde{H}_3(z_1, z_2)\tilde{H}_2^*(z_1, z_2)\tilde{H}_4^*(z_1, z_2) + \tilde{H}_1^*(z_1, z_2)\tilde{H}_3^*(z_1, z_2)\tilde{H}_2(z_1, z_2)\tilde{H}_4(z_1, z_2) \quad (4.180)$$

where $\tilde{H}^*$ denotes the filter with coefficients which are the complex conjugate of those of $\tilde{H}_1$, etc.

The frequency characteristic of the obtained filter, H_{11}, corresponds to the filter of specification (4.177); H_{22} has a frequency characteristic which is a clockwise 90°-rotated version of H_{11}.

Although both terms on the right-hand side of equations (4.179) and (4.180) individually represent zero-phase filters with complex coefficients, the coefficients of the resulting filters $H_{11}(z_1, z_2)$ and $H_{22}(z_1, z_2)$ are real because each term is the complex conjugate of the other. During the implementation step, it can also be shown that filter functions H_{11} and H_{22} are functions of complex variables z_1 and z_2 with integer powers (terms with fraction powers of z_1 and z_2 will be cancelled).

4.11.3.2. Quadrant Fan-Filter Design

The frequency characteristic of a quadrant fan filter is specified as

$$H_f(e^{j\omega_1 T}, e^{j\omega_2 T}) = \begin{cases} 1 & \text{for } \omega_1\omega_2 \geq 0 \\ 0 & \text{for } \omega_1\omega_2 < 0 \end{cases} \quad (4.181)$$

We consider the same ideal prototype filter. Then the transformed filters $\tilde{H}_{12}(z_1, z_2)$, $\tilde{H}_{14}(z_1, z_2)$, $\tilde{H}_{34}(z_1, z_2)$, and $\tilde{H}_{23}(z_1, z_2)$ are obtained via equation (4.178) with $(\alpha_1/\beta_1, \alpha_2/\beta_2)$ equal to $(1, 0)$, $(0, 1)$, $(-1, 0)$, and $(0, -1)$, respectively, and $\phi = \pi/2$. The subscript on $\tilde{H}$ refers to quadrants to which the low-pass characteristics have been shifted. Figure 4.29 shows the amplitude response of these transformed filters.

Next, one can construct the filter characteristics $H_{13}(z_1, z_2)$ and $H_{24}(z_1, z_2)$ in a similar manner to equations (4.179) and (4.180). The frequency characteristics of the resulting filter H_{13} correspond to the quadrant fan filter specified in relationship (4.181). Function H_{24} has a frequency characteristic which is a clockwise 90°-rotated version of H_{13}. Both H_{13} and H_{24} are zero phase.

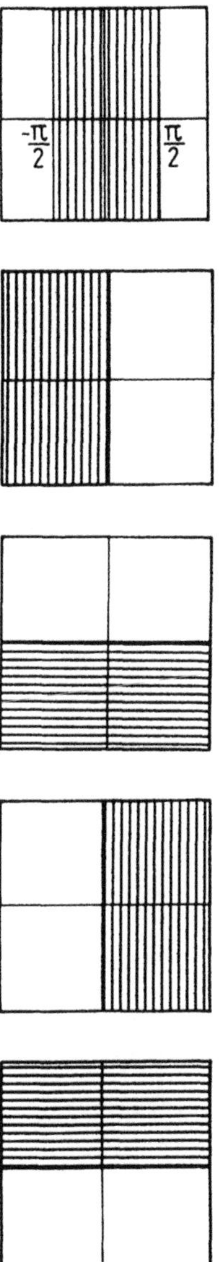

Figure 4.29. Basic building blocks for quadrant fan-filter design.

EXAMPLE 4.10. We consider a low-pass 1-D filter function

$$H(z) = (0.03112) \prod_{k=1}^{3} \frac{z^{-2} + a_k z^{-1} + b_k}{z^{-2} + c_k z^{-1} + d_k} \tag{4.182}$$

The coefficients of this function are given in Table 4.8. Filter function (4.182)

Table 4.8. Coefficient Values of the One-Dimensional Filter

k	a_k	b_k	c_k	d_k
1	−0.72968	1.0	−0.94645	0.93267
2	1.17818	1.0	−1.09725	0.38569
3	−0.38510	1.0	−0.99086	0.71690

has the following specifications for the magnitude characteristics:

$$|H(e^{j\omega T})| = \begin{cases} 1 \pm \delta_p & \omega \leq 0.3333333\pi \\ \delta_s & 0.376111\pi \leq \omega \leq \pi \end{cases}$$

Since the filter does not have a cutoff frequency at $\omega_c = \pi/2$, a low-pass to low-pass frequency transformation (4.113) is used to obtain a filter with a cutoff frequency at $\omega_c = \pi/2$. Then it is readily applied to equation (4.179) to yield a symmetric fan filter. Figure 4.30 shows the magnitude plot of the resulting fan filter. Similarly, a quadrant filter can be obtained as discussed

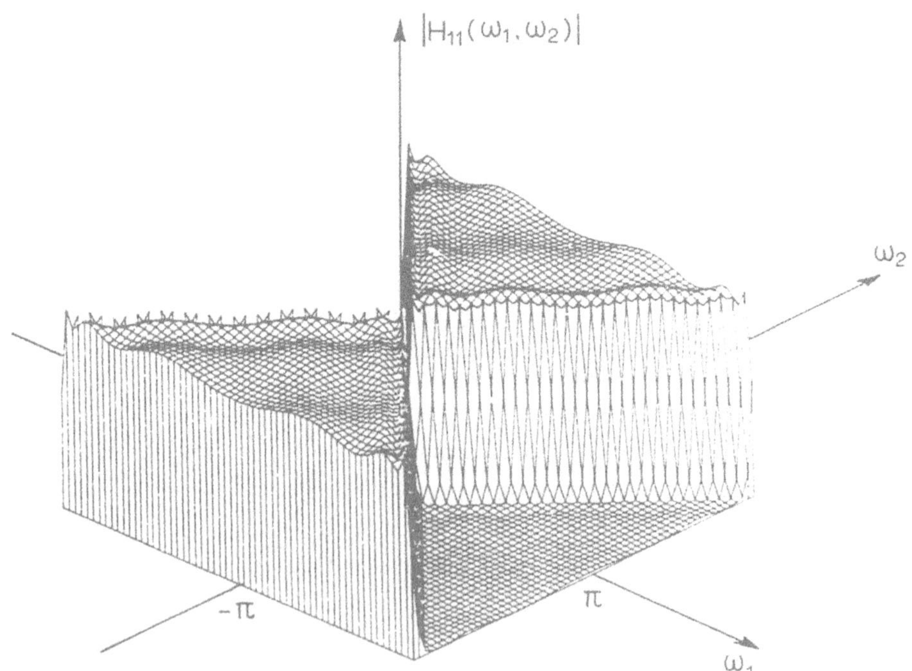

Figure 4.30. Magnitude characteristics of a symmetrical fan filter.

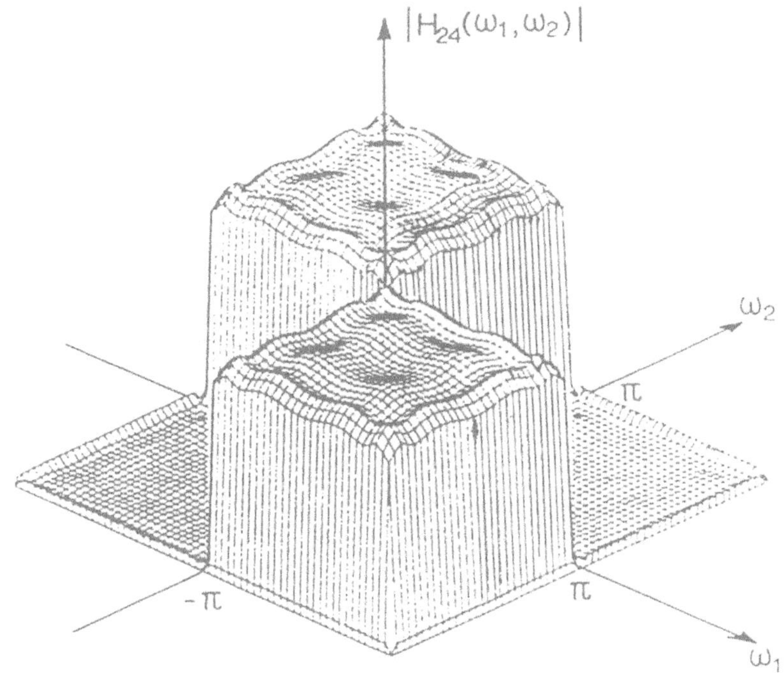

Figure 4.31. Magnitude plot of a quadrant fan filter.

earlier. Figure 4.31 shows the magnitude plot of the resulting quadrant fan filter.

4.12. ITERATIVE APPROACH TO THE DESIGN OF TWO-DIMENSIONAL FILTERS

4.12.1. Linear Programming Approach

The method to be presented here is an extension to two dimensions of the technique discussed in Section 4.3.2 and due to Chottera and Jullien.[21] Their approach consists of first specifying the desired magnitude and linear-phase characteristics. The magnitude specification is straightforward; however, the linear phase is specified in terms of the desired group delay. Based on these specifications, a linear program is then formulated such that the variables of the linear program are expressed in terms of the desired filter coefficients. The linear program is then solved using the standard solution procedures to obtain filter coefficients directly from the linear program.

4.12.1.1. Linear Phase Specification

Linear phase is specified in terms of the desired constant group delays in the respective spatial directions. Letting τ_1 and τ_2 be the constant group delays with respect to frequency variables ω_1 and ω_2 respectively, we have the phase specification

$$\phi(\omega_1, \omega_2) = -(\tau_1\omega_1 + \tau_2\omega_2) \tag{4.183}$$

where ω_1 and ω_2 are the normalized frequency variables. We simplify the specification by setting $\tau_1 = \tau_2 = \tau$ and therefore the desired phase specification becomes

$$\phi(\omega_1, \omega_2) = -\tau(\omega_1 + \omega_2) \tag{4.184}$$

The desired linear phase is specified up to the Nyquist frequency. However, the resulting approximation of phase is good only in the pass and transition bands. The stop-band characteristics are of no consequence, since the magnitude is insignificant.

4.12.1.2. Approximation Procedure

The transfer function $H(z_1, z_2)$ of the 2-D recursive digital filter is written in the form (4.123) with the orders of the numerator and denominator equal,

$$H(z_1, z_2) = \frac{N(z_1, z_2)}{D(z_1, z_2)} = \frac{\sum_{m=0}^{M} \sum_{n=0}^{M} a(m, n) z_1^{-m} z_2^{-n}}{\sum_{m=0}^{M} \sum_{n=0}^{M} b(m, n) z_1^{-m} z_2^{-n}} \tag{4.185}$$

where $z_1 = e^{j\omega_1 T}$ and $z_2 = e^{j\omega_2 T}$. The term $b(0, 0)$ and quantity T can be assumed equal to 1.0 without any loss of generality.

Now, given an arbitrary magnitude and linear-phase specification, it is desired to formulate the approximation as a linear programming problem, such that the constraints of the problem are in terms of the coefficients of the transfer function (4.185). This is carried out as follows.

We let $M(\omega_{1m}, \omega_{2n})$ and $\phi(\omega_{1m}, \omega_{2n})$ be the given magnitude and linear-phase specifications, respectively, where $\phi(\omega_{1m}, \omega_{2n})$ is as in equation (4.184), specified at a discrete grid of frequency points ω_{1m}, $m = 1, 2, \ldots, L_1$; ω_{2n}, $n = 1, 2, \ldots, L_2$. Given the above, the real and imaginary components $Y_R(\omega_{1m}, \omega_{2n})$ and $Y_I(\omega_{1m}, \omega_{2n})$, respectively, of the frequency domain specifications can be expressed in the forms

$$\begin{aligned} Y_R(\omega_{1m}, \omega_{2n}) &= M(\omega_{1m}, \omega_{2n}) \cos[\phi(\omega_{1m}, \omega_{2n})] \\ &= M(\omega_{1m}, \omega_{2n}) \cos[-\tau(\omega_{1m} + \omega_{2n})] \end{aligned} \tag{4.186}$$

and

$$\begin{aligned} Y_I(\omega_{1m}, \omega_{2n}) &= M(\omega_{1m}, \omega_{2n}) \sin[\phi(\omega_{1m}, \omega_{2n})] \\ &= M(\omega_{1m}, \omega_{2n}) \sin[-\tau(\omega_{1m} + \omega_{2n})] \end{aligned} \tag{4.187}$$

A complex error is now defined as

$$E(\omega_{1m}, \omega_{2n}) = Y_R(\omega_{1m}, \omega_{2n}) + \mathrm{j}Y_I(\omega_{1m}, \omega_{2n}) - \frac{N(\omega_{1m}, \omega_{2n})}{D(\omega_{1m}, \omega_{2n})} \tag{4.188}$$

where

$$H(e^{-j\omega_{1m}}, e^{-j\omega_{2n}}) = \frac{N(\omega_{1m}, \omega_{2n})}{D(\omega_{1m}, \omega_{2n})}$$

Multiplication of equation (4.188) by the denominator D yields the real and imaginary components, E_R and E_I, of the complex weighted error as follows:

$$\begin{aligned} E_R(\omega_{1m}, \omega_{2n}) &= \mathrm{Re}[E(\omega_{1m}, \omega_{2n})D(\omega_{1m}, \omega_{2n}) \\ &= Y_R(\omega_{1m}, \omega_{2n})D_R(\omega_{1m}, \omega_{2n}) \\ &\quad + Y_I(\omega_{1m}, \omega_{2n})D_I(\omega_{1m}, \omega_{2n}) - N_R(\omega_{1m}, \omega_{2n}) \end{aligned} \tag{4.189}$$

and

$$\begin{aligned} E_I(\omega_{1m}, \omega_{2n}) &= \mathrm{Im}[E(\omega_{1m}, \omega_{2n})D(\omega_{1m}, \omega_{2n}) \\ &= Y_I(\omega_{1m}, \omega_{2n})D_R(\omega_{1m}, \omega_{2n}) \\ &\quad - Y_R(\omega_{1m}, \omega_{2n})D_I(\omega_{1m}, \omega_{2n}) + N_I(\omega_{1m}, \omega_{2n}) \end{aligned} \tag{4.190}$$

where N and D are given by

$$D(\omega_{1m}, \omega_{2n}) = D_R(\omega_{1m}, \omega_{2n}) - \mathrm{j}D_I(\omega_{1m}, \omega_{2n}) \tag{4.191}$$

and

$$N(\omega_{1m}, \omega_{2n}) = N_R(\omega_{1m}, \omega_{2n}) - \mathrm{j}N_I(\omega_{1m}, \omega_{2n}) \tag{4.192}$$

A good approximation to the desired characteristics can now be obtained by minimizing a positive variable ε such that

$$|E_R(\omega_{1m}, \omega_{2n})| \leq \varepsilon \tag{4.193}$$

and

$$|E_I(\omega_{1m}, \omega_{2n})| \leq \varepsilon \tag{4.194}$$

for all m and n. The following inequalities can be derived from the latter two relationships:

$$E_R(\omega_{1m}, \omega_{2n}) - \varepsilon \leq 0 \tag{4.195}$$

$$-E_R(\omega_{1m}, \omega_{2n}) - \varepsilon \leq 0 \tag{4.196}$$

$$E_I(\omega_{1m}, \omega_{2n}) - \varepsilon \leq 0 \tag{4.197}$$

$$-E_I(\omega_{1m}, \omega_{2n}) - \varepsilon \leq 0 \tag{4.198}$$

for $m = 0, 1, \ldots, L_1$; $n = 0, 1, \ldots, L_2$.

The above approximation problem is amenable to linear programming, since $E_R(\omega_{1m}, \omega_{2n})$ and $E_I(\omega_{1m}, \omega_{2n})$ are linear in terms of the 2-D recursive filter coefficients **a** and **b**. Therefore letting $\delta = -\varepsilon$, substituting for E_R and E_I from equations (4.189) and (4.190), and expressing D_R, D_I, N_R, and N_I in terms of coefficients of the filter, the linear programming approximation problem can be expressed as

$$\text{Maximize } g = \delta \tag{4.199}$$

subject to

$$\begin{aligned}\sum_{(k,l)}\sum b(k, l)\{M(\omega_{1m}, \omega_{2n}) \cos[(k+\tau)\omega_{1m} + (l+\tau)\omega_{2n}]\} \\ + \sum_{(k,l)}\sum a(k, l)\{-\cos[k\omega_{1m} + l\omega_{2n}]\} + \delta \leq 0\end{aligned} \tag{4.200}$$

$$\begin{aligned}\sum_{(k,l)}\sum b(k, l)\{-M(\omega_{1m}, \omega_{2n}) \sin[(k+\tau)\omega_{1m} + (l+\tau)\omega_{2n}]\} \\ + \sum_{(k,l)}\sum a(k, l)\{\sin[k\omega_{1m} + l\omega_{2n}]\} + \delta \leq 0\end{aligned} \tag{4.201}$$

$$\begin{aligned}-\sum_{(k,l)}\sum b(k, l)\{M(\omega_{1m}, \omega_{2n}) \cos[(k+\tau)\omega_{1m} + (l+\tau)\omega_{2n}]\} \\ - \sum_{(k,l)}\sum a(k, l)\{\cos[k\omega_{1m} + l\omega_{2n}]\} + \delta \leq 0\end{aligned} \tag{4.202}$$

$$-\sum_{(k,l)}\sum b(k,l)\{-M(\omega_{1m},\omega_{2n})\sin[(k+\tau)\omega_{1m}+(l+\tau)\omega_{2n}]\}$$

$$-\sum_{(k,l)}\sum a(k,l)\{\sin[k\omega_{1m}+l\omega_{2n}]\}+\delta\leq 0 \tag{4.203}$$

for all $m = 0, 1, \ldots, L_1$; $n = 0, 1, \ldots, L_2$. In the above constraints $b(0,0)$ is set equal to unity.

It is clear that the upper bound on g is zero, and maximizing g minimizes the weighted real and imaginary components, E_R and E_I respectively, of the complex weighted error. If g is greater than zero, the solution is meaningless and therefore it is to be neglected. It should be noted that the constraints given above just suffice for the purpose of approximation, but they do not ensure stability of the resulting filter. Therefore additional constraints, which are linear in form, are required to be incorporated into the linear program for stable filter design.

4.12.1.3. Stability Constraints

As indicated above, the stability constraint must be linear in form since the approximation procedure involves linear programming. The authors propose a constraint of the form given by

$$\text{Re}[D(z_1, z_2)] > 0 \qquad \text{for } |z_1| = 1 \text{ and } |z_2| = 1 \tag{4.204}$$

This constraint is linear in terms of filter coefficients and can therefore easily be incorporated into a linear program.

It should be noted, however, that constraint (4.204) is only a sufficient condition for stability, and the filters designed incorporating the above constraint would therefore belong to a subclass of stable filters.

The actual form of the constraint that is incorporated into the linear programming design is a slightly modified form of condition (4.204) and is given by

$$\text{Re}\, D(z_1, z_2) \geq \xi \qquad \text{for } |z_1| = 1 \text{ and } |z_2| = 1 \tag{4.205}$$

where ξ is a small positive quantity. With $b(0,0) = 1$, the above constraint can be rewritten as

$$-\sum_{\substack{(k,l)\\ k+l\neq 0}}\sum b(k,l)\cos(k\omega_1+l\omega_2)\leq 1-\xi \qquad \text{for } -\pi\leq\omega_1\leq\pi \text{ and } 0\leq\omega_2\leq\pi \tag{4.206}$$

Thus constraint (4.206) subject to conditions (4.200)-(4.203) completely defines the linear programming design of a 2-D recursive digital filter for a specified magnitude and linear phase.

The above procedure can be summarized as follows:

1. Specify the desired magnitude characteristics.
2. Choose the desired filter order.
3. Choose the desired group delay τ and generate the linear-phase characteristics.
4. Solve the linear programming problem (4.200)–(4.206) to obtain the filter coefficients and compute the error measures.

In step (4), if a solution to the linear programming does not exist, then decrease τ and repeat steps (3) and (4). In step (3), the value of τ can be chosen equal to the order of the filter.

4.12.2. Nonlinear Programming Approach

The general approach to the design of a stable 2-D recursive digital filter is to assign a two-variable VSHP, which can be generated by any of the techniques discussed in Section 3.13, to the denominator of a 2-D analogue filter expressed as

$$H_a(s_1, s_2) = \frac{N(s_1, s_2)}{D(s_1, s_2)} = \frac{\sum_{i=0}^{M} \sum_{j=0}^{M} n(i,j) s_1^i s_2^j}{\sum_{i=0}^{M} \sum_{j=0}^{M} d(i,j) s_1^i s_2^j} \tag{4.207}$$

while the numerator is left unchanged. The analogue 2-D transfer function is discretized by applying the double bilinear transformation. Now the error of the magnitude response, calculated using equation (4.150), is given by

$$E_{\text{Mag}}(j\omega_{1m}, j\omega_{2n}) = M_{\text{I}}(\omega_{1m}, \omega_{2n}) - M_{\text{D}}(\omega_{1m}, \omega_{2n}) \tag{4.208}$$

where E_{Mag} is the error of the magnitude response while M_{I} and M_{D} are the magnitude responses of the ideal and designed filter, respectively. Also, to calculate the group-delay responses E_{τ_i} we use equation (4.151), which yields

$$E_{\tau_i}(j\omega_{1m}, j\omega_{2n}) = \tau_{\text{I}} T - \tau_i(\omega_{1m}, \omega_{2n}) \tag{4.209}$$

where τ_{I} is a constant representing the ideal group-delay response of the filter and τ_i $(i = 1, 2)$ is the group-delay response of the designed filter. The design problems for simultaneous approximation of the magnitude and group-delay response of the filter can be formulated by choosing the general mean square as in equation (4.152) and using equations (4.208) and (4.209). This gives

$$E_G(j\omega_{1m}, j\omega_{2n}, \boldsymbol{\psi}) = \sum_{m,n \in I_{ps}}\sum E^2_{\text{Mag}}(j\omega_{1m}, j\omega_{2n}, \boldsymbol{\psi}) + \sum_{m,n \in I_p}\sum E^2_{\tau_1}(j\omega_{1m}, j\omega_{2n}, \boldsymbol{\psi})$$
$$+ \sum_{m,n \in I_p}\sum E^2_{\tau_2}(j\omega_{1m}, j\omega_{2n}, \boldsymbol{\psi}) \quad (4.210)$$

where in this equation $\boldsymbol{\psi}$ is the coefficient vector to be calculated, I_{ps} is a set of all discrete frequency pairs along the ω_1 and ω_2 axes covering the pass band and stop band of the filter, while I_p is a set of all discrete frequency pairs along the ω_1 and ω_2 axes covering only the pass band of the filter. To design a stable 2-D filter satisfying a prescribed magnitude and constant group-delay response, $\boldsymbol{\psi}$ should be calculated in such a way that quantity E_G in equation (4.210) is minimized subject to constraints needed for the denominator polynomial to be a VSHP. It should be noted that, if magnitude-only specification is required, calculation of the sum of squares of group-delay errors in the last two terms of equation (4.210) will be dropped. This is a very simple nonlinear optimization problem, which can be solved easily by using the Fletcher and Powell technique.[22] For further details of the nonlinear programming approach, readers are referred elsewhere.[23-28]

4.13. SPATIAL DESIGN TECHNIQUE

In the previous sections, it was assumed that the filter characteristics are always given in the frequency domain, and that is why most of the previous design techniques are termed frequency domain design techniques. In many instances, the designer is provided with the impulse response of the desired filter, and therefore it is necessary to have design approaches accommodating this situation. We term this approach "spatial design technique."

4.13.1. The Shanks, Treitel, and Justice Method[16]

Given the desired impulse response $h(m, n)$ of a 2-D recursive filter for $m = 0, 1, \ldots, M$ and $n = 0, 1, \ldots, N$, its z transform is then

$$H(z_1, z_2) = \sum_{m=0}^{M} \sum_{n=0}^{N} h(m, n) z_1^{-m} z_2^{-n} \quad (4.211)$$

while the approximating 2-D recursive filter will always be of the form

$$F(z_1, z_2) = \frac{A(z_1, z_2)}{B(z_1, z_2)} = \frac{\sum_{k=0}^{K_1} \sum_{l=0}^{L_1} a(k, l) z_1^{-k} z_2^{-l}}{\sum_{k=0}^{K_2} \sum_{l=0}^{L_2} b(k, l) z_1^{-k} z_2^{-l}} \tag{4.212}$$

where K_1, L_1, K_2, and L_2 are arbitrary (but fixed) parameters. Equation (4.212) can be rewritten as

$$F(z_1, z_2)B(z_1, z_2) = A(z_1, z_2) \tag{4.213}$$

and since multiplication of the z polynomials is equivalent to convolution of the arrays in the space domain, it follows that

$$a(k, l) = \sum_{m=0}^{K_2} \sum_{n=0}^{L_2} b(m, n) f(k - m, l - n) \tag{4.214}$$

The coefficients $a(k, l)$ are defined by equation (4.214) over the integers $k = 0, 1, \ldots, K_1$; $l = 0, 1, \ldots, L_1$.

For all other values of k and l, coefficients $a(k, l)$ are zero. Therefore, if a set of integer pairs S_a is defined as

$$S_a = \{(k, l): 0 \leq k \leq K_1, 0 \leq l \leq L_1\} \tag{4.215}$$

and another set $\hat{S}_a$ defined as the set of all other values of (k, l) greater than zero:

$$\hat{S}_a = \{(k, l): k > 0, l > 0, (k, l) \notin S_a\} \tag{4.216}$$

then equation (4.214) can be written in the alternative form

$$f(k, l) = -\sum_{\substack{m=0 \\ (m,n)\neq 0}}^{K_2} \sum_{n=0}^{L_2} b(m, n) f(k - m, l - n) \qquad \text{for } (k, l) \in \hat{S}_a \tag{4.217}$$

and when $b(m, n)$ are suitably chosen in equation (4.217), $f(k, l)$ approximates the desired impulse response $h(m, n)$. Thus

$$h(k, l) \simeq -\sum_{m=0}^{K_2} \sum_{n=0}^{L_2} b(m, n) h(k - m, l - n) \qquad \text{for } (k, l) \in \{\hat{S}_a \cap S_d\} \tag{4.218}$$

where S_d, the set of integer pairs over which the desired filter is defined, is given by

$$S_d = \{(k, l): 0 \leq k \leq M, 0 \leq l \leq N\} \tag{4.219}$$

An error $e(k, l)$ can be added to the right-hand side of equation (4.218) to produce the equality

$$h(k, l) = e(k, l) - \sum_{\substack{m=0 \\ m+n\neq 0}}^{K_2} \sum_{n=0}^{L_2} b(m, n)h(k-m, l-n) \qquad \text{for } (k, l) \in \{\hat{S}_a \cap S_d\} \tag{4.220}$$

which, with $h(k, l)$ moved inside the summation, can be written as

$$e(k, l) = \sum_{m=0}^{K_2} \sum_{n=0}^{L_2} b(m, n)h(k-m, l-n) \qquad \text{for } (k, l) \in \{\hat{S}_a \cap S_d\} \tag{4.221}$$

The obvious choice of $b(m, n)$ is that based on minimization of the mean-square error:

$$e^2 = \sum_k \sum_l \left[\sum_{m=0}^{K_2} \sum_{n=0}^{L_2} b(m, n)h(k-m, l-n) \right]^2 \qquad \text{for } (k, l) \in \{\hat{S}_a \cap S_d\} \tag{4.222}$$

so that, by differentiating equation (4.222) with respect to $b(m, n)$ and setting the resultant equations equal to zero, the $b(m, n)$ that minimizes e^2 above can be found. This involves solving $K_2 \times L_2$ equations of the type

$$\sum_{\substack{m=0 \\ m+n\neq 0}}^{K_2} \sum_{n=0}^{L_2} b(m, n)\phi(i, j, m, n) = \phi(i, j) \tag{4.223}$$

for $i = 0, 1, \ldots, K_2$; $j = 0, 1, \ldots, L_2$, but $i + j \neq 0$, where

$$\phi(i, j, m, n) = \sum_k \sum_l h(k-m, l-n)h(k-i, l-j) \tag{4.224}$$

and

$$\phi(i, j) = -\sum_k \sum_l h(k, l)h(k-i, l-j) \qquad \text{for } (k, l) \in \{\hat{S}_a \cap S_d\} \tag{4.225}$$

With the denominator $B(z_1, z_2)$ of the equation (4.212) computed, the numerator $A(z_1, z_2)$ must next be determined. One way of doing this is to compute those coefficients of $A(z_1, z_2)$ that minimize the mean-square difference between the coefficients of $F(z_1, z_2) = A(z_1, z_2)/B(z_1, z_2)$ and the coefficients of the desired response $H(z_1, z_2)$. This is a Wiener filtering problem in two dimensions. It consists of finding the optimum filter $A(z_1, z_2)$, given an input $1/B(z_1, z_2)$ and the desired output $H(z_1, z_2)$.

In a simpler but less accurate method, the array B can be convolved with H to give the A array. Since $A(z_1, z_2)/B(z_1, z_2) \simeq H(z_1, z_2)$, the coefficients $\mathbf{a}$ are computed from $A(z_1, z_2) = B(z_1, z_2)H(z_1, z_2)$ for $(k, l) \in S_a$. This technique is generally very easy to implement, but has the disadvantage that the stability of the resulting filter cannot be guaranteed.

4.13.2. The Bordner Technique[29]

This is a modified form of the above technique due to Shanks *et al.*[16] To ensure the stability of the designed filter, the finite 2-D sequence of equation (4.211) representing the desired impulse response can be augmented with an infinite sequence. It is intuitively apparent that the best such sequence is that which would represent the most natural extension of the original impulse response sequence $h(m, n)$ for $m = 0, 1, \ldots, M$ and $n = 0, 1, 2, \ldots, N$, and which is square-summable over the region $(0, \infty)$. In other words, given the impulse response of the filter $h(m, n)$, a square-summable tail $g(m, n)$ should be added to it, yielding a new 2-D sequence $\hat{h}(m, n)$ where

$$\hat{h}(m, n) = \begin{cases} h(m, n) & \text{for } 0 \le m \le M \text{ and } 0 \le n \le N \\ g(m, n) & M + 1 \le m \le \infty,\ 0 \le n \text{ and} \\ & 0 \le m,\ N + 1 \le n \le \infty \end{cases} \tag{4.226}$$

The square-summability constraint on the augmented sequence implies stability, since

$$\sum_m \sum_n |\hat{h}(m, n)|^2 < \infty \tag{4.227}$$

In this proposed method, the transfer function of the filter is restricted to the sum of 2-D second-order all-pole sections of the form

$$F(z_1, z_2) = \sum_{i=0}^{N} \frac{a_i}{1 + b_i z_1 + c_i z_2 + d_i z_1 z_2} \tag{4.228}$$

The parameter set (a_i, b_i, c_i, d_i) must be calculated so as to minimize the error function e in a least-squares sense:

$$e = \sum_{m=0}^{M} \sum_{n=0}^{N} [h(m, n) - f(m, n)]^2$$

$$+ \left\{ \sum_{m=0}^{\infty} \sum_{n=0}^{\infty} [g(m, n) - f(m, n)]^2 - \sum_{m=0}^{M} \sum_{n=0}^{N} [g(m, n) - f(m, n)]^2 \right\}$$

The only restriction in this method is that the infinite array $g(m, n)$ must be square-summable.

It can be seen that the second spatial design technique discussed in this section is based on some modification applied to the technique of Shanks, Treitel, and Justice(16) to overcome the stability problem encountered by the signal technique. Guaranteeing the stability of the designed filter through the above modification of course increases the computational costs.

REFERENCES

1. L. R. Rabiner, N. Y. Graham, and M. D. Helms, Linear programming design of IIR digital filters with arbitrary magnitude function, *IEEE Trans. Acoust., Speech, Signal Process.* **ASSP-22,** 117-123 (1974).
2. L. R. Rabiner and B. Gold, *Theory and Application of Digital Signal Processing,* Prentice-Hall, Englewood Cliffs, NJ (1975).
3. A. Chottera and G. A. Jullien, Designing Near Linear Phase Recursive Filters Using Linear Programming, Proc. IEEE Int. Conf. on Acoustics, Speech and Signal Processing, 88-92 (May 1977).
4. E. I. Jury, *Theory and Application of the z-Transform Method,* John Wiley and Sons, New York (1964).
5. E. Robinson, *Statistical Communication and Detection,* Hafner, New York (1967).
6. F. H. Borphy and A. C. Salazar, Two design techniques for digital phase networks, *Bell Syst. Tech. J.* **54,** 767-781 (1975).
7. V. Ramachandran and C. S. Gargour, Implementation of a stability test of 1-D discrete system based on Schussler's theorem and some consequent coefficient conditions, *J. Franklin Inst.* **317,** 341-358 (1984).
8. V. Ramachandran, C. S. Gargour, M. Ahmadi, and M. T. Boraie, Direct design of recursive digital filters based on a new stability test, *J. Franklin Inst.* **318,** 407-413 (1984).
9. Lonnie C. Ludeman, *Fundamentals of Digital Signal Processing,* Harper and Row, New York (1986).
10. A. Antoniou, *Digital Filters: Analysis and Design,* McGraw-Hill, New York (1979).
11. A. Antoniou, M. Ahmadi, and C. Charalambous, Design of factorable lowpass 2-dimensional digital filters satisfying prescribed specifications, *IEE Proc.* **128,** Part G, No. 2, 53-60 (1981).
12. C. Charalambous, Design of 2-dimensional circularly-symmetric digital filters, *IEE Proc.* **129,** Part G, No. 2, 47-54 (1982).

13. K. Rajan and M. N. S. Swamy, Design of separable denominator 2-dimensional digital filters possessing real circularly symmetric frequency responses, *IEE Proc.* **129**, Part G, No. 5, 235–240 (1982).
14. M. Ahmadi, M. T. Boraie, V. Ramachandran, and C. S. Gargour, Design of 2-D recursive digital filters with constant group delay characteristics using separable denominator transfer function and a new stability test, *IEEE Trans. Acoust., Speech, Signal Process.* **ASSP-33,** 1316–1318 (1985).
15. T. S. Huang, J. W. Burnett, and A. G. Deczky, The importance of phase in image processing filters, *IEEE Trans. Acoust., Speech, Signal Process.* **ASSP-23,** 529–542 (1975).
16. J. L. Shanks, S. Treitel, and J. M. Justice, Stability and synthesis of two-dimensional recursive filters, *IEEE Trans. Audio Electroacoust.* **AU-20,** 115–128 (1982).
17. J. M. Costa and A. N. Venetsanopoulos, Design of circularly symmetric two-dimensional recursive filters, *IEEE Trans. Acoust., Speech, Signal Process.* **ASSP-22,** 432–443 (1974).
18. M. Ahmadi, A. G. Constantinides, and R. A. King, Design Technique for a Class of Stable 2-Dimensional Recursive Digital Filters, Proc. IEEE Int. Conf. on Acoustics, Speech and Signal Processing, Philadelphia, USA, 145–147 (April 1976).
19. R. A. King and A. H. Kayran, A New Transformation Technique for the Design of 2-Dimensional Stable Recursive Digital Filters, Proc. IEEE Int. Symp. on Circuits and Systems, Chicago, 196–199 (April 1981).
20. A. H. Kayran and R. A. King, Design of recursive and nonrecursive fan filters with complex transformations, *IEEE Trans. Circuits Syst.* **CAS-30,** 849–857 (1983).
21. A. Chottera and G. A. Jullien, Design of 2-dimensional recursive digital filters using linear programming, *IEEE Trans. Circuits Syst.* **CAS-29**, 817–826 (1982).
22. R. Fletcher and M. J. D. Powell, A Rapid descent method for minimization, *Comput. J.* **6,** 163–168 (1963).
23. S. A. H. Aly and M. M. Fahmy, Design of two-dimensional recursive digital filters with specified magnitude and group delay characteristics, *IEEE Trans. Circuits Syst.* **CAS-25,** 908–916 (1978).
24. P. A. Ramamoorthy and L. T. Bruton, Design of stable two-dimensional analogue and digital filters with applications in image processing, *Int. J. Circuit Theory Appl.* **7,** 229–245 (1979).
25. S. Golikeri, M. Ahmadi, and V. Ramachandran, Design of 2-D recursive digital filters satisfying prescribed magnitude and constant group delay response, *Electron. Lett.* **19,** 9–11 (1983).
26. V. Ramachandran and M. Ahmadi, Design of 2-D stable analog and recursive digital filters using properties of the derivative of even or odd parts of Hurwitz polynomials, *J. Franklin Inst.* **315,** 259–267 (1983).
27. M. Ahmadi and V. Ramachandran, New method for generating two-variable VSHPs and its application in the design of two-dimensional recursive digital filters with prescribed magnitude and constant group delay responses, *IEE Proc.* **131,** Part G, No. 4, 151–155 (1984).
28. M. Ahmadi, M. O. Ahmad, and V. Ramachandran, Transfer function realization of a class of doubly-terminated two-variable lossless networks and their application in linear phase 2-dimensional filter design, *J. Franklin Inst.* **321,** 147–153 (1986).
29. G. W. Bordner, *Time Domain Design of Stable Recursive Digital Filters*, Ph.D. Dissertation, State University of New York at Buffalo (1974).

5

Quantization and Roundoff Errors

5.1. SOURCES OF ERRORS IN DIGITAL FILTERS

A one-dimensional (1-D) digital filter, as noted in Section 1.3, is generally defined by

$$y_n = \sum_{i=0}^{M} a_i u_{n-i} - \sum_{i=1}^{N} b_i y_{n-i} \tag{5.1}$$

where $\{u_n\}$ is the input sequence, $\{y_n\}$ is the output sequence, and a_i and b_i are some constants.

Since the aim of realizing a digital filter is to integrate it within a digital network, two obvious advantages should consequently be achieved. These are: flexibility, due to the simplicity of changing the filter characteristics (coefficients a_i and b_i); and reliability, due to the computational algorithm. There exists, however, an inherent limitation on accuracy, because of the finite word length of the computer in use.

There are five main factors responsible for the accuracy problem in digital filters, and they are all related to the finite word length of the computer in some way or another. These factors are[1-5]:

1. The input quantization error caused by the quantization step of the input signal $\{u_n\}$ into a finite number of bits.
2. The representation of coefficients a_i and b_i by a finite number of bits.
3. Accumulation of roundoff errors at each computational operation.
4. The form of realization of the digital filter, i.e., direct, canonic, parallel, or cascade form.
5. The type of number system to be considered.

5.2. INPUT QUANTIZATION EFFECTS

The preliminary process of digitizing the continuous input signal $u(t)$ into the amplitude samples $u(nT)$, where T is the sampling interval, involves a quantization into one of several levels at intervals of E_0, which normally introduces an error into the system. This error will be denoted by $e(nT)$ and is analyzed by considering it as additive noise superimposed at the input of the filter. The input sequence may thus be expressed as the sum of two sequences in the form

$$u_i(nT) = u(nT) + e(nT) \tag{5.2}$$

where $u(nT)$ is the noiseless input signal and $e(nT)$ is the error sequence.

In our analysis of this problem, it will be assumed that the errors $e(nT)$ and $e(mT)$ are uncorrelated for any integers m and n, $m \neq n$, and that the error signal $e(nT)$ and the input signal are also uncorrelated.[6] Bennett[7] has shown that these assumptions hold for most practical signals which are rich in variation. Consequently, the input quantization error associated with each sample is uniformly distributed in the range

$$-E_0/2 \leq e \leq E_0/2 \qquad \text{for rounding}$$

and

$$-E_0 \leq e \leq 0 \qquad \text{for truncation}$$

The effect of quantization and rounding may be regarded as an additive white random noise with zero mean and a variance σ^2 given by

$$p(e) = 1/E_0, \qquad -E_0/2 \leq e \leq E_0/2 \tag{5.3}$$

and

$$\sigma^2 = \int_{-E_0/2}^{E_0/2} e^2 p(e)\, de = E_0^2/12 \tag{5.4}$$

where $p(e)$ is the probability density of the error (Figure 5.1).

Let us consider the linear system shown in Figure 5.2 with an impulse response $h(nT)$; the output noise sequence $e_o(nT)$ is obtained by convolving $h(nT)$ with the input noise sequence $e_i(nT)$. Thus

$$e_o(nT) = \sum_{m=0}^{\infty} h(mT) e_i[(n-m)T] \tag{5.5}$$

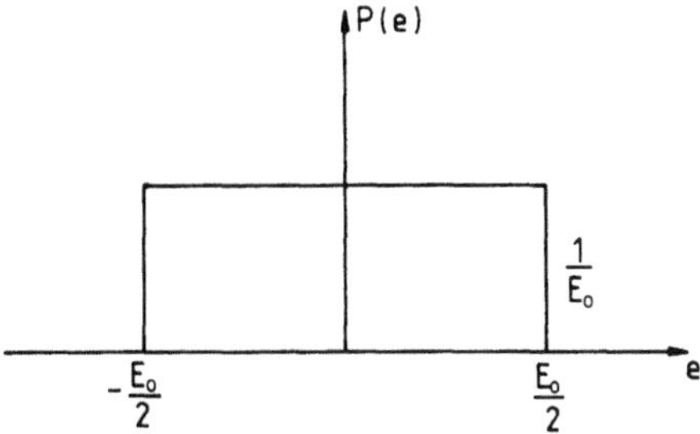

Figure 5.1. Probability density function for roundoff error in the case of rounding.

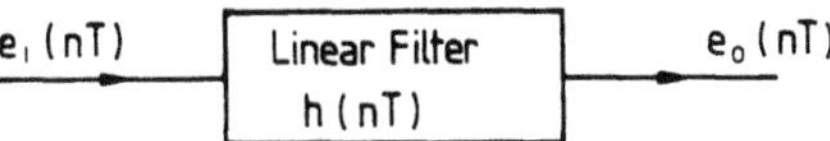

Figure 5.2. Representation of the error function as a linear, discrete-time system.

The sequence $e_o(nT)$ in expression (5.5) can be viewed as a weighted sum of individual random variables with an autocorrelation function given by

$$\begin{aligned} R_{oo}(kT) &= \sum_{n=0}^{\infty} e_o(nT)e_o[(n-k)T] \\ &= \sum_{n=0}^{\infty} \sum_{m=0}^{\infty} h(mT)e_i[(n-m)T]h[(m-k)T]e_i[(n-m+k)T] \\ &= R_{ii}(kT) \sum_{m=0}^{\infty} h(mT)h[(m-k)T] \end{aligned} \tag{5.6}$$

where $R_{ii}(kT)$ is the autocorrelation of $e_i(nT)$ given by

$$R_{ii}(kT) = \sum_{n=0}^{\infty} e_i(nT)e_i[(n-k)T] \tag{5.6a}$$

The variance is given when $k = 0$ in equation (5.6); thus

$$\sigma_o^2 = R_{oo}(0) = R_{ii}(0) \sum_{m=0}^{\infty} h^2(mT) \tag{5.7}$$

and with $R_{ii}(0) = \sigma_i^2 = E_0^2/12$, equation (5.7) can be expressed as

$$\sigma_o^2 = E_0^2/12 \sum_{m=0}^{\infty} h^2(mT) \tag{5.8}$$

When the impulse response $h(mT)$ is infinite, it is simpler to evaluate the output variance in the Z domain from the discrete Parseval theorem,[8] by calculating the residues of a contour integral. Thus

$$\sigma_o^2 = \frac{E_0^2}{12} \cdot \frac{1}{2\pi j} \oint H(z)H^*(z)\frac{dz}{z} \tag{5.9}$$

where $H(z)$ is the transfer function of the filter and $H^*(z)$ its complex conjugate.

5.3. TYPES OF ARITHMETIC

The implementation of a digital filter can be carried out using different types of arithmetic. The two common modes of computation, however, are the fixed-point and floating-point number systems.

In fixed-point arithmetic, every state or coefficient variable x is normalized so as to satisfy the inequality

$$-1 \le x \le 1 \tag{5.10}$$

and is represented in binary 2's-complement form as

$$f_0 \cdot f_1 f_2 \cdots f_t \tag{5.11}$$

where f_i are 1 or 0 and $(\cdot)$ represents the binary point. Expression (5.11) represents the number

$$\sum_{i=1}^{t} 2^{-i} f_i \qquad \text{when } f_0 = 0 \tag{5.12}$$

or

$$-\left(\sum_{i=1}^{t} 2^{-i} \bar{f}_i + 2^{-t}\right) \qquad \text{when } f_0 = 1 \tag{5.13}$$

where $\bar{f}_i$ is the complement of f_i.

This representation of a real number corresponds to a quantization step size of 2^{-t}. Thus, if x and y are quantized in this manner, the addition of x and y will not generate any errors, assuming that there is no overflow:

$$(x+y)_q = x + y + e, \qquad e = 0 \tag{5.14}$$

where $(\cdot)_q$ represents the quantized result of the operation $(\cdot)$.

However, the multiplication of x and y consisting of $(t+1)$ bits each (including the sign bit) yields a $(2t+1)$-bit number, and to represent the result in a finite word length machine, the least significant t bits must be removed, either by rounding or truncation. In the case of rounding, $2^{-(t+1)}$ is added to the result of the multiplication and the least significant t-bits dropped. Consequently, the magnitude error in the rounded multiplication is $\leq 2^{-(t+1)}$; thus

$$(xy)_q = xy + e, \qquad |e| \leq 2^{-(t+1)} \tag{5.15}$$

On the other hand, in the case of truncation, the magnitude error is bounded by

$$|e| \leq 2^{-t}$$

The type of arithmetic used in the filter implementation not only determines the amount of coefficient quantization noise, but also influences the dynamic range of the filter. While fixed-point arithmetic filters are easier to analyze, they suffer from the disadvantages of having a fixed binary point and more rigid limitations on their dynamic range.[4,9]

The dynamic range (DR) of a filter is defined as the ratio between the largest positive representable value and the smallest positive representable value. In the case of fixed-point arithmetic, this is given by

$$(\mathrm{DR})_{\text{fixed}} = (1 - 2^{-t})/2^{-t} = 2^t - 1 \tag{5.16}$$

We consider now a floating point number defined by

$$m_0 \cdot m_1 m_2 \cdots m_t g_0 g_1 \cdots g_k \tag{5.17}$$

where m_i and g_i are 1 or 0, and the m part is the fractional part or mantissa and the g part is the exponent. Expression (5.17) always has to be normalized so that the mantissa is constrained by

$$\tfrac{1}{2} \leq |\text{mantissa}| \leq 1 \tag{5.18}$$

Since the base of the exponent is 2 and the exponent is an integer, expression (5.17) represents the number

$$\left(\sum_{i=1}^{t} 2^{-i} m_i\right) \cdot 2^{G_k} \qquad \text{when } m_0 = 0 \tag{5.19}$$

or

$$-\left(\sum_{i=1}^{t} 2^{-i} \bar{m}_i + 2^{-t}\right) \cdot 2^{G_k} \qquad \text{when } m_0 = 1 \tag{5.20}$$

where

$$G_k = \begin{cases} \sum_{i=1}^{k} 2^{k-i} g_i & \text{when } g_0 = 0 \quad (5.21) \\ -\left(\sum_{i=1}^{k} 2^{k-i} \bar{g}_i + 1 \right) & \text{when } g_0 = 1 \quad (5.22) \end{cases}$$

$\bar{m}_i$ and $\bar{g}_i$ being the complements of m_i and g_i, respectively.

The addition operation in floating-point systems is more complex than in fixed-point systems. This is shown by considering the two numbers $x_1 = m_1 \cdot 2^{g_1}$ and $x_2 = m_2 \cdot 2^{g_2}$ where $g_1 > g_2$. In this case x_2 is divided successively by 2 (a shift-right operation) until it is expressed as

$$x_2 = \hat{m}_2 \cdot 2^{g_1} \tag{5.23}$$

The two mantissas m_1 and $\hat{m}_2$ are then added and scaled up or down to conform to the inequality constraint given in relation (5.18). The mantissa is rounded to $(t+1)$ bits and g_1 is adjusted accordingly. If the summation result is $m_3 \cdot 2^{g_3}$ exactly, then the modulus of the error is bounded by

$$|e| \leq 2^{-(t+1)} \cdot 2^{g_3} \tag{5.24}$$

In general, the error bound is expressed by

$$(x_1 + x_2)_q = (x_1 + x_2)(1 + e), \qquad -2^{-t} \leq e \leq 2^{-t} \tag{5.25}$$

The floating-point multiplication of two numbers x_1 and x_2 is relatively simple, since it involves the addition of the exponents g_1 and g_2 and the multiplication of the mantissas; the product is then scaled and rounded to give a $(t+1)$-bit mantissa. The result of a floating-point multiplication is given by

$$(x_1 x_2)_q = x_1 x_2 (1 + e), \qquad -2^{-t+1} \leq e \leq 0 \tag{5.26}$$

When considering the dynamic range of a floating-point system, this is expressed by

$$\begin{aligned} (\mathrm{DR})_{\text{floating}} &= \frac{(1 - 2^{-t}) 2^{(2^k - 1)}}{2^{-2^k}/2} \\ &= (1 - 2^{-t}) 2^{2^{k+1}} \end{aligned} \tag{5.27}$$

5.4. ROUNDOFF NOISE

It is clear from Section 5.3 that the type of arithmetic selected to implement a digital filter has a direct bearing on the element of error in the implementation. Furthermore, although these errors, in general, were separately classified in Section 5.1, they nevertheless all interact together to produce noise effects in the final design of the filter.

By analogy to the input quantization effects discussed in Section 5.2, there exists a similar effect due to the quantization of the filter coefficients into a finite number of bits.

The ideal output sequence of a digital filter is given by the difference equation (5.1). This equation, however, is altered if the coefficients a_i and b_i are quantized due to their machine representation by a finite number of bits. Equation (5.1) thus becomes

$$[y_n]_q = \sum_{i=0}^{M} [a_i]_q u_{n-i} - \sum_{i=1}^{N} [b_i]_q [y_{n-i}]_q \tag{5.28}$$

where

$$[a_i]_q = a_i + \varepsilon \qquad \text{and} \qquad [b_i]_q = b_i + \delta \tag{5.29}$$

ε and δ being the quantization error occurring in a_i and b_i, respectively. Although the input quantization effect is not considered in equation (5.28), the output elements y_i are shown to suffer from the effects of coefficient quantization and, consequently, become themselves sources of additional errors.

Furthermore, the quantization of coefficients a_i and b_i alters the original pole and zero positions of the synthesized filter and thus leads to the alteration of its amplitude and phase spectra.

Owing to the fact that the output y_n suffers from additive noise with each recurrence of equation (5.28), an overall roundoff noise is generated. This additive roundoff noise has the same properties as the quantization noise, with the exception that the transfer function is considered from the stage where noise is introduced to the output, and the output mean-square error due to quantization and roundoff noises is obtained by summing the individual contributions at the output. This is possible, since the analysis relies on the statistical independence of the various sources of errors. Equation (5.1) may thus be rewritten as

$$w_n = \sum_{i=0}^{M} a_i u_{n-i} - \sum_{i=1}^{N} b_i w_{n-i} + \varepsilon_n \tag{5.30}$$

where ε_n is the roundoff noise contribution of sequence element n.

The complete error sequence e_n is defined as[10]

$$e_n = w_n - y_n \tag{5.31}$$

and from equations (5.1), (5.30), and (5.31) we obtain

$$e_n = -\sum_{i=1}^{N} b_i(w_{n-i} - y_{n-i}) + \varepsilon_n \tag{5.32}$$

or

$$e_n = \varepsilon_n - \sum_{i=1}^{N} b_i e_{n-i} \tag{5.33}$$

This problem is hence reduced to the same earlier problem but, in this case, the input is ε_n and the output e_n. If the transfer function corresponding to equation (5.33) is denoted by $1/D(z)$, and as $\{\varepsilon_n\}$ is a zero-mean wide-sense stationary process (since it is assumed that the filter is excited by a wide-sense stationary input), then, in terms of the power spectral density $\phi_{\varepsilon\varepsilon}(z)$ of $\{\varepsilon_n\}$, it can be shown[3] that $\{e_n\}$ is also a zero-mean wide-sense stationary process with power spectral density $\phi_{ee}(z)$ given by

$$\phi_{ee}(z) = \frac{1}{D(z)} \frac{1}{D(1/z)} \phi_{\varepsilon\varepsilon}(z) \tag{5.34}$$

and the mean-square error of e_n can be expressed as

$$E\{e_n^2\} = \frac{1}{2\pi j} \oint \phi_{ee}(z) \frac{dz}{z} \tag{5.35}$$

Figure 5.3 shows how the roundoff-noise contribution ε_n relates to the input u_n, resulting in w_n. In Figure 5.4 we show a detailed model of the noise effects corresponding to the direct realization of filter (5.1).

When considering the effect of rounding in floating-point arithmetic, the same approach could be adopted.[5,6,11-14] The direct realization of filter

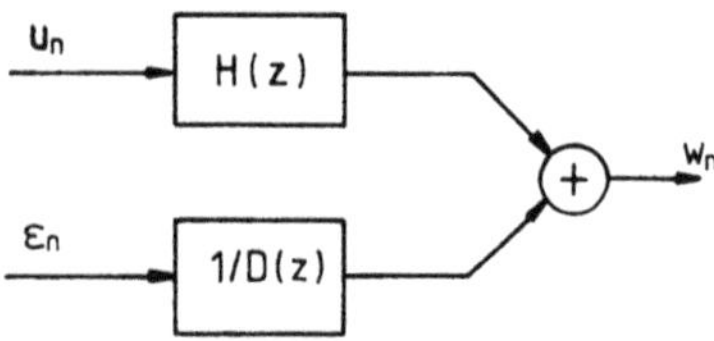

Figure 5.3. Model for the additive property of the overall roundoff-noise contribution.

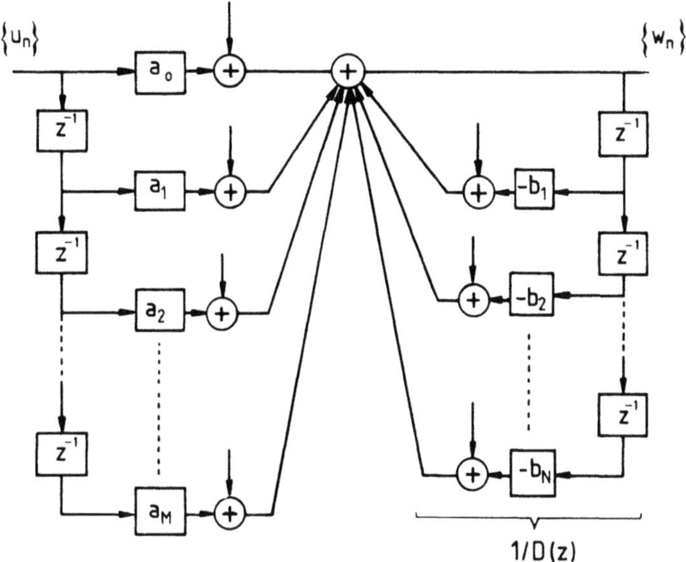

Figure 5.4. Model for the noise effects in a direct realization.

(5.1) under floating-point condition yields

$$w_n = \left(\sum_{i=0}^{M} a_i u_{n-i} - \sum_{i=1}^{N} b_i w_{n-i} \right)_{\text{fl}} \tag{5.36}$$

Relationships (5.25) and (5.26) can be used to draw a flow diagram (Figure 5.5), where the quantities $\delta_{n,i}$, $\varepsilon_{n,i}$, $\zeta_{n,i}$, $\eta_{n,i}$, and ψ_n represent the errors which are caused by roundoff at each arithmetic step and which satisfy the previous statistical requirements of independence, randomness, and identical distribution.

The actual output sequence is thus given by

$$w_n = \sum_{i=0}^{M} a_i \theta_{n,i} u_{n-i} - \sum_{i=1}^{N} b_i \phi_{n,i} w_{n-i} \tag{5.37}$$

where

$$\theta_{n,0} = (1 + \psi_n)(1 + \delta_{n,0}) \prod_{i=1}^{M} (1 + \zeta_{n,i})$$

$$\theta_{n,j} = (1 + \psi_n)(1 + \delta_{n,j}) \prod_{i=j}^{M} (1 + \zeta_{n,i}) \qquad \text{for } j = 1, 2, \ldots, M \tag{5.38}$$

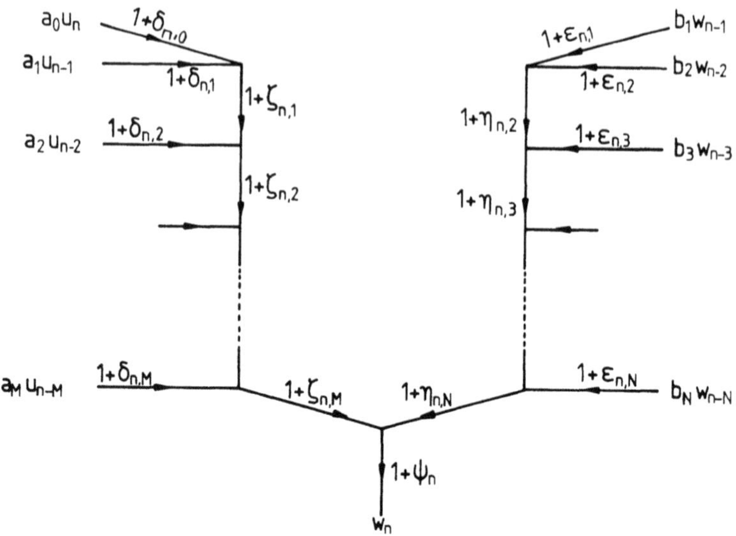

Figure 5.5. Floating-point quantization in an Nth-order system (after Liu and Kaneko[5]).

and

$$\phi_{n,1} = (1+\psi_n)(1+\varepsilon_{n,1})\prod_{i=2}^{N}(1+\eta_{n,i})$$

$$\phi_{n,j} = (1+\psi_n)(1+\varepsilon_{n,j})\prod_{i=j}^{N}(1+\eta_{n,i}) \qquad \text{for } j = 2, 3, \ldots, N \tag{5.39}$$

These random variables, according to the discussion in Section 5.3, are uniformly distributed over the interval $[-2^{-t}, 2^{-t}]$ or $[-2^{-t+1}, 0]$ according to whether rounding or truncation is used in the floating-point arithmetic implementation.

By defining $b_0 = 1$ and $\phi_{n,0} = 1$, equation (5.37) may be expressed in the form

$$\sum_{i=0}^{N} b_i\phi_{n,i}w_{n-i} = \sum_{i=0}^{M} a_i\theta_{n,i}u_{n-i} \tag{5.40}$$

and since coefficients $\phi_{n,i}$ and $\theta_{n,i}$ are time-varying random coefficients, the solution of equation (5.40) involves the computation of w'_n, w''_n, $w'''_n, \ldots, w_n^{(k)}, \ldots$, which are defined by[5]

$$\sum_{i=0}^{N} b_i\bar{\phi}_i w'_{n-i} = \sum_{i=0}^{M} a_i\bar{\theta}_i u_{n-i} \tag{5.41}$$

$$\sum_{i=0}^{N} b_i \bar{\phi}_i w''_{n-i} = \sum_{i=0}^{M} a_i(\theta_{n,i} - \bar{\theta}_i)u_{n-i} - \sum_{i=0}^{N} b_i(\phi_{n,i} - \bar{\phi}_i)w'_{n-i} \tag{5.42}$$

and

$$\sum_{i=0}^{N} b_i \bar{\phi}_i w^{(k)}_{n-i} = -\sum_{i=0}^{N} b_i(\phi_{n,i} - \bar{\phi}_i)w^{(k-1)}_{n-i}, \qquad k = 3, 4, 5, \ldots \tag{5.43}$$

where

$$\bar{\phi}_i = E\{\phi_{n,i}\} \qquad \text{and} \qquad \bar{\theta}_i = E\{\theta_{n,i}\}$$

With the initial assumption that the input $\{u_n\}$ is zero-mean wide-sense stationary, a relationship could be derived[5] for the error power spectral density $\phi_{ee}(z)$ in terms of the power spectral densities $\phi_{y'y'}(z)$, $\phi_{y'',y''}(z), \ldots$ of the sequences $\{y'\}, \{y''\}, \ldots$, respectively. Each of the $y^{(k)}$ is itself zero-mean wide-sense stationary, with its corresponding $\phi_{y^{(k)}y^{(k)}}(z)$ being a function of $\phi_{uu}(z)$, the power spectral density of the input sequence.

5.5. EFFECT OF FILTER REALIZATION ON THE NOISE MODEL

In Section 5.4, the roundoff-noise effects were discussed for the direct form of realization of a given transfer function. However, the cascade (with its more recent transpose configuration) and parallel forms of realization (see Section 6.2) are of equal importance when considering the design specifications of a filter.

Many researchers have offered alternative methods of error analyses for such realizations,[2,15-17] in both arithmetic implementations. The statistical analyses carried out for such structures follow the same lines as for the direct realization and are beyond the scope of this book. However, we shall briefly consider each form and give the general formula for its overall output noise variance.

The cascade structure of a digital filter is shown in Figure 5.6, where the transfer function $H(z)$ of the filter is initially decomposed into first- or

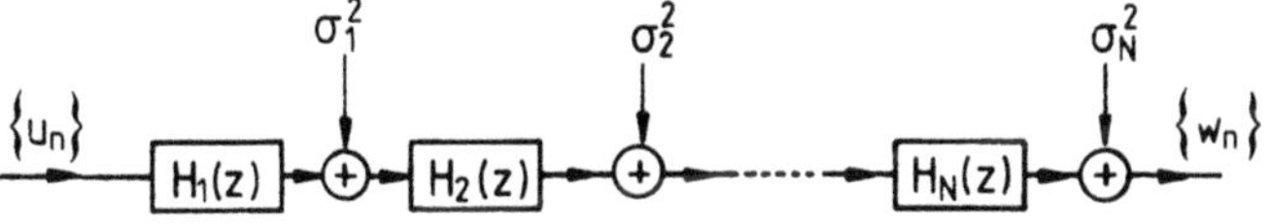

Figure 5.6. Noise model for an N-stage cascade structure.

second-order structures such that, for N stages in cascade,

$$H(z) = H_1(z) \bullet H_2(z) \bullet \cdots \bullet H_N(z)$$
$$= \frac{z - \alpha_1}{1 - \beta_1 z} \bullet \frac{z - \alpha_2}{1 - \beta_2 z} \bullet \cdots \bullet \frac{z^2 - \gamma z + \delta}{\pi z^2 + \zeta z + 1} \tag{5.44}$$

The arrows at the different summation points represent the input noise at each section, which includes the noise at the output of the previous section together with the noise generated within the current section. The output noise from the first stage therefore passes through the zeros and poles of the $N - 1$ remaining stages. In general, the average output noise power at the ith stage passes through the $(i + 1)$th and subsequent stages of the system.

Again, using Parseval's theorem, the total average noise power σ_c^2 can be expressed as

$$\sigma_c^2 = \sum_{i=1}^{N} \sigma_i^2 \frac{1}{2\pi j} \oint \prod_{k=i+1}^{N} [H_k(z) \cdot H_k^*(z)] \frac{dz}{z} \tag{5.45}$$

where σ_i^2 is the noise variance associated with stage i.

The noise model for the parallel form of realization, on the other hand, is shown in Figure 5.7 where, again, the filter transfer function is broken down into N separate stages of first- or second-order functions which are linearly combined:

$$H(z) = H_1(z) + H_2(z) + \cdots + H_N(z)$$
$$= \frac{\alpha_1}{1 - \beta_1 z} + \frac{\alpha_2}{1 - \beta_2 z} + \cdots + \frac{\gamma \delta z}{\pi + \zeta z + \eta z^2} \tag{5.46}$$

In this case, the overall output noise variance is the sum of the individual

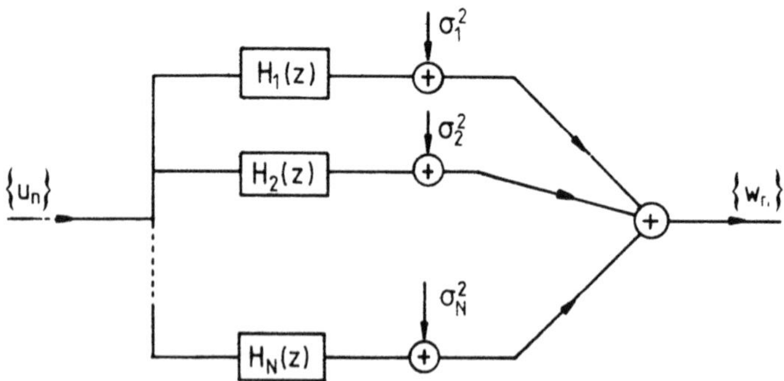

Figure 5.7. Noise model for an N-stage parallel structure.

average noise power at the output of each section; thus

$$\sigma_{\mathrm{p}}^2 = \sum_{i=1}^{N} \sigma_i^2 \tag{5.47}$$

The detailed comparison of various digital filter implementations executed in either fixed or floating point yields interesting results. Some realizations result in better signal-to-noise ratios than others for a certain range of parameters. However, the basic question as to which realization yields the best signal-to-noise ratio based on input quantization and roundoff noises, and given the type of arithmetic and the number of bits, is not yet fully answered.

5.6. LIMIT-CYCLE OSCILLATIONS IN DIGITAL FILTERS

Owing to the finite word length used in the implementation of a recursive digital filter, a nonzero periodic output is possible under zero-input conditions. Called "limit cycles," these nonlinearities can be produced by[18]

1. internal register overflow,
2. internal product quantization.

Overflow limit cycles can always be eliminated by using "saturation arithmetic,"[19] and only the second limit cycle type will be examined here, specifically those generated by second-order sections using fixed-point, 2's-complement arithmetic with internal product rounding. The choice of a second-order section reflects the fact that early research[20] discovered that it was preferable to construct higher-order filters from first- or second-order configurations.

A second-order digital filter with zero-input and fixed-point arithmetic may be modeled by

$$y(n) = -[ay(n-1)]_{\mathrm{R}} - [by(n-2)]_{\mathrm{R}} \tag{5.48}$$

where $y(n)$, without loss of generality, is assumed to be an integer, and $[\cdot]_{\mathrm{R}}$ denotes rounding to the nearest integer.

A zero-input, self-sustaining oscillation generated by equation (5.48) is called a limit cycle, and constitutes a periodic sequence (period N) of integers, $\{y(n)\}$. Such a limit cycle may be uniquely represented by any N-element sequence $\{y_i\}_{i=0}^{N-1}$ selected from $\{y(n)\}$.

It is not possible to predict exactly which specific limit cycle frequencies a given filter can support, or what is the maximum limit cycle amplitude,[21-23] but Jackson's effective-value model provides good estimates.[18]

When $b \geq 0.5$, limit cycles with a frequency of oscillation (f_0) such that

$$0 < f_0/f_s < \tfrac{1}{2} \tag{5.49}$$

are of special interest. Based upon his effective-value model, Jackson's estimate for an amplitude bound for such limit cycles is[18]

$$\max_i |y_i| \leq J, \qquad J = \left[\frac{0.5}{1-b}\right]_{TR} \tag{5.50a}$$

where $[\cdot]_{TR}$ denotes truncation of the fractional part.

5.6.1. Reverse Limit Cycle (RLC)

If a limit cycle $\{y(n)\}$ exists and is represented by $\{y_i\}_{i=0}^{N-1}$, then $\{\hat{y}(n)\}$ is the reverse limit cycle if

$$\hat{y}_i = y_{N-1-i} \qquad \forall 0 \leq i \leq N-1 \tag{5.50b}$$

where $\{\hat{y}_i\}_{i=0}^{N-1}$ is one of the N-element sequences representing $\{\hat{y}(n)\}$. For example, $\{2, -21, 7, 18, -15, -11, 20\}$ is the RLC of $\{20, -11, -15, 18, 7, -21, 2\}$.

5.6.2. Phase-Plane Symmetry (PPS)

Limit cycles may be displayed in the phase plane, with axes $y(n)$ and $y(n-1)$, by plotting successive states $(y(n), y(n-1))\ \forall n$.[21]

Phase-plane symmetry is said to exist if a straight line can be drawn through the origin in the phase plane, splitting the state trajectory into two mirror images, as in Figure 5.8, where PPS exists about two axes.

Alternatively, if $\{y_i\}_{i=0}^{N-1}$ is equally spaced around the circumference of a circle, then PPS is equivalent to finding a diameter (which may pass through some elements of $\{y_i\}_{i=0}^{N-1}$) about which even or odd symmetry exists.

By either definition, only the first two of the following observed limit cycles have PPS:

$$\{1, -3, 4, -3, 1\}$$

$$\{4, -2, -1, 4, -5, 4, -1, -2, 4, -4, 2, 1, -4, 5, -4, 1, 2, -4\}$$

$$\{-8, 7, -2, -4, 8, -7, 2, 4\}$$

$$\{25, -8, -19, 21, 4, -24, 13, 15, -24, 2, 23, -18, -10\}$$

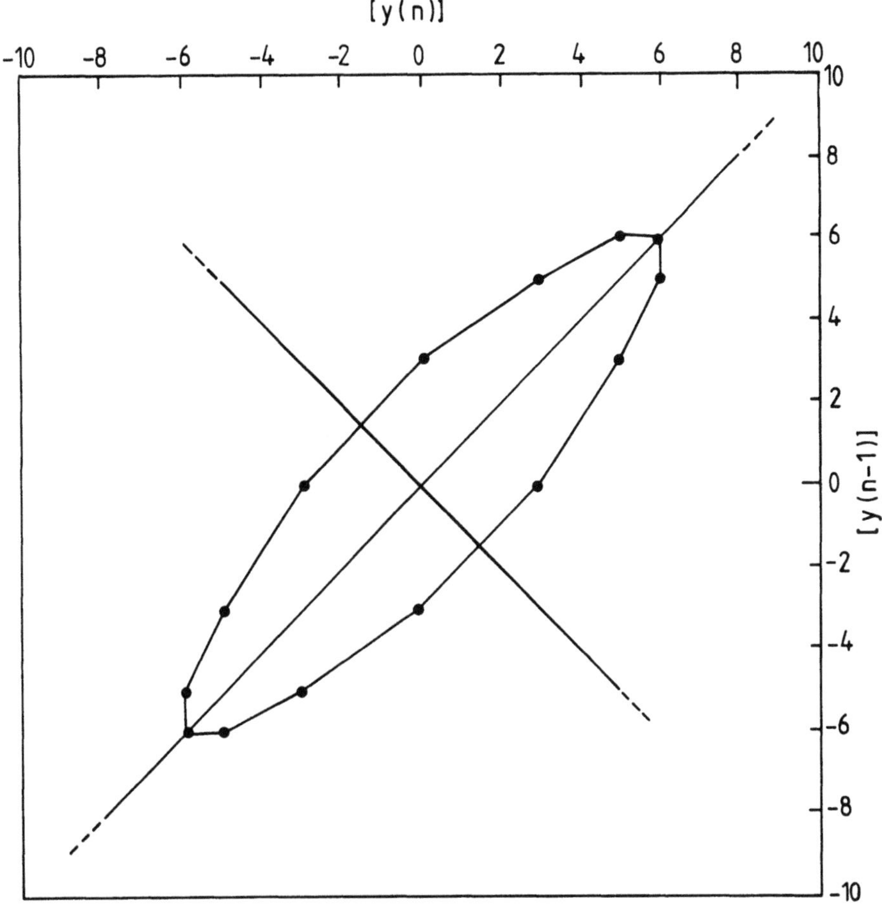

Figure 5.8. Phase-plane symmetry about two axes, $\{y_{ij}\} = \{-6, -5, -3, 0, 3, 5, 6, 6, 5, 3, 0, -3, -5, -6\}$.

5.6.3. Limit Cycle Properties

Jackson's[18] heuristic approach to limit cycles employed an effective-value linear model. From this model, estimates for the limit cycle amplitude bound and the frequency of oscillation were derived. Three further properties may be deduced from his original effective-value model.[24]

Property 1. For any limit cycle of length $N > 1$ with oscillation frequency $f_0 = C/N$, where C is the number of complete cycles of oscillation within an $(N + 1)$-length sequence, C and N are mutually prime.

For example, $a = 0.703125$ and $b = 0.98046875$ support

$$\{25, -8, -19, 21, 4, -24, 13, 15, -24, 2, 23, -18, -10\}$$

with $f_0 = 4/13$.

Property 2. For any limit cycle of length N (N even),

$$y(n) = -y(n + N/2) \ \forall n \tag{5.51}$$

For example, $a = 1.36328125$ and $b = 0.9296875$ support $\{-7, 8, -4, -2, 7, -8, 4, 2\}$.

Property 3. The mean value of a limit cycle is zero. That is

$$\sum_{i=0}^{N-1} y_i = 0 \tag{5.52}$$

for a limit cycle of length N.

For example, $a = 1.796875$ and $b = 0.98828125$ support $\{-3, -16, 32, -42, 43, -35, 21\}$.

Justification for each property derives from the assumption (based upon the effective-value model) that the limit cycle is taken to be N consecutive samples from an integer number (C) of complete cycles of a sinusoid.

So with reference to property 1, C and N must therefore have no common factor because, if they do, any limit cycle obtained by the method just stated will actually be one unique limit cycle (of length $[N/(\text{highest common factor of } C \text{ and } N)]$) concatenated with itself a number of times to produce a sequence of length N.

In property 2, such limit cycles are called type A.[25] Those not exhibiting this half-wave symmetry, but with N still even, are called type B2.[25] But, from the previously assumed model for limit-cycle generation, all limit cycles with N even must have half-wave symmetry, and therefore type B2 cannot exist.

Finally, property 3 derives from the fact that the average value of an integer number of cycles from a sinusoid is zero.

Using the technique of Munson *et al.*,[26] which may be employed as a computationally efficient method for discovering all the limit cycles that a second-order section will support, hundreds of different limit cycles have been recorded for many complex poles within the unit circle. As the effective-value model for limit-cycle generation is not exact,[18,21,22] exceptions to the previous three properties were expected. In each case these amounted to only between 4% and 10% of the total number of limit cycles examined.

5.7. DIGITAL FILTERING USING THE LOGARITHMIC NUMBER SYSTEM

So far in our analysis, we have only considered digital filters which are implemented in either fixed-point or floating-point arithmetic, these being the two most common and widely used systems. However, recently there has been considerable interest in the application of a third type of arithmetic, the logarithmic arithmetic, to digital filter implementation.[27-31]

The logarithmic number system is defined by

$$sd_0d_1 \cdots d_t \cdot e_1e_2 \cdots e_k \tag{5.53}$$

where s, d_i, and e_i are either 0 or 1; s is the sign value and the d and e parts combined represent the exponent of the number; the base is assumed to be a positive constant, b; $(\cdot)$ represents the binary point of the exponent. Thus system (5.53) represents the number

$$\pm b^{\pm[d \text{ part} \cdot e \text{ part}]} \tag{5.54}$$

or, in a more rigorous form,

$$\begin{cases} b^{L_{t,k}} & \text{when } s = 0 \\ -b^{L_{t,k}} & \text{when } s = 1 \end{cases} \tag{5.55}$$

where $L_{t,k}$ is defined as

$$L_{t,k} = \sum_{i=1}^{t} 2^{t-i}d_i + \sum_{i=1}^{k} 2^{-i}e_i \qquad \text{when } d_0 = 0 \tag{5.56}$$

$$L_{t,k} = -\left(\sum_{i=1}^{t} 2^{t-i}\bar{d}_i + \sum_{i=1}^{k} 2^{-i}\bar{e}_i + 2^{-k}\right) \qquad \text{when } d_0 = 1 \tag{5.57}$$

$\bar{d}_i$ and $\bar{e}_i$ being the complements of d_i and e_i, respectively.

The logarithmic number system has a dynamic range given by

$$(\text{DR})_{\log} = \frac{b^{(2^t - 2^{-k})}}{b^{-2^t}} = b^{(2^{t+1} - 2^{-k})} \tag{5.58}$$

A comparison between the dynamic ranges of the three number systems would obviously depend on the number of bits assigned to each part of the numbers (5.11), (5.17), and (5.53). However, for an 8-bit machine, $t = 7$ in the fixed-point system, and for the floating-point system $t + k = 6$; we assume $t = 3$ and $k = 3$. In the logarithmic system, for a base $b = 2$, $t + k = 6$ such that $t = 3$ and $k = 3$.

On comparing the dynamic ranges it is seen that

$$(\mathrm{DR})_{\mathrm{fixed}} = 2^7 - 1 = 127$$

$$(\mathrm{DR})_{\mathrm{floating}} = (1 - 2^{-3})2^{2^4} = 57344$$

$$(\mathrm{DR})_{\mathrm{log}} = 2^{(2^4 - 2^{-3})} \cong 60097$$

Thus the logarithmic number system can provide the largest range. Furthermore, the usual assignment for the floating-point case is that t is 3 or 4 times larger than k, and consequently a much larger dynamic range can be achieved in the case of the logarithmic system.

In terms of logarithmic arithmetic, the product of a multiplication operation is obtained by a fixed-point addition of the two exponents. Hence, if $x = 2^m$ and $y = 2^n$, then

$$xy = 2^m \cdot 2^n = 2^{m+n} \tag{5.59}$$

Addition, however, is more complicated and involves the use of look-up tables. For the two numbers x and y to be added, we can write

$$x + y = 2^m + 2^n = 2^m(1 + 2^{n-m}) = 2^{m + \log_2(1 + 2^{n-m})} \tag{5.60}$$

where $\log_2(1 + 2^{n-m})$ is the machine representation and is given by a precomputed look-up table. In this respect, the addition operation gives rise to roundoff errors. Consider the number Z to be the true result of addition and Z_{q} its quantized machine version. The absolute error e_{abs} can be written in the form

$$\begin{aligned} e_{\mathrm{abs}} &= Z_{\mathrm{q}} - Z \\ &= Z[(Z_{\mathrm{q}}/Z) - 1] \\ &= Ze \end{aligned} \tag{5.61}$$

and e is defined as the relative error

$$e = (Z_{\mathrm{q}}/Z) - 1 \tag{5.62}$$

If Z is uniformly distributed over the range $[Z_1, Z_2)$ and $0 < Z_1 < Z_2$, where Z_1 and Z_2 are defined by

$$Z_1 = Z_q b^{-2^{-k-1}} \quad \text{and} \quad Z_2 = Z_q b^{2^{-k-1}} \tag{5.63}$$

and if we define the probability density function $f(Z)$ of Z by

$$f(Z) = \begin{cases} 1/(Z_2 - Z_1) & \text{for } Z_1 \leq Z < Z_2 \\ 0 & \text{otherwise} \end{cases} \tag{5.64}$$

and if equation (5.62) is rewritten in the form

$$Z = w(e) = Z_q/(e+1) \tag{5.65}$$

then, by using the rule of transformation of variables,[32] the error probability density function $g(e)$ can be obtained by[29]

$$g(e) = f[w(e)]|w'(e)| = \begin{cases} \dfrac{1}{Z_2 - Z_1} \dfrac{Z_q}{(e+1)^2} & \text{for } (Z_q/Z_2 - 1 \leq e \\ & \quad \leq Z_q/Z_1 - 1) \\ 0 & \text{otherwise} \end{cases} \tag{5.66}$$

Alternatively,

$$g(e) = \begin{cases} \dfrac{1}{b^{2^{-k-1}} - b^{-2^{-k-1}}} \cdot \dfrac{1}{(e+1)^2} & \text{for } (b^{-2^{-k-1}} - 1 \leq e \leq b^{2^{-k-1}} - 1) \\ 0 & \text{otherwise} \end{cases} \tag{5.67}$$

This error probability density function has been tested for uniformity both statistically and experimentally by Kurokawa.[29] The mean (μ) and variance (σ^2) of the relative error e are given by

$$\mu = (b^{2^{-k-1}} + b^{-2^{-k-1}} - 2)/2 \tag{5.68}$$

and

$$\sigma^2 = (b^{2^{-k-1}} - b^{-2^{-k-1}})^2/12 \tag{5.69}$$

while the expected values of the relative error + 1 and of its square are

$$E\{e+1\} = (b^{2^{-k-1}} + b^{-2^{-k-1}})/2 \tag{5.70}$$

and

$$E\{(e+1)^2\} = E^2\{e+1\} + \sigma^2 \tag{5.71}$$

As we mentioned earlier, multiplication is exact in the logarithmic number system and error consideration has to be made only for addition. Thus, if $Z = x + y$, then by equation (5.61) we have

$$(x + y)_q = (x + y)(1 + e) \tag{5.72}$$

and $(x + y)_q$ is the machine version of $x + y$.

The error spectrum analysis performed in floating-point number systems can be carried out similarly for logarithmic systems. It is shown[29,30] that, for the case of stochastic input, the theoretical noise-to-signal ratio is less using logarithmic number systems. Since the dynamic range in these systems is larger than in other number systems, the logarithmic number system yields a filtering performance superior to that of a floating-point system of equivalent word length. Furthermore, when considering the hardware implementation of such filters, the logarithmic number system provides very high speed arithmetic in limited hardware.[31]

REFERENCES

1. L. R. Rabiner and B. Gold, *Theory and Application of Digital Signal Processing*, Prentice-Hall, Englewood Cliffs, NJ (1975).
2. L. B. Jackson, *An Analysis of Roundoff Noise in Digital Filters*, ScD Thesis, Stevens Institute of Technology, Hoboken, New Jersey (1969).
3. B. Liu, Effect of finite word length on the accuracy of digital filters, *IEEE Trans. Circuit Theory* **CT-18,** 670–677 (1971).
4. V. B. Lawrence, *Use of Orthogonal Functions in the Design of Digital Filters*, PhD Thesis, London University (1972).
5. B. Liu and T. Kaneko, Error analysis of digital filters realized with floating-point arithmetic, *Proc. IEEE* **57,** 1735–1747 (1969).
6. J. K. Aggarwal, Input Quantization and Arithmetic Roundoff in Digital Filters—A Review, Proc. NATO Advanced Study Institute on Network and Signal Theory, PPL Conf. Publ. No. 12, 315–343 (1972).
7. W. R. Bennett, Spectra of quantized signals, *Bell Syst. Tech. J.* **27,** 446–472 (1948).
8. A. Papoulis, *Probability, Random Variables and Stochastic Processes*, McGraw-Hill, New York (1965).
9. L. Jackson, On the interaction of roundoff noise and dynamic range in digital filters, *Bell Syst. Tech. J.* **49,** 159–184 (1970).
10. B. Liu and M. E. Van Valkenburg, On roundoff error of fixed-point digital filters using sign-magnitude truncation, *IEEE Trans. Circuit Theory* **CT-19,** 536–537 (1972).
11. I. W. Sandberg, Floating-point-roundoff accumulation in digital-filter realizations, *Bell Syst. Tech. J.* **46,** 1775–1791 (1967).
12. T. Kaneko and B. Liu, Effect of coefficient rounding in floating-point digital filters, *IEEE Trans. Aerosp. Electron. Syst.* **AES-7,** 995–1003 (1971).
13. E. P. F. Kan and J. K. Aggarwal, Error analysis of digital filters employing floating-point arithmetic, *IEEE Trans. Circuit Theory* **CT-18,** 678–686 (1971).

14. A. Fettweis, Some Statistical Properties of Floating-Point Roundoff Noise, Proc. 16th Midwest Symp. Circuit Theory, II.2.1-II.2.10 (1973).
15. D. S. K. Chan, *Roundoff Noise in Cascade Realization of Finite Impulse Response Digital Filters*, SB and SM Thesis, Department of Electrical Engineering, Massachusetts Institute of Technology, Cambridge, Mass. (1972).
16. L. B. Jackson, Roundoff-noise analysis for fixed-point digital filters realized in cascade or parallel form, *IEEE Trans. Audio Electroacoust.* **AU-18,** 107-122 (1970).
17. D. S. K. Chan and L. R. Rabiner, Theory of roundoff noise in cascade realizations of finite impulse response digital filters, *Bell Syst. Tech. J.* **52,** 329-345 (1973).
18. L. B. Jackson, An Analysis of Limit Cycles due to Multiplication Rounding in Recursive Digital (Sub) Filters, Proc. 7th Annual Allerton Conference on Circuit and System Theory, 69-78 (October 1969).
19. P. M. Ebert, J. E. Mazo, and M. G. Taylor, Overflow oscillations in digital filters, *Bell Syst. Tech. J.* **48,** 2990-3020 (1969).
20. J. F. Kaiser, Some Practical Considerations in the Realization of Linear Digital Filters, Proc. 3rd Allerton Annual Conference on Circuit and System Theory, 621-633 (October 1965).
21. S. R. Parker and S. F. Hess, Limit cycle oscillations in digital filters, *IEEE Trans. Circuit Theory* **CT-18,** 687-697 (1971).
22. K. P. Prasad and P. S. Reddy, Limit cycles in second-order digital filters, *J. Inst. Electron. Telecommun. Eng.* (*India*) **26,** 85-86 (1980).
23. V. B. Lawrence and K. V. Mina, A new and interesting class of limit cycles in recursive digital filters, *Bell Syst. Tech. J.* **58,** 379-408 (1979).
24. D. C. McLernon and R. A. King, Additional properties of one-dimensional limit cycles, *IEE Proc.* **133,** Part G, No. 3, 140-144 (1986).
25. T. A. C. M. Claasen, W. F. G. Mecklenbrauker, and J. B. H. Peek, Some remarks on the classification of limit cycles in digital filters, *Philips Res. Rep.* **28,** 297-305 (1973).
26. D. C. Munson, Jr., J. H. Strickland, Jr., and T. P. Walker, Maximum amplitude zero-input limit cycles in digital filters, *IEEE Trans. Circuits Syst.* **CAS-31,** 266-275 (1984).
27. N. G. Kingsbury and P. J. W. Rayner, Digital filtering using logarithmic arithmetic, *Electron. Lett.* **7,** 56-58 (1971).
28. E. E. Swartzlander, Jr. and A. G. Alexopoulos, The Sign/logarithm number system, *IEEE Trans. Comput.* **C-24,** 1238-1242 (1975).
29. T. Kurokawa, *Error Analysis of Digital Filters with Logarithmic Number System*, Ph.D. Dissertation, Univ. Oklahoma, Norman (1978).
30. T. Kurokawa, J. A. Payne, and S. C. Lee, Error analysis of recursive digital filters implemented with logarithmic number systems, *IEEE Trans. Acoust., Speech, Signal Process.* **ASSP-28,** 706-715 (1980).
31. F. J. Taylor, Logarithmic Arithmetic Unit for Signal Processing, IEEE Int. Conf. on Acoustics, Speech and Signal Process., Vol. 3, 44.10/1-4 (1984).
32. R. V. Hogg and T. Craig, *Introduction to Mathematical Statistics*, The MacMillan Company, London (1970).

6

Direct Realization and Implementation Using Linear Transformations

6.1. INTRODUCTION

Having obtained a network transfer function in either recursive or nonrecursive form to give an appropriate approximation to the desired specification using some of the techniques outlined in the preceding chapters, we must now implement this transfer function using an interconnection of delays, digital multipliers, and adders. The delays may be realized using shift registers, and the multipliers and adders by appropriate digital circuitry; thus we may obtain a purpose-built system to implement the desired specification. Alternatively, the complete implementation of the IIR or FIR digital equations may be realized in a general-purpose digital computer. In either case, we need to choose an appropriate interconnection of these circuit elements, the choice being determined by factors such as dynamic range, multiplier roundoff error, and sensitivity to component variation.

6.2. CLASSICAL CANONIC IMPLEMENTATION

The simplest form of implementation is that derived from the recursion equation

$$y(n) = \sum_{i=0}^{M} a_i u(n-i) - \sum_{i=1}^{N} b_i y(n-i) \tag{6.1}$$

or, in two dimensions,

$$y(m, n) = \sum_{j=0}^{M_1} \sum_{k=0}^{M_2} a_{j,k} u(m-j, n-k) - \sum_{\substack{j=0 \\ j+k\neq 0}}^{N_1} \sum_{k=0}^{N_2} b_{j,k} y(m-j, n-k) \tag{6.2}$$

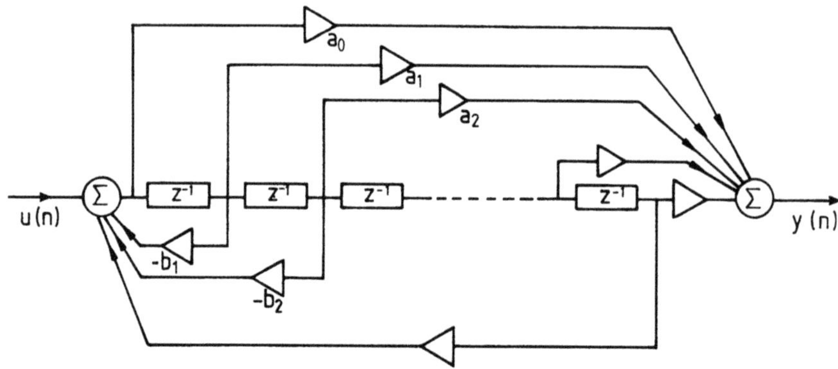

Figure 6.1. Direct implementation of a one-dimensional recursive filter.

6.2.1. Direct Canonic Form

Equation (6.1) may be directly implemented using K delay elements, where K is the greater of M and N. This is shown in Figure 6.1. It may also be noted that the total number of multipliers is $M + N + 1$, which is equal to the total number of coefficients a_i and b_i. This implementation is thus not only canonic in the number of delay elements, but also in the number of multipliers.

In two dimensions,[1] equation (6.2) may be implemented by the circuit diagram of Figure 6.2 (drawn for $M_1 = N_1 = 3$, $M_2 = 2$, $N_2 = 1$). In this it may be seen that the number of delay elements is greater than the sum of the highest powers of z_1 and z_2 in the numerator or denominator. In Figure 6.2, the number of delays in the z_1 dimension is equal to the greater of M_1 and N_1. In addition, in the z_2 dimension, M_2 delay elements are needed to implement the nonrecursive part of the output and N_2 delay elements to realize the recursive part. The number of multipliers in this realization is equal to the number of coefficients in the discrete equation.

6.2.2. Parallel Realization

In one dimension it is possible to expand the transfer function into a set of partial fractions having real or complex linear denominators. The complex elements will occur in conjugate pairs and thus may be combined to form a quadratic function with real coefficients. Thus the transfer function $H(z)$ may be broken up into the sum of a number of terms of the form $H_1(z)$ and $H_2(z)$, where

$$H_1(z) = \frac{a_0}{1 + b_1 z^{-1}} \tag{6.3a}$$

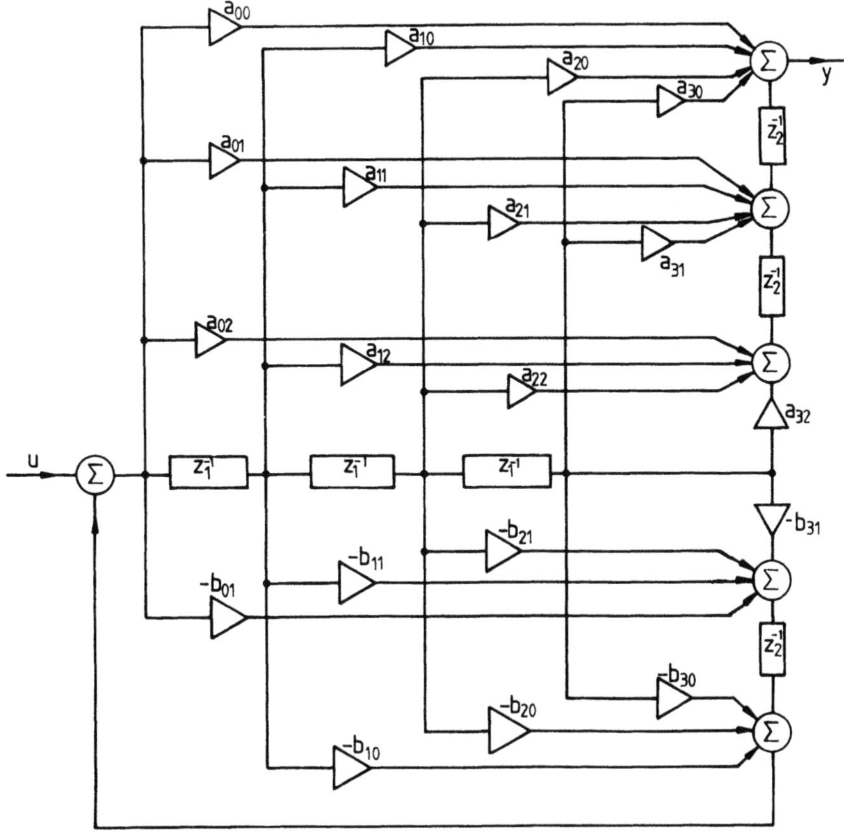

Figure 6.2. Direct implementation of a two-dimensional recursive filter.

and

$$H_2(z) = \frac{a_0 + a_1 z^{-1}}{1 + b_1 z^{-1} + b_2 z^{-2}} \tag{6.3b}$$

Equations (6.3a) and (6.3b) may be realized by the circuits of Figure 6.3a and b. Again this is a canonical realization, both in delay elements and multipliers. The overall transfer function $H(z)$ is thus implemented by a parallel connection of elementary networks like $H_1(z)$ and $H_2(z)$.

This form of realization is not generally possible for two-dimensional (2-D) systems, since the fundamental theorem of algebra does not apply. Thus, in the general case, it is not possible to factorize the denominator into linear and quadratic factors with real coefficients. Where factorization of the denominator is possible, a limited form of parallel realization may be used.

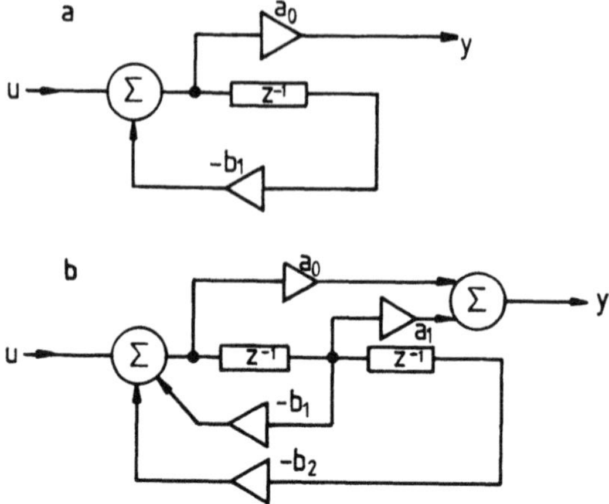

Figure 6.3. Parallel realization: (a) first-order element; (b) second-order element.

6.2.3. Cascade Realization

Again, in one dimension it is always possible to factorize both numerator and denominator into linear and quadratic factors. The numerator factors may now be paired with the denominator factors, resulting in a set of unit filter elements having transfer functions $H_3(z)$ and $H_4(z)$ given by

$$H_3(z) = \frac{a_0 + a_1 z^{-1}}{1 + b_1 z^{-1}} \tag{6.4a}$$

and

$$H_4(z) = \frac{a_0 + a_1 z^{-1} + a_2 z^{-2}}{1 + b_1 z^{-1} + b_2 z^{-2}} \tag{6.4b}$$

which may similarly be realized by networks of the form of Figure 6.3a, with an additional multiplier a_1 from the output of the delay element to the output adder, and Figure 6.3b, with a multiplier a_2 from the output of the second delay element. The complete realization consists of an appropriate number of these elements connected in cascade.

6.3. IMPLEMENTATION BASED ON LADDER NETWORKS

The advantages of passive ladder networks in the realization of analogue filters is well known. The particular configuration of interest in the present context is the doubly terminated ladder network. It has the merit that the magnitude of the transfer function has very low sensitivity to variations in the values of the elements in the network; this low-sensitivity property, however, does not extend to the phase of the transfer function. Again, as a result of the ladder topology, the filter is also insensitive to component variations in the pass band.

The implementation technique, which will be described in this section, depends on the representation of the analogue elements, inductors, and capacitors by digital elements that simulate the integrations involved. Considerable flexibility in the design results from the representation of the analogue variables of voltage and current at the ports of the network by linear transformations of these; one specific example of such transformation parameters is the scattering parameters, where the variables represent incident and reflected voltage and current.[2-7] However, the dimension of the variables is irrelevant in a digital filter, and so the analogue prototype will be represented by a signal flow graph.

6.3.1. Flow Graph of Analogue Prototype Filter

A general two-port network, shown in Figure 6.4, may be represented by its transmission (chain) matrix A relating the input to the output variables via

$$\begin{bmatrix} V_1 \\ I_1 \end{bmatrix} = \begin{bmatrix} a_{11} & a_{12} \\ a_{21} & a_{22} \end{bmatrix} \begin{bmatrix} V_2 \\ I_2 \end{bmatrix} \tag{6.5a}$$

$$= A \begin{bmatrix} V_2 \\ I_2 \end{bmatrix} \tag{6.5b}$$

where A is the modified chain matrix with the output current I_2 having reversed polarity from the conventional form.

It is now possible to apply a linear transformation P to the input voltage and current, to create new input variables

$$\begin{bmatrix} X_1 \\ Y_1 \end{bmatrix} = P \begin{bmatrix} V_1 \\ I_1 \end{bmatrix} \tag{6.6}$$

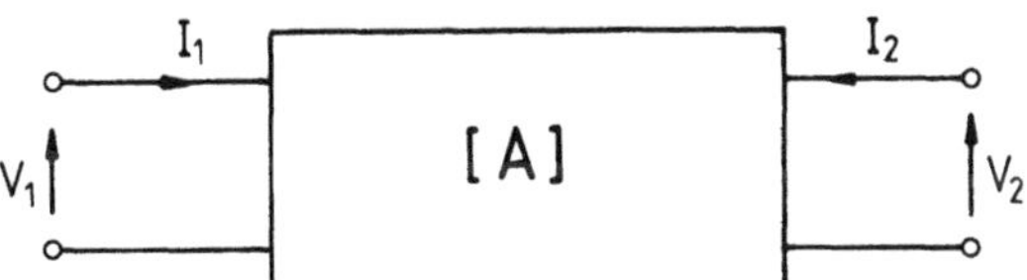

Figure 6.4. General two-port network.

Similarly, a linear transformation Q may be applied at the output to create new output variables at port 2,

$$\begin{bmatrix} X_2 \\ Y_2 \end{bmatrix} = Q \begin{bmatrix} V_2 \\ I_2 \end{bmatrix} \tag{6.7}$$

The network matrix Φ relating input to output is given by

$$\begin{bmatrix} X_1 \\ Y_1 \end{bmatrix} = \Phi \begin{bmatrix} X_2 \\ Y_2 \end{bmatrix} \tag{6.8a}$$

where, using equations (6.5)-(6.7),

$$\Phi = PAQ^{-1} \tag{6.8b}$$

$$\Phi = \begin{bmatrix} \phi_{11} & \phi_{12} \\ \phi_{21} & \phi_{22} \end{bmatrix} \tag{6.8c}$$

Although there are no formal constraints on the dimensions of the elements of the matrix Φ, a simple form of implementation results if they are all dimensionless parameters. This may be achieved if we define the P and Q matrices in terms of two scalar quantities R_1 and R_2, representing the resistances of ports 1 and 2 respectively, such that

$$P = \begin{bmatrix} p_{11} & p_{12}R_1 \\ p_{21} & p_{22}R_1 \end{bmatrix} \tag{6.9a}$$

$$Q = \begin{bmatrix} q_{11} & q_{12}R_2 \\ q_{21} & q_{22}R_2 \end{bmatrix} \tag{6.9b}$$

It will be seen that, with this choice of P and Q, the port variables have the dimensions of voltage.

An alternative formulation for this would be

$$P = \begin{bmatrix} p_{11}G_1 & p_{12} \\ p_{21}G_1 & p_{22} \end{bmatrix} \tag{6.9c}$$

$$Q = \begin{bmatrix} q_{11}G_2 & q_{12} \\ q_{21}G_2 & q_{22} \end{bmatrix} \tag{6.9d}$$

where G_1 is the conductance of port 1 and G_2 the conductance of port 2, while $G_1 = 1/R_1$ and $G_2 = 1/R_2$. Here the port variables have dimensions of current.

In the digital implementation, it will be assumed that X_i $(i = 1, 2)$ are the input variables and Y_i $(i = 1, 2)$ the output. It is therefore desirable to write the output vector in terms of the input vector according to

$$\begin{bmatrix} Y_1 \\ Y_2 \end{bmatrix} = \sigma \begin{bmatrix} X_1 \\ X_2 \end{bmatrix} \tag{6.10a}$$

where

$$\sigma = \begin{bmatrix} \sigma_{11} & \sigma_{12} \\ \sigma_{21} & \sigma_{22} \end{bmatrix} \tag{6.10b}$$

and

$$\begin{aligned} \sigma_{11} &= \phi_{22}/\phi_{12} \\ \sigma_{12} &= (\phi_{12}\phi_{21} - \phi_{11}\phi_{22})/\phi_{12} \\ \sigma_{21} &= 1/\phi_{12} \\ \sigma_{22} &= -\phi_{11}/\phi_{12} \end{aligned} \tag{6.10c}$$

It is now possible to draw a flow diagram for the transformation given by equation (6.10); this is shown in Figure 6.5.

6.3.1.1. Cascade of Elementary Networks[8]

A ladder structure consists of a cascade of elementary alternate shunt and series branches, each of which may be represented by the flow diagram of Figure 6.5 and the transmission equation (6.10). It is now necessary to consider the constraints imposed by interconnecting two successive elementary networks. As a consequence of the direction of current flow in Figure 6.4, the output current from the ith section $I_{2(i)} = -I_{1(i+1)}$, the input current of the $(i + 1)$th section; hence

$$\begin{bmatrix} V_2 \\ I_2 \end{bmatrix}_{(i)} = \theta \begin{bmatrix} V_1 \\ I_1 \end{bmatrix}_{(i+1)} \tag{6.11a}$$

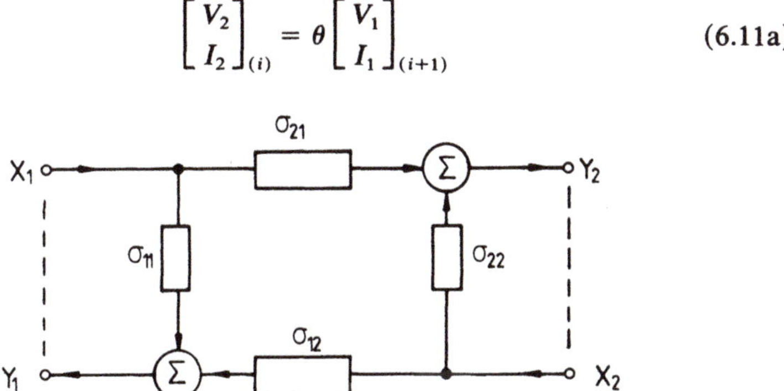

Figure 6.5. Flow diagram for general network transformation.

where

$$\theta = \begin{bmatrix} 1 & 0 \\ 0 & -1 \end{bmatrix} \tag{6.11b}$$

Thus, using equations (6.6) and (6.7) for the $(i+1)$th and ith network respectively, we have

$$\begin{bmatrix} X_2 \\ Y_2 \end{bmatrix}_{(i)} = M \begin{bmatrix} X_1 \\ Y_1 \end{bmatrix}_{(i+1)} \tag{6.12a}$$

where

$$M = Q_{(i)} \theta P_{(i+1)}^{-1} \tag{6.12b}$$

In order to minimize the number of multipliers and adders, M must be of the form

$$M = \begin{bmatrix} 0 & m_1 \\ m_2 & 0 \end{bmatrix} \tag{6.12c}$$

and for the simplest possible implementation

$$M = \begin{bmatrix} 0 & 1 \\ 1 & 0 \end{bmatrix} \tag{6.12d}$$

As a consequence, a constraint exists between the output transformation of one section $Q_{(i)}$ and the input transformation $P_{(i+1)}$ of the subsequent section given by

$$\begin{bmatrix} 0 & 1 \\ 1 & 0 \end{bmatrix} \begin{bmatrix} p_{11} & p_{12}R_1 \\ p_{21} & p_{22}R_1 \end{bmatrix}_{(i+1)} = \begin{bmatrix} q_{11} & q_{12}R_2 \\ q_{21} & q_{22}R_2 \end{bmatrix}_{(i)} \begin{bmatrix} 1 & 0 \\ 0 & -1 \end{bmatrix} \tag{6.13a}$$

Also, since $R_{1\,(i+1)} = R_{2\,(i)}$ for matching between sections, we have

$$p_{11(i)} = q_{21(i+1)}$$

$$p_{12(i)} = -q_{22(i+1)}$$

$$p_{21(i)} = q_{11(i+1)}$$

$$p_{22(i)} = -q_{12(i+1)} \tag{6.13b}$$

These are termed the compatibility conditions.

By successive application of equations (6.6)-(6.8) for an n-section system, we obtain

$$\begin{bmatrix} X_1 \\ Y_1 \end{bmatrix}_{(1)} = P_1 A_1 Q_1^{-1} M P_2 A_2 Q_2^{-1} \cdots M P_n A_n Q_n^{-1} \begin{bmatrix} X_2 \\ Y_2 \end{bmatrix}_{(n)} \quad (6.14a)$$

$$= P_1 A_1 \theta A_2 \theta \cdots A_n Q_n^{-1} \begin{bmatrix} X_2 \\ Y_2 \end{bmatrix}_{(n)} \quad (6.14b)$$

Now the modified chain matrix of the complete system, A, is given by

$$A\theta = (A_1\theta)(A_2\theta) \cdots (A_n\theta)$$

and hence

$$\begin{bmatrix} X_1 \\ Y_1 \end{bmatrix}_{(1)} = P_1 A Q_n^{-1} \begin{bmatrix} X_2 \\ Y_2 \end{bmatrix}_{(n)} \quad (6.15)$$

This shows that the input-output relationship of the equivalent structure is independent of the P and Q matrices at the junctions of the various sections and only dependent on the two transformation matrices at the terminations. It is therefore obvious that the matrices P and Q at each junction may be chosen independently, provided that each pair is compatible.

The above property permits considerable flexibility in design. The simplest arrangement is to make all the pairs Q_i, P_{i+1} identical; this is considered as having a cycle length of one. Another useful option is to use two such pairs and alternate them throughout the ladder structure, a cycle length of two. A third possibility, which only uses one topology of digital structure to implement a ladder filter whose series and shunt elements are duals, uses a set of all different pairs Q_i, P_{i+1} at the various section junctions. We shall consider some of these possibilities subsequently.

When all the (P_i, Q_i) pairs are identical, the compatibility condition is further restricted. Assuming $R_1 = R_2$, each section is constrained by equations (6.13b), so that

$$p_{11} = q_{21}$$

$$p_{12} = -q_{22}$$

$$p_{21} = q_{11}$$

$$p_{22} = -q_{12}$$

6.3.1.2. *Digital Realization of Ladder Filter*

In order to obtain a digital implementation of a ladder structure, we replace each of the elements of the σ matrix in Figure 6.5 by the appropriate z-transfer function by using the bilinear transformation

$$s = \frac{1 - z^{-1}}{1 + z^{-1}}$$

on the corresponding s-transfer function. In order to carry out computations, there must be no situations in which the value of any variable is directly proportional to itself via a loop in which there are no delays. The only place where a loop may occur in a cascade of such sections is at the junction between two adjacent sections, where a loop is formed of $\sigma_{22(i)}$ in series with $\sigma_{11(i+1)}$.

Now on transformation to the z plane, the coefficient of z^0 must be zero in the product of σ_{11} and σ_{22}. This may be expressed by saying that

$$\sigma_{11}|_{z^{-1}=0} = 0 \qquad \text{or} \qquad \sigma_{22}|_{z^{-1}=0} = 0$$

or alternatively, by transformation to the s plane,

$$\sigma_{11}|_{s=1} = 0 \tag{6.16a}$$

or

$$\sigma_{22}|_{s=1} = 0 \tag{6.16b}$$

and this must hold for all the junctions between sections, including those at the source and the load. The above realizability identities may be forced to hold either on σ_{11}, when the structures will be referred to as derived from the load end, or on $\sigma_{22} = 0$ throughout, when the structures will be derived from the source end.

6.3.1.3. *Analogue Transformations*

We shall now obtain the s matrix for the two basic components of a ladder structure, namely, a shunt admittance Y and a series impedance Z.

The modified chain matrices for these two elements are

$$A_Y = \begin{bmatrix} 1 & 0 \\ Y & -1 \end{bmatrix} \tag{6.17a}$$

and

$$A_Z = \begin{bmatrix} 1 & -Z \\ 0 & -1 \end{bmatrix} \tag{6.17b}$$

6.3.1.3a. Shunt Arm. Substituting matrices (6.9c) and (6.9d) and (6.17a) into equation (6.8), and then using relations (6.10c), we have

$$\begin{aligned}
\sigma_{11} &= (p_{21}q_{12}G_1 + p_{22}q_{11}G_2 + p_{22}q_{12}Y)/D \\
\sigma_{12} &= -\Delta_P/D \\
\sigma_{21} &= -\Delta_Q/D \\
\sigma_{22} &= (p_{11}q_{22}G_1 + p_{12}q_{21}G_2 + p_{12}q_{22}Y)/D
\end{aligned} \tag{6.18a}$$

where Δ_P and Δ_Q are the determinants of matrices P and Q, and

$$D = p_{11}q_{12}G_1 + p_{12}q_{11}G_2 + p_{12}q_{12}Y \tag{6.18b}$$

6.3.1.3b. Series Arm. In a similar manner, using matrices (6.9c), (6.9d), and (6.17b) with equation (6.8) and relations (6.10c), we obtain for the series arm

$$\begin{aligned}
\sigma_{11} &= (p_{21}q_{12}R_2 + p_{22}q_{11}R_1 + p_{21}q_{11}Z)/D \\
\sigma_{12} &= -R_1R_2\Delta_P/D \\
\sigma_{21} &= -R_1R_2\Delta_Q/D \\
\sigma_{22} &= (p_{11}q_{22}R_2 + p_{12}q_{21}R_1 + p_{11}q_{21}Z)/D
\end{aligned} \tag{6.19a}$$

where

$$D = p_{11}q_{12}R_2 + p_{12}q_{11}R_1 + p_{11}q_{11}Z \tag{6.19b}$$

For realizations of cycle length one, where all P and Q matrices for all junctions are the same, further simplifications may be made to the above using relations (6.13b).

6.3.1.3c. Load Resistance. The load transformation can be carried out in the same manner. For a load resistance R_L with current and voltage polarities as shown in Figure 6.6a, we have

$$V = R_L I \tag{6.20}$$

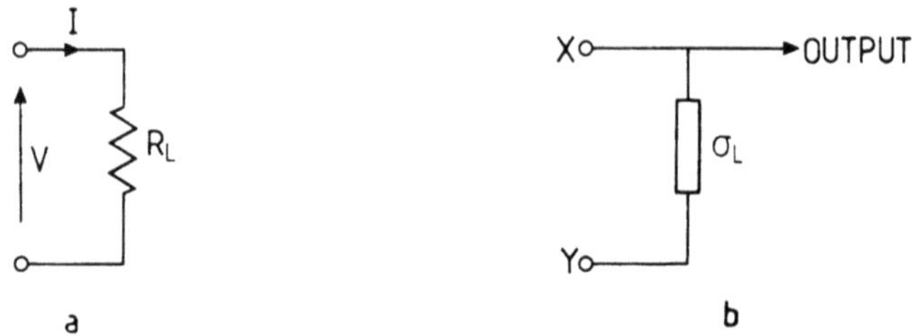

Figure 6.6. Load element: (a) network diagram; (b) flow graph.

Substitution of matrix (6.9c) in equation (6.6) with relation (6.20) yields the ratio $\sigma_L = Y/X$ for the load resistance, where

$$\sigma_L = (p_{21}G_1R_L + p_{22})/(p_{11}G_1R_L + p_{12}) \tag{6.21}$$

The flow graph is shown in Figure 6.6b.

6.3.1.3d. Source Relationship. The voltage–current relationship for a source voltage V_S in series with a resistance R_S, as in Figure 6.7a, is

$$V = V_S + R_SI \tag{6.22}$$

Again combining this with equations (6.9b) and (6.7), we obtain

$$Y = \sigma_SX + U$$

where

$$\sigma_S = (q_{21}R_S + q_{22}R_2)/(q_{11}R_S + q_{12}R_2) \tag{6.23a}$$

and

$$U = -V_S(q_{11}q_{22} - q_{12}q_{21})/(q_{11}R_S + q_{12}R_2) \tag{6.23b}$$

The flow graph for this is shown in Figure 6.7b.

Figure 6.7. Source element: (a) network diagram; (b) flow graph.

6.4. IMPLEMENTATION OF ONE-DIMENSIONAL DIGITAL FILTERS

We shall now consider the problem of the implementation of the various sections of a ladder-type filter represented by the σ matrix. It is obvious that there is a very wide choice for the transformation matrices P and Q, but we shall restrict the discussion to a small set of these of rather simple form. One simplification would follow from setting one or more of the elements of P (or Q) to zero. It is not difficult to show that only one of these parameters may be zero without violating the realizability condition.

If we choose transformation sets with one zero element, we may further simplify the implementation by setting the remaining values of p_{ij} and q_{ij} to be identical; and for convenience we choose a value of unity. This may be shown by considering, for example, the value of the matrix σ for a shunt element when $p_{12} = 0$ and P and Q are constrained by relations (6.13b):

$$\begin{aligned}
\sigma_{11} &= \frac{p_{21}}{p_{11}} R_1(G_1 - G_2) + \frac{p_{22}}{p_{11}} R_1 Y \\
\sigma_{12} &= 1 \\
\sigma_{21} &= G_2/G_1 \\
\sigma_{22} &= 0
\end{aligned} \tag{6.24}$$

Now the factors p_{21}/p_{11} and p_{22}/p_{11} represent additional multipliers required in the implementation, and these may be eliminated if they are all made equal.

In obtaining the one-dimensional (1-D) digital flow graph from the analogue form, we use the bilinear substitution $s = (1 - z^{-1})/(1 + z^{-1})$.

6.4.1. Transformation Set with Unity Cycle Length

The condition for all the transformation sets (P_{i+1}, Q_i) to be identical is given by equations (6.13b). Considering the case where one of the elements of P (and Q) is zero, we have four possible cases, each of which will give rise to a different implementation. We consider one specific example to illustrate the realization technique:

$$P = \begin{bmatrix} G_1 & 0 \\ G_1 & 1 \end{bmatrix} \quad \text{and} \quad Q = \begin{bmatrix} G_2 & -1 \\ G_2 & 0 \end{bmatrix} \tag{6.25}$$

We may now derive appropriate digital realizations for some specific elements in an analogue filter network. We shall assume that the prototype is a low-pass network with series elements of inductors and shunt-capacitor elements, terminated by resistors.

6.4.1.1. *Shunt Capacitor*

The analogue representation of this network, where $Y = sC$ and $R_1 = 1/G_1$, is obtained from equations (6.18) in the form

$$\sigma_{11} = R_1(G_1 - G_2 + Y)$$

$$\sigma_{12} = 1$$

$$\sigma_{21} = G_2 R_1$$

$$\sigma_{22} = 0$$

In order to avoid delay-free loops, we need to apply the realizability condition (6.16a), giving

$$G_1 = G_2 - C$$

Finally, applying the bilinear relationship, we obtain

$$\sigma_{11} = -2az^{-1}/(1 + z^{-1})$$

$$\sigma_{12} = 1$$

$$\sigma_{21} = 1 + a$$

$$\sigma_{22} = 0 \tag{6.26}$$

where $a = R_1 C$, and the equivalent digital structure has the form of Figure 6.8a.

6.4.1.2. *Series Inductor*

In this case, we implement the series inductor $Z = sL$ using equation (6.19) with the appropriate forms of P and Q:

$$P = \begin{bmatrix} 1 & 0 \\ 1 & R_1 \end{bmatrix} \quad \text{and} \quad Q = \begin{bmatrix} 1 & -R_2 \\ 1 & 0 \end{bmatrix} \tag{6.27}$$

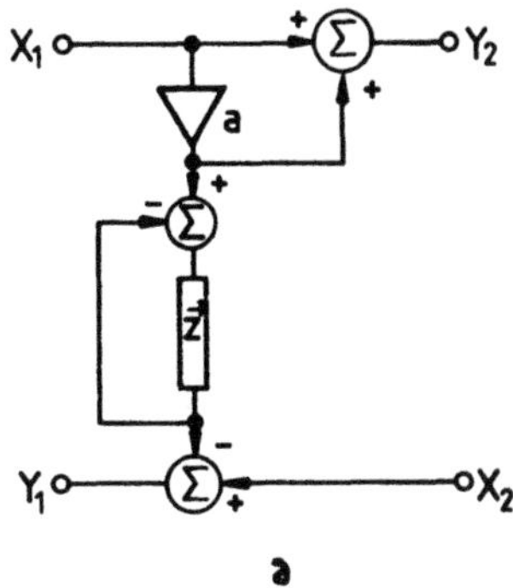

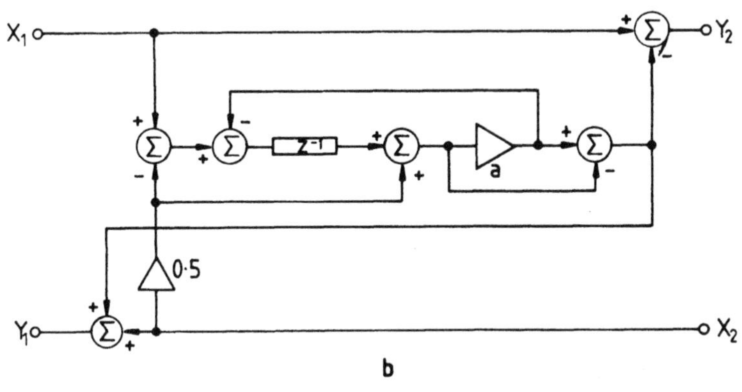

Figure 6.8. Flow graph for (a) shunt capacitor, (b) series inductor, (c) load termination, (d) source network.

We obtain

$$\sigma_{11} = (R_2 - R_1 - Z)/(R_2 - Z)$$

$$\sigma_{12} = R_2/(R_2 - Z)$$

$$\sigma_{21} = R_1/(R_2 - Z)$$

$$\sigma_{22} = -Z/(R_2 - Z)$$

In order to satisfy the realizability constraint (6.16a), we require

$$R_1 = R_2 - L$$

and, using the bilinear relationship, the final implementation equations are

$$\sigma_{11} = (a-1)z^{-1}/(1+az^{-1})$$

$$\sigma_{22} = (1-a)(1-z^{-1})/2(1+az^{-1})$$

$$\sigma_{12} = 1 - \sigma_{22}$$

$$\sigma_{21} = 1 - \sigma_{11} \tag{6.28}$$

where $a = 1 + 2L/R_1$.

A digital structure representing this is given in Figure 6.8b.

6.4.1.3. *Load Resistance*

As developed in equation (6.21), and applying the transformation P of equations (6.25), we obtain

$$\sigma_L = (G_1 R_L + 1)/G_1 R_L$$
$$= 1 + 1/G_1 R_L$$

In order to satisfy the realizability condition and avoid delay-free loops, $1 + 1/G_1 R_L = 0$. Hence

$$G_1 = -1/R_L \tag{6.29}$$

and the digital implementation is simply a sink connected to X, as illustrated in Figure 6.8c, since $Y = 0$.

6.4.1.4. *Source Transformation*

From equation (6.23), and applying the transformation (6.26),

$$\sigma_S = R_S/(R_S - R_2) \qquad \text{and} \qquad U = -V_S R_2/(R_S - R_2) \tag{6.30}$$

and the implementation is shown in Figure 6.8d.

6.4.1.5. *Realization of Other Circuit Elements*

The only analogue circuit elements considered so far have been series inductors and shunt capacitors. These are the elements from which ladder-type low-pass networks are constructed. In high-pass circuits, the basic elements are series capacitors and shunt inductors; band-pass circuits are composed of series elements consisting of parallel *LC* combinations and shunt elements which are series *LC* combinations.

These could all be implemented individually, but it is simpler to realize such high-pass and band-pass networks by frequency transformations from a low-pass prototype.

The high-pass transformation involves the replacement of s by $1/s$. After application of the bilinear transformation, it is seen that substitution of z by $-z$ in a low-pass realization results in a corresponding high-pass network.

Again, the low-pass to band-pass transformation is given by the replacement

$$s \to s + 1/s$$

which corresponds, when using the bilinear transformation, to

$$z \to -z^2$$

Thus, any filter specification may be transformed to its corresponding normalized low-pass form, designed according to the preceding or subsequently described methods, and then transformed as above in the z domain.

6.4.1.6. *Transformation of General Reactance Elements*

A valuable extension to the transformation of ladder-filter sections comprising an arbitrary reactance function has been proposed by Erfani *et al.*[11]

We consider a general shunt reactance element

$$Y(s) = K \prod_i \frac{s^2 + \omega_i^2}{s(s^2 + \omega_{i+1}^2)} \tag{6.31}$$

which is transformed via the bilinear transformation into

$$Y(z) = H \prod_i \frac{1 + z^{-1}}{1 - z^{-1}} \frac{1 - 2\rho_i z^{-1} + z^{-2}}{1 + 2\rho_{i+1} z^{-1} + z^{-2}} \tag{6.32a}$$

where

$$H = Y(z)|_{z^{-1}=0} = Y(s)|_{s=1} = \prod_i K \frac{1+\omega_i^2}{1+\omega_{i+1}^2} \tag{6.32b}$$

and

$$\rho_i = \frac{1-\omega_i^2}{1+\omega_i^2} \tag{6.32c}$$

Now function $Y(z)$ may be written in the form

$$Y(z) = H\frac{N(z)}{D(z)} \tag{6.32d}$$

where $N(z)$ and $D(z)$ are polynomials in z, both having coefficients of z^0 equal to unity.

As an illustration we shall use the (P, Q) matrices of unity cycle length,

$$P = \begin{bmatrix} 1 & R_1 \\ 1 & -R_1 \end{bmatrix} \quad \text{and} \quad Q = \begin{bmatrix} 1 & R_2 \\ 1 & -R_2 \end{bmatrix} \tag{6.33}$$

Thus, equations (6.32), (6.33), and (6.17) enable us to obtain

$$\sigma_{11} = \frac{(G_1 - G_2)D(z) - HN(z)}{(G_1 + G_2)D(z) + HN(z)}$$

$$\sigma_{12} = \frac{2G_2D(z)}{(G_1 + G_2)D(z) + HN(z)}$$

$$\sigma_{21} = \frac{2G_1D(z)}{(G_1 + G_2)D(z) + HN(z)}$$

$$\sigma_{22} = \frac{(G_1 - G_2)D(z) + HN(z)}{(G_1 + G_2)D(z) + HN(z)} \tag{6.34}$$

In order to satisfy the realizability constraints on σ_{22}, we need

$$G_2 = G_1D(z) + HN(z)|_{z^{-1}=0}$$

and, since the coefficient of z^0 in $D(z)$ and $N(z)$ is unity, the realizability constraint is

$$G_2 = G_1 + H \tag{6.35}$$

With this constraint, the scattering parameters (6.34) may be written

$$\sigma_{11} = (a_1 - 1)/[1 + a_1 G(z)]$$

$$\sigma_{12} = [1 + G(z)]/[1 + a_1 G(z)]$$

$$\sigma_{21} = a_1[1 + G(z)]/[1 + a_1 G(z)]$$

$$\sigma_{22} = (1 - a_1)G(z)/[1 + a_1 G(z)] \tag{6.36a}$$

where

$$a_1 = G_1/G_2 \tag{6.36b}$$

and

$$G(z) = [D(z) - N(z)]/[D(z) + N(z)] \tag{6.36c}$$

Similarly, for a series impedance Z, we obtain

$$\sigma_{11} = (1 - a_2)/[1 + a_2 G(z)]$$

$$\sigma_{12} = a_2[1 + G(z)]/[1 + a_2 G(z)]$$

$$\sigma_{21} = [1 + G(z)]/[1 + a_2 G(z)]$$

$$\sigma_{22} = (a_2 - 1)G(z)/[1 + a_2 G(z)] \tag{6.37a}$$

where

$$a_2 = R_1/R_2 \tag{6.37b}$$

and $G(z)$ is given by equation (6.36c).

The structure realizing a shunt element, represented by system (6.36), is given in Figure 6.9a, and for a series element in Figure 6.9b. It may be seen that the function $G(z)$ plays the same role as the delay element z^{-1} performs in transformations of single-element sections. It may thus be regarded as a generalized delay element whose functional form is related to the structure of the elements.

As an example, we may consider a filter having a shunt element consisting of a series LC network. The admittance of this is

$$Y(s) = \frac{s}{L(s^2 + 1/LC)}$$

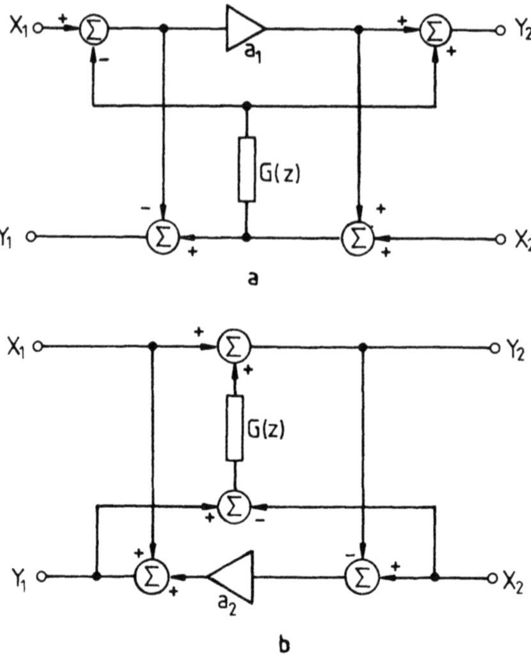

Figure 6.9. Flow graph for a generalized element: (a) shunt element; (b) series element.

Using the bilinear transformation, we obtain

$$Y(z) = \frac{C}{1 + LC} \cdot \frac{1 - z^{-2}}{1 - 2\rho z^{-1} + z^{-2}}$$

where $\rho = (LC - 1)/(LC + 1)$.

The generalized delay function becomes

$$G(z) = \frac{z^{-1}(z^{-1} - \rho)}{1 - \rho z^{-1}}$$

and this is represented in Figure 6.10. The complete implementation of this

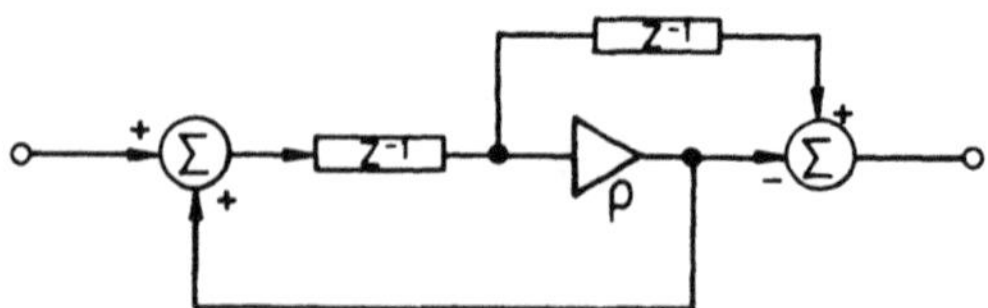

Figure 6.10. Implementation of generalized element comprising a series LC network.

section is obtained by replacing the delay element $G(z)$ in Figure 6.9a by the flow graph of Figure 6.10.

6.4.2. Transformation Set with Cycle Length of Two

As an alternative to the above implementation with all transformation sets identical, it is possible to implement a digital representation of an analogue ladder filter by employing two sets of transformation matrices used alternately on the series and shunt arms of the ladder.

It will be seen that such a set of transformations results in an implementation which has simpler structures. The number of delays is identical with the unity cycle length transformation; the number of multipliers is essentially the same (the factor of two appearing in these structures may be performed by a single shift), and the number of adders is never greater than these.

As an illustration of this, we shall apply the transformation

$$P_1 = \begin{bmatrix} G_{1\,(1)} & 0 \\ G_{1\,(1)} & 1 \end{bmatrix} \quad \text{and} \quad Q_1 = \begin{bmatrix} -G_{2\,(1)} & -1 \\ 0 & -1 \end{bmatrix} \tag{6.38a}$$

to the shunt elements, and the transformation

$$P_2 = \begin{bmatrix} 0 & 1 \\ -G_{1\,(2)} & 1 \end{bmatrix} \quad \text{and} \quad Q_2 = \begin{bmatrix} G_{2\,(2)} & -1 \\ G_{2\,(2)} & 0 \end{bmatrix} \tag{6.38b}$$

to the series sections.

Transformation of a shunt capacitor by means of the pair (P_1, Q_1) results in the σ parameters defined in the z plane,

$$\sigma_{11} = 2az^{-1}/(1 + z^{-1})$$

$$\sigma_{12} = 1$$

$$\sigma_{21} = a - 1$$

$$\sigma_{22} = 1 \tag{6.39a}$$

where $a = -R_1C$, and the realizability constraint is $G_1 = -C - G_2$. This may be implemented by the flow graph of Figure 6.11a.

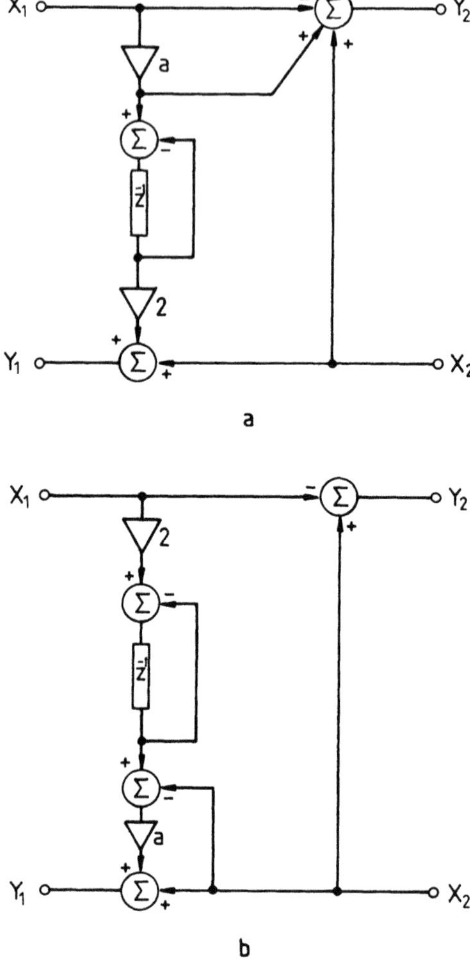

Figure 6.11. Transformation with cycle length of two: (a) shunt capacitor; (b) series inductor.

Transformation of the series inductor by means of the pair (P_2, Q_2) results in the σ parameters

$$\sigma_{11} = 2az^{-1}/(1 + z^{-1})$$

$$\sigma_{12} = 1 - a$$

$$\sigma_{21} = -1$$

$$\sigma_{22} = 1 \tag{6.39b}$$

where $a = G_1L$, and the realizability constraint is $R_1 = L - R_2$. The implementation of this is shown in Figure 6.11b.

The load and source terminations are derived in the same manner as in Sections 6.4.1.3 and 6.4.1.4, the transformation matrix being the one appropriate to the position in the system; the matrix Q_2 will be used on the source and, for the load transformation, either P_1 if the preceding section is even, or P_2 if it is odd.

Transformations for other analogue elements in series or shunt branches are derived as in Section 6.4.1.5.

6.4.3. Full Length Cycle Transformations

It is interesting to consider a specific group of transformations in which all pairs are different. If appropriately chosen, it may be shown that the structure for implementing a series element is identical to that for a shunt element whose analogue form is the topological dual of the series element.

The set of such transformations is as follows. For odd values of i (assumed to be a shunt arm),

$$P_{(i)} = \begin{bmatrix} -\mu_{(i)}G_{1\,(i)} & 0 \\ -\mu_{(i)}G_{1\,(i)} & \mu_{(i)} \end{bmatrix} \quad \text{and} \quad Q_{(i)} = \begin{bmatrix} \mu_{(i)}G_{1\,(i)} & \mu_{(i)}G_{1\,(i)}R_{2\,(i)} \\ 0 & \mu_{(i)}G_{1\,(i)}R_{2\,(i)} \end{bmatrix} \tag{6.40a}$$

where

$$\mu_{(i)} = \begin{cases} -R_{1\,(i)} & \text{for } i = 1 \\ -R_{1\,(i)} \prod_{j=1}^{(i-1)/2} R_{1\,(2j)}G_{1\,(2j+1)} & \text{otherwise} \end{cases}$$

For even values of i (assumed to be a series arm),

$$P_{(i)} = \begin{bmatrix} 0 & -\mu_{(i)}R_{1\,(i)} \\ \mu_{(i)} & -\mu_{(i)}R_{1\,(i)} \end{bmatrix} \quad \text{and} \quad Q_{(i)} = \begin{bmatrix} -\mu_{(i)}R_{1\,(i)}G_{2\,(i)} & -\mu_{(i)}R_{1\,(i)} \\ -\mu_{(i)}R_{1\,(i)}G_{2\,(i)} & 0 \end{bmatrix} \tag{6.40b}$$

where

$$\mu_{(i)} = \begin{cases} -1 & \text{for } i = 2 \\ -\prod_{j=1}^{(i-2)/2} R_{1\,(2j)}G_{1\,(2j+1)} & \text{otherwise} \end{cases}$$

If, however, the odd sections are series elements and the even sections are shunt elements, equation (6.40b) is applied to odd sections and equation

(6.40a) to even sections, with appropriate modifications of the upper limit on the products.

For a shunt capacitor, we obtain

$$\sigma_{11} = 2a_i z^{-1}/(1 + z^{-1})$$

$$\sigma_{12} = a_i - 1$$

$$\sigma_{21} = 1$$

$$\sigma_{22} = 1$$

where $a_i = R_{1\,(i)}C_i$, with realizability constraint $G_{1\,(i)} = G_{2\,(i)} + C_i$.

For a series inductor,

$$\sigma_{11} = 2a_i z^{-1}/(1 + z^{-1})$$

$$\sigma_{12} = a_i - 1$$

$$\sigma_{21} = 1$$

$$\sigma_{22} = 1$$

where $a_i = L_i G_{1\,(i)}$, and the realizability constraint is $R_{1\,(i)} = R_{2\,(i)} + L_i$.

These realizations are identical for series and shunt elements, and are also independent of the position of the section in the network. They may all therefore be implemented using the structure of Figure 6.12. The intercon-

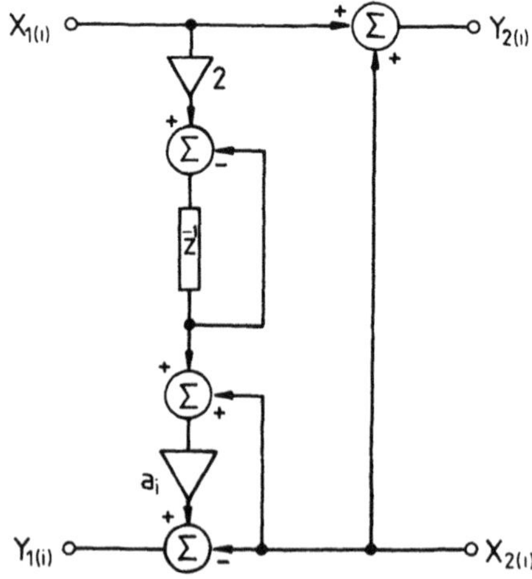

Figure 6.12. Full length cycle transformations; implementation of all sections.

nection of this series of structures is permissible, since it may easily be shown from equation (6.13b) that the $Q_{(i)}$ matrix for a shunt element is compatible with the $P_{(i+1)}$ matrix for a series element.

There is a large number of other structural transformations which have the same modular form as the one considered, but they mostly involve higher complexity of implementation.[8]

6.5. TWO-DIMENSIONAL IMPLEMENTATION

Two-dimensional filters may be implemented in a similar manner to those in one dimension. The advantages in this case stem mainly from the ability to obtain filters whose sensitivity to multiplier coefficient variation is small. Further advantages lie in the fact that, in certain situations, the realization is of lower complexity, or that there is a higher degree of modularity, as was evident in the 1-D implementation of Section 6.4.3.

The procedure for deriving 2-D filters from a ladder prototype involves the following steps:

1. Select an appropriate passive LC ladder prototype filter.
2. Apply an appropriate set of transformation matrices (P_i, Q_i) to the prototype to obtain the elements of the relevant 1-D flow diagram.
3. Use one of the transformations from the 1-D digital domain to the 2-D digital domain, appropriate to the form of frequency response required.

Thus, all the procedures for 1-D realization may be extended to two dimensions. Shunt admittances Y are replaced by blocks having transfer functions $Y[G(z_1, z_2)]$, and series impedances Z by $Z[G(z_1, z_2)]$, both of which are implemented by the appropriate flow graph dependent upon the choice of matrices P and Q.

Since the load and source sections are independent of frequency, the transformations are identical to those of the 1-D system.

Again, as in the 1-D situation, all delay-free loops must be eliminated. This results in the following realizability constraints on implementation.

Two-dimensional structures are realizable when derived by linear transformation from one-dimensional LC ladder networks if either

$$\sigma_{11}(s_1, s_2)|_{s_1=s_2=1} = \sigma_{11}(z_1, z_2)|_{z_1^{-1}=z_2^{-1}=0} = 0$$

or

$$\sigma_{22}(s_1, s_2)|_{s_1=s_2=1} = \sigma_{22}(z_1, z_2)|_{z_1^{-1}=z_2^{-1}=0} = 0 \tag{6.41}$$

holds for all sections including load and source.

6.5.1. Realization using Unity Cycle Length

We may consider here the implementation of a 2-D filter derived via the (P, Q) matrix pair given by equations (6.33), used on all sections of the prototype analogue filter; this leads to the wave digital representation.[9-12]

By considering a shunt branch admittance $Y(s)$, we obtain

$$\sigma_{11} = (G_1 - G_2 - Y)/(G_1 + G_2 + Y)$$

$$\sigma_{12} = 2G_2/(G_1 + G_2 + Y)$$

$$\sigma_{21} = 2G_1/(G_1 + G_2 + Y)$$

$$\sigma_{22} = (G_1 - G_2 + Y)/(G_1 + G_2 + Y) \tag{6.42a}$$

The realizability condition for this realization is

$$G_2 = G_1 + Y(s_1, s_2)|_{s_1=1, s_2=1} \tag{6.42b}$$

where $Y(s_1, s_2)$ is the 2-D analogue admittance derived from the 1-D reactance function via the chosen transformation function.

6.5.1.1. *Circularly Symmetric Filter*

To obtain a 2-D circularly symmetric filter, we may use the transformation of Section 4.11.2 in which

$$s = \frac{a_1 s_1 + a_2 s_2}{1 + b s_1 s_2} \tag{6.43}$$

and then apply the bilinear transform to each of the variables s_1 and s_2 to obtain a 2-D admittance function

$$Y(z_1, z_2) = H\frac{N(z_1, z_2)}{D(z_1, z_2)} \tag{6.44}$$

As in Section 6.4.1.6, we generate a 2-D generalized delay element

$$G(z_1, z_2) = \frac{D(z_1, z_2) - N(z_1, z_2)}{D(z_1, z_2) + N(z_1, z_2)} \tag{6.45}$$

which may then be used to obtain the scattering matrix for the 2-D admittance.

As an example, we may consider a simple shunt capacitance having an admittance $Y = sC$. From equation (6.43) we have

$$Y(s_1, s_2) = C\frac{a_1 s_1 + a_2 s_2}{1 + b s_1 s_2}$$

and, using the bilinear transformation,

$$Y(z_1, z_2) = C\frac{a_1 + a_2}{1 + b} \cdot \frac{1 - p_1 z_1^{-1} + p_1 z_2^{-1} - z_1^{-1} z_2^{-1}}{1 + p_2 z_1^{-1} + p_2 z_2^{-1} + z_1^{-1} z_2^{-1}} \tag{6.46}$$

where

$$p_1 = (a_1 - a_2)/(a_1 + a_2) \qquad \text{and} \qquad p_2 = (1 - b)/(1 + b) \tag{6.47}$$

Thus

$$H = C(a_1 + a_2)/(1 + b)$$

$$D(z_1, z_2) = 1 + p_2 z_1^{-1} + p_2 z_2^{-1} + z_1^{-1} z_2^{-1}$$

$$N(z_1, z_2) = 1 - p_1 z_1^{-1} + p_1 z_2^{-1} - z_1^{-1} z_2^{-1}$$

and hence

$$G(z_1, z_2) = \frac{z_1^{-1}(p_1 + p_2)/2 + z_2^{-1}(p_2 - p_1)/2 + z_1^{-1} z_2^{-1}}{1 + z_1^{-1}(p_2 - p_1)/2 + z_2^{-1}(p_1 + p_2)/2} \tag{6.48}$$

which is an all-pass network function.

Following a similar procedure to that used in Section 6.4.1.6, we have by analogy with equations (6.36a)

$$\sigma_{11} = (a_1 - 1)/[1 + a_1 G(z_1, z_2)]$$

$$\sigma_{12} = [1 + G(z_1, z_2)]/[1 + a_1 G(z_1, z_2)]$$

$$\sigma_{21} = a_1[1 + G(z_1, z_2)]/[1 + a_1 G(z_1, z_2)]$$

$$\sigma_{22} = (1 - a_1) G(z_1, z_2)/[1 + a_1 G(z_1, z_2)] \tag{6.49}$$

where $a_1 = G_1/G_2$.

The final implementation of a shunt capacitance therefore has the flow graph of Figure 6.9a, with $G(z)$ replaced by $G(z_1, z_2)$. This function $G(z_1, z_2)$, given by equation (6.48), is realized by the flow graph of Figure 6.13, which is canonical in the number of multipliers involved.

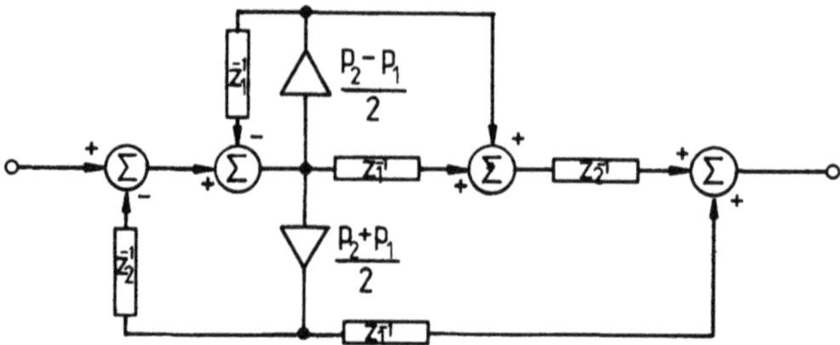

Figure 6.13. Two-dimensional implementation. Generalized network for a circular-symmetric realization.

A similar realization of a series inductor is obtained with the flow graph of Figure 6.9b, the generalized delay unit $G(z_1, z_2)$ having the same form as above and the wave parameters given by equations (6.37a) modified for two dimensions.

6.5.1.2. *Fan Filter*

A fan filter may be designed[13] by transforming a 1-D reference filter into a 2-D fan filter using the transformation

$$z = (z_1 z_2)^{1/2}$$

This will generally lead to a transfer function in which terms involving $z_1^{-1/2}$ and $z_2^{-1/2}$ will appear. We may realize the operator $[G(z)]^{1/2}$ by means of a continued fraction expansion.[14]

We may write the identity in the form

$$[G(z)]^{1/2} = 1 + \frac{G(z) - 1}{1 + [G(z)]^{1/2}} \tag{6.50}$$

and by successive iterations of this equation we obtain the infinite continued-fraction expansion

$$[G(z)]^{1/2} = 1 + \cfrac{G(z) - 1}{2 + \cfrac{G(z) - 1}{2 + \cfrac{G(z) - 1}{2 + \cdots}}} \tag{6.51}$$

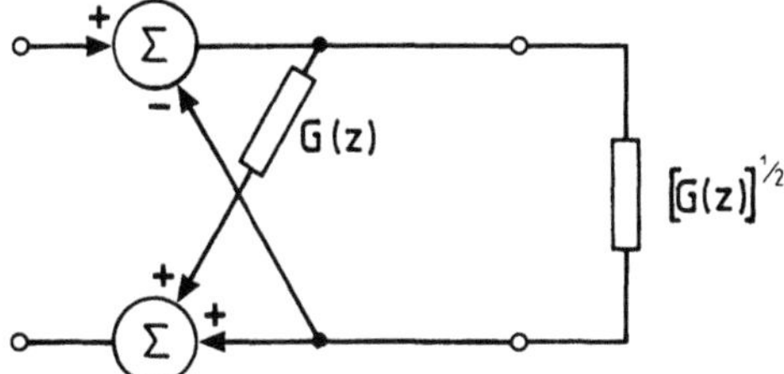

Figure 6.14. Implementation of function $G(z^{1/2})$ by a lattice network.

If we consider the particular case where $G(z) = z^{-1}$, we have

$$z^{-1/2} = 1 + \cfrac{z^{-1} - 1}{2 + \cfrac{z^{-1} - 1}{2 + \cfrac{z^{-1} - 1}{2 + \cdots}}} \tag{6.52}$$

Now equation (6.50) may be implemented by the network of Figure 6.14, giving an input function $[G(z)]^{1/2}$. Thus, an infinite cascade of such networks will realize the network function $[G(z)]^{1/2}$. As an example with $G(z) = z^{-1}$, we may realize the function $z^{-1/2}$. Since an infinite network is not practicable, we may truncate the infinite series of equation (6.52) to a finite number of terms. For example, considering only eight terms, we have

$$\begin{aligned} z^{-1/2} &\cong G_8(z) \\ &= \frac{8z^{-4} + 56z^{-3} + 56z^{-2} + 8z^{-1}}{z^{-4} + 18z^{-3} + 70z^{-2} + 28z^{-1} + 1} \end{aligned} \tag{6.53}$$

and this gives a very close approximation to the linear phase and constant magnitude, as required.

The extension to two dimensions follows directly in a manner similar to that discussed above.

REFERENCES

1. S. Erfani and M. Ahmadi, Direct form realization of 2-D digital systems, Internal Report, University of Windsor (1986).
2. A. Fettweis, Digital filter structures related to classical filter networks, *Arch. Elektron. Uebertragungstech.* **25**, 79–89 (1971).
3. A. Fettweis, Some principles of designing digital filters imitating classical filter structures, *IEEE Trans. Circuit Theory* **CT-18**, 314–316 (1971).

4. J. Bingham, A new type of digital filter with a reciprocal ladder configuration, IEEE Symp. on Circuit Theory, 129–130 (December 1970).
5. A. Chu and R. E. Crochiere, Comments and experimental results on optimal digital ladder structures, *IEEE Trans. Audio Electroacoust.* **AU-20,** 217–218 (1972).
6. K. Renner and S. Gupta, On the design of wave digital filters wih low sensitivity properties, *IEEE Trans. Circuit Theory* **CT-20,** 555–566 (1973).
7. R. E. Crochiere, Digital ladder structures and coefficient sensitivity, *IEEE Trans. Audio Electroacoust.* **AU-20,** 240–246 (1972).
8. A. Ali, *Design and Realisation of Low Sensitivity Active and Digital Filters in One and Two Dimensions,* PhD. Thesis, University of London (1980).
9. S. Erfani, B. Peikari, and M. Ahmadi, Realization of a class of two-dimensional zero-phase recursive digital filters with variable cut-off boundary, *Can. Electron. Eng. Journal* **11,** 18–22 (1986).
10. S. Erfani, M. Ahmadi, and V. Ramachandran, Modified realization technique for digital filters derived from the analog counterparts, 28th Midwest Symp. on Circuits and Systems, 445–448 (August 1985).
11. S. Erfani and B. Peikari, Digital design of general *LC* structures, *IEEE Trans. Circuits Syst.* **CAS-25,** 269–273 (1978).
12. S. Erfani and B. Peikari, Digital design of two-dimensional *LC* structures, *IEEE Trans. Circuits Syst.* **CAS-28,** 75–77 (1981).
13. A. H. Kayran and R. A. King, Design of recursive and non-recursive fan filters with complex transformations, *IEEE Trans. Circuits Syst.* **CAS-30,** 849–857 (1983).
14. S. Erfani, M. Ahmadi, V. Ramachandran, and M. O. Ahmad, A unified approach to the synthesis of the irrational immitance ($Z(s)$) and the fractional-step delay operator $z^{-1/2}$, *J. Franklin Inst.* **321,** 219–232 (1986).

7

State-Space Description

In the preceding chapters, we have regarded the system as a closed box which converts an input signal into an output. The system has been described directly in terms of the relationship between input and output signals, either in the time or space domains, or alternatively in the Fourier-transform domain of either of these.

The system may be described in greater detail if we also consider the internal structure. This may be achieved by specifying a number of internal variables termed the *state variables.* The state-space equations relating these variables to each other and to the input and output are always first-order recursive equations, which can be combined into a first-order vector-matrix equation. Using the state-space equation the information about the state or behavior of a dynamic system at any given point in time or space in the future may be determined by knowing the past and present states and the corresponding input point. For systems governed by differential or difference equations the state is simply a sufficient set of initial conditions.

In addition to the computational benefits obtained by writing the system equation in state-space form, several other advantages can also be expected. These include:

1. Applications to nonlinear and time-varying systems.
2. More insight into the internal structure of the system.
3. Extension to multi-input, multi-output systems.

In this chapter state-space formulations for one-dimensional (1-D) and two-dimensional (2-D) systems are introduced. In this connection, the conditions for stability, controllability, and observability of 1-D and 2-D systems are investigated. Methods for realizing these state-space descriptions are also discussed.

7.1. ONE-DIMENSIONAL STATE–SPACE MODELS

From equation (1.2) it may be seen that a linear 1-D discrete-time system may be represented by its finite-difference equation

$$\sum_{i=0}^{N} b_i y(n-i) = \sum_{i=0}^{N} a_i u(n-i) \tag{7.1}$$

with $b_0 = 1$, where $u(n)$ is the input signal and $y(n)$ the output signal. In the above, it is assumed that the number of input and output lags are identical; if this is not the case, the appropriate coefficients a_i or b_i may be set to zero.

By taking the Z transform we obtain the alternative representation, namely, the transfer function

$$\begin{aligned} H(z) &= \frac{Y(z)}{U(z)} = \frac{\sum_{i=0}^{N} a_i z^{-i}}{\sum_{i=0}^{N} b_i z^{-i}} = \frac{A(z)}{B(z)} \\ &= H_1(z)H_2(z) \end{aligned} \tag{7.2}$$

where $H_1(z)$ represents the FIR filter

$$H_1(z) = \sum_{i=0}^{N} a_i z^{-i} = \frac{Y(z)}{W(z)} = A(z) \tag{7.3}$$

and $H_2(z)$ represents the all-pole filter

$$H_2(z) = 1 \Big/ \sum_{i=0}^{N} b_i z^{-i} = \frac{W(z)}{U(z)} = \frac{1}{B(z)} \tag{7.4}$$

The inverse transform of equation (7.4) yields the recursion equation

$$\sum_{i=0}^{N} b_i w(n-i) = u(n) \tag{7.5}$$

The state variables $x_i(n)$ are defined by

$$x_{N-i+1}(n) = w(n-i), \qquad i = 1, 2, \ldots, N \tag{7.6}$$

The state equations comprise the set of N first-order recursive equations relating the state of the system to a linear combination of its states at the preceding instant and are given by

$$x_1(n+1) = x_2(n)$$

$$x_2(n+1) = x_3(n)$$

$$\vdots$$

$$x_{N-1}(n+1) = x_N(n)$$

Since $b_0 = 1$, we have from the recursion equation (7.5)

$$x_N(n+1) = -\sum_{i=1}^{N} b_i x_{N-i+1}(n) + u(n) \tag{7.7}$$

These equations may be represented graphically by the state diagram of Figure 7.1. In matrix form they may be written as

$$x(n+1) = Ax(n) + Bu(n) \tag{7.8}$$

where $x(n) = [x_1(n), x_2(n), \ldots, x_N(n)]^t$ is the state vector of dimension N; the state matrix

$$A = \begin{bmatrix} 0 & 1 & \cdots & & \cdots & 0 \\ & 0 & 1 & & & \\ & & 0 & 1 & \ddots & \\ 0 & & & \ddots & \ddots & \\ & & & & \ddots & 1 \\ -b_N & -b_{N-1} & \cdots & & \cdots & -b_1 \end{bmatrix} \tag{7.9a}$$

is of dimension $N \times N$, relating the present state of the system to the past

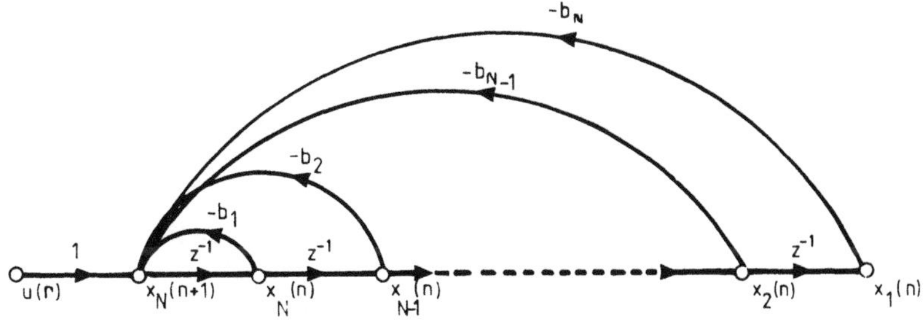

Figure 7.1. State-space flow graph of an all-pole IIR filter.

state, and the input matrix

$$B = [0\ 0\ 0\ \cdots\ 1]^t \tag{7.9b}$$

is a vector of dimension N, relating the present state to the present input.

The FIR part of the transfer function (7.3) may be represented by the discrete-time convolutional equation

$$\begin{aligned} y(n) &= \sum_{i=0}^{N} a_i w(n-i) \\ &= \sum_{i=1}^{N} a_i w(n-i) + a_0 w(n) \end{aligned}$$

On combining this with state variables (7.6), we obtain

$$y(n) = \sum_{i=1}^{N} (a_i - a_0 b_i) x_{N-i+1}(n) + a_0 u(n) \tag{7.10}$$

In matrix form this becomes

$$y(n) = Cx(n) + Du(n) \tag{7.11}$$

where

$$C = [c_1\ c_2\ \cdots\ c_N] \tag{7.12a}$$

is an $N \times 1$ vector in which $c_i = a_{N-i+1} - a_0 b_{N-i+1}$, relating the output of the system to the present state, and

$$D = a_0 \tag{7.12b}$$

which relates the output directly to the input. The diagram representing these output equations is shown in Figure 7.2.

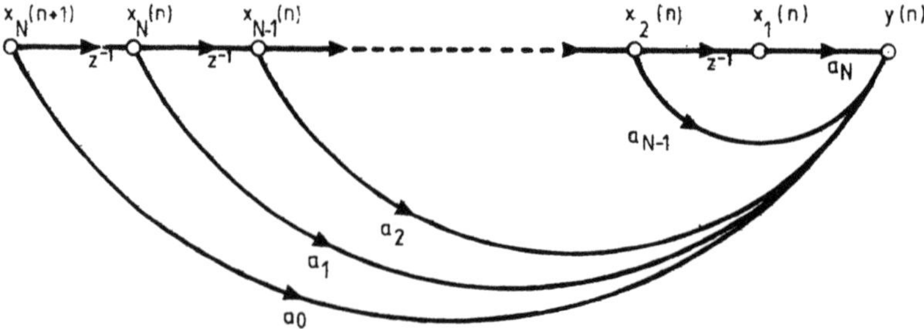

Figure 7.2. State-space flow graph of a FIR filter.

In the Z domain, assuming zero initial state, the state equations become

$$zX(z) = AX(z) + BU(z) \tag{7.13a}$$

and

$$Y(z) = CX(z) + DU(z) \tag{7.13b}$$

The transfer function $H(z)$ is obtained directly by eliminating $X(z)$. Hence

$$H(z) = Y(z)/U(z) = C(zI - A)^{-1}B + D \tag{7.14}$$

In the completely general case, where a multi-input, multi-output (MIMO) system is excited by P independent inputs and the output may be obtained from Q outputs, equations (7.8) and (7.11) take the form

$$x(n+1) = Ax(n) + B\tilde{u}(n) \tag{7.15a}$$

and

$$\tilde{y}(n) = Cx(n) + D\tilde{u}(n) \tag{7.15b}$$

respectively, where $x(n)$ and A have dimensions as previously, but $\tilde{u}(n)$ is a P length vector, $\tilde{y}(n)$ is a Q length vector, B is an $(N \times P)$ matrix, C a $(Q \times N)$ matrix, and D a $(Q \times P)$ matrix.

In this case equation (7.14) gives the transfer-function matrix $\tilde{H}(z)$ of the MIMO system, the general form of which is

$$\tilde{H}(z) = \begin{bmatrix} H_{11}(z) & H_{12}(z) & \cdots & H_{1P}(z) \\ H_{21}(z) & H_{22}(z) & & H_{2P}(z) \\ \vdots & & & \vdots \\ H_{Q1}(z) & H_{Q2}(z) & \cdots & H_{QP}(z) \end{bmatrix} \tag{7.16}$$

where

$$H_{ij}(z) = \left.\frac{Y_i(z)}{U_j(z)}\right|_{U_k(z)=0} \qquad k \neq j, \quad \begin{matrix} i \in [1, Q] \\ j \in [1, P] \end{matrix}$$

Thus $H_{ij}(z)$ represents the scalar transfer function of the single-input, single-output (SISO) system between the jth input and the ith output when all the other channels are deactivated. Further information on multivariable can be obtained elsewhere.[1]

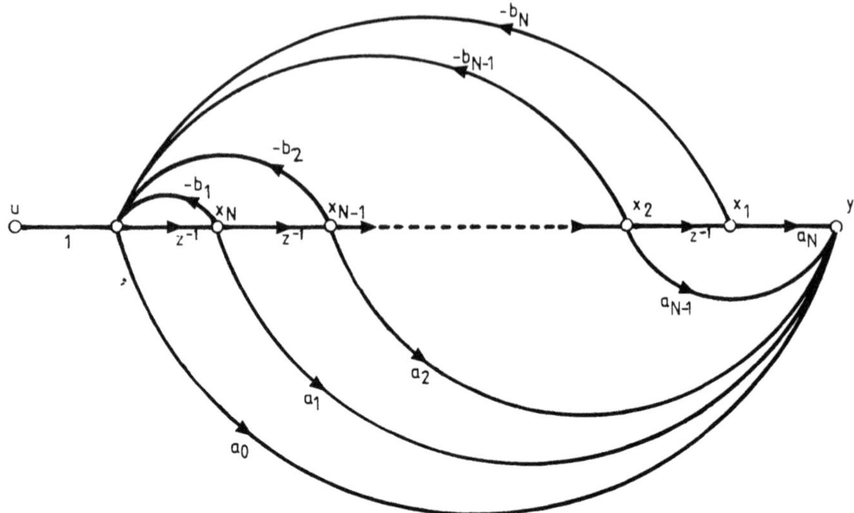

Figure 7.3. State-space flow graph of a complete IIR filter.

7.1.1. Realization of State–Space Model

The complete state diagram defined by equations (7.13) may be realized by the combination of Figures 7.1 and 7.2, shown in Figure 7.3, from which it is seen that the number of multipliers is $2N + 1$ and the number of delays is N. It is thus a canonical structure.

7.2. ONE-DIMENSIONAL STRUCTURE TRANSFORMATION

It has already been remarked that the state vector is not unique; many alternative state vectors are possible, all of which may be obtained by linear combinations of the elements of $x(n)$.

The original vector, represented by $x(n)$, may be transformed linearly to $x'(n)$ via

$$x'(n) = Tx(n) + Ku(n-1) \tag{7.17}$$

The $N \times N$ nonsingular matrix T represents scaling and rotation of the state vector, and K, an $N \times 1$ vector, gives the translational transformation.

7.2.1. Effect of Translation

We consider the simplified form of equation (7.17) involving the translational term only:

$$x'(n) = x(n) + Ku(n-1) \tag{7.18}$$

Substitution of this expression in equations (7.8) and (7.11) yields

$$x'(n+1) = Ax'(n) + (B+K)u(n) - AKu(n-1) \tag{7.19a}$$

and

$$y(n) = Cx'(n) + Du(n) - CKu(n-1) \tag{7.19b}$$

which gives an alternative form of the state equations. However, the only effect is in the way in which the input is introduced into the equations, the state matrix remaining unaffected.

7.2.2. Effect of Rotation and Scaling

The more important aspect of rotation and scaling will now be examined while neglecting any translation effect. Equation (7.17) reduces to

$$x'(n) = Tx(n) \tag{7.20}$$

and substitution into system (7.15) leads to a new set of state equations, which may be represented in standard form by introducing

$$A' = TAT^{-1} \tag{7.21a}$$

$$B' = TB \tag{7.21b}$$

$$C' = CT^{-1} \tag{7.21c}$$

$$D' = D \tag{7.21d}$$

We note that as both A and T are nonsingular, any new state matrix obtained by this transformation is also nonsingular. An infinite number of state vectors may therefore be obtained by variation of T, leading to infinite state-space models.

It may be seen that, if T is a diagonal matrix, this represents a scaling of all state variables with zero rotation, while off-diagonal elements are responsible for rotation.

7.2.3. Fully Decoupled States

These transformations on the state vector allow the creation of any of the possible implementations of a given transfer function. Most of these implementations are noncanonical and involve considerably more multipliers and adders than the direct form from which they were developed.

One transformation of particular interest is that which allows the various states to be completely decoupled in the output. The state matrix A'_D in this case is diagonal and the transformation T_D creating this is the normalizing transformation obtained from the eigenvectors. Thus, if a state matrix A has distinct real eigenvalues,

$$A'_D = T_D A T_D^{-1} \tag{7.22a}$$

where

$$A'_D = \begin{bmatrix} \lambda_1 & & & & 0 \\ & \lambda_2 & & & \\ & & \ddots & & \\ & & & \ddots & \\ 0 & & & & \lambda_N \end{bmatrix}$$

where λ_i are the distinct eigenvalues of A, and $T_D = [V_1 \ V_2 \ \cdots \ V_N]$ is a matrix whose columns are the set of eigenvectors V_i corresponding to the eigenvalues λ_i for $i = 1, 2, \ldots, N$. The other matrices in the transformed model are

$$B'_D = [b_{D1} \ b_{D2} \ \cdots \ b_{DN}]^t = T_D B \tag{7.22b}$$

and

$$C'_D = [c_{D1} \ c_{D2} \ \cdots \ c_{DN}] = C T_D^{-1} \tag{7.22c}$$

The signal flow diagram for this state matrix is given in Figure 7.4.

In situations where the eigenvalues are real but not distinct, the state matrix cannot be diagonalized. The procedure will result in a state matrix

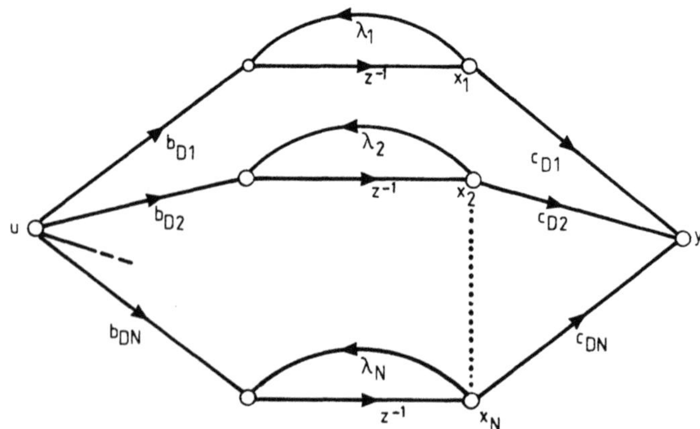

Figure 7.4. Flow graph of a filter implemented with fully decoupled states.

having the form

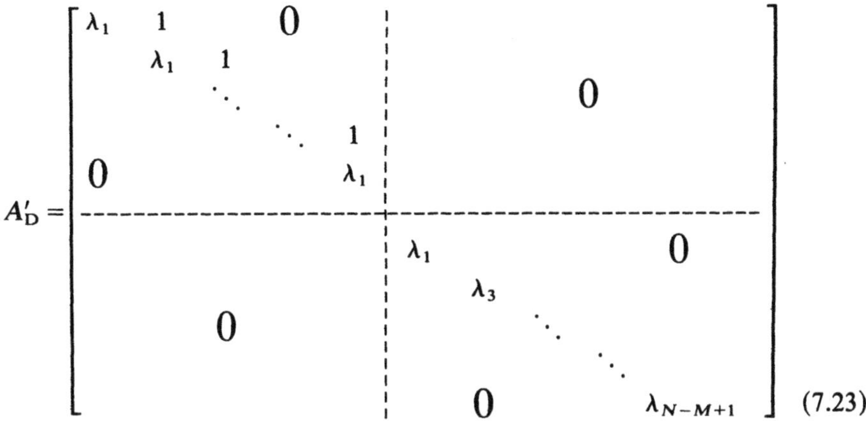

$$A'_{\mathrm{D}} = \left[\begin{array}{ccccc|cccc} \lambda_1 & 1 & & 0 & & & & & \\ & \lambda_1 & 1 & & & & & 0 & \\ & & \ddots & \ddots & 1 & & & & \\ 0 & & & & \lambda_1 & & & & \\ \hline & & & & & \lambda_1 & & & 0 \\ & & & & & & \lambda_3 & & \\ & & 0 & & & & & \ddots & \\ & & & & & 0 & & & \lambda_{N-M+1} \end{array}\right] \qquad (7.23)$$

where the eigenvalue λ_1 is of multiplicity M and the other eigenvalues are distinct. The state matrix in this case is known as the Jordan canonical form. The realization of this state matrix for the case $M = 3$ is shown in Figure 7.5.

For systems in which the characteristic equation has complex roots, it is possible to arrive at a real state matrix A from the direct form of realization. However, any attempt to decouple the variables by diagonalizing A will result in a normalized matrix with complex elements which cannot be

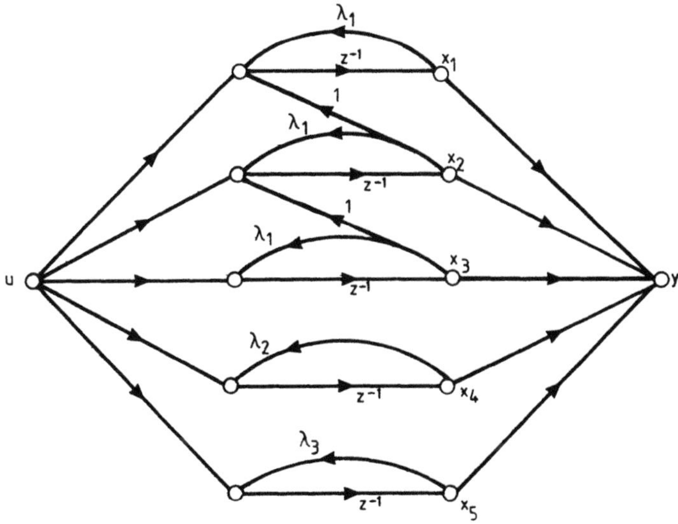

Figure 7.5. Flow graph for a filter with repeated real eigenvalues.

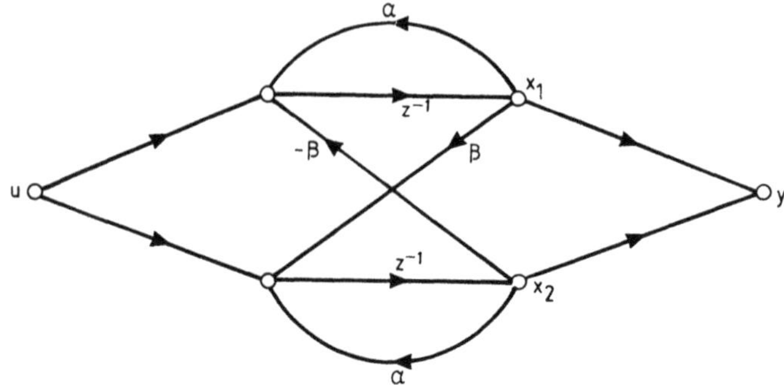

Figure 7.6. Flow graph for a filter with complex eigenvalues.

realized. In this case it is is possible to perform a transformation which, although not diagonalizing the matrix, does introduce partial decoupling.

As an example, a second-order system may be considered with conjugate eigenvalues λ and λ^*, where $\lambda = \alpha + j\beta$. In this case the partially decoupled state matrix will be the skew-symmetric matrix

$$A'_{\mathrm{D}} = \begin{bmatrix} \alpha & -\beta \\ \beta & \alpha \end{bmatrix} \tag{7.24}$$

The flow diagram for its implementation is shown in Figure 7.6. In the general case, the partially decoupled matrix will contain diagonal elements corresponding to the real eigenvalues and, in addition, 2×2 submatrices on the diagonal for each pair of complex eigenvalues.

7.3. SOLUTION OF ONE-DIMENSIONAL STATE EQUATIONS

This may be approached both in the time domain and in the z domain leading, naturally, to identical results.

7.3.1. Time-Domain Solution

The solution in the time domain may be obtained by superposition of the zero-state and zero-input responses.

The zero-input response, namely, the solution of

$$x(n+1) = Ax(n) \tag{7.25}$$

is obtained by successive iteration, assuming that the first nonzero state is $x(k)$:

$$x(n) = A^{n-k}x(k), \qquad 0 \le k \le n \tag{7.26a}$$

$$x(n) = 0, \qquad n < k \tag{7.26b}$$

If $x(0)$ is the initial state vector, this leads to the zero-input solution

$$x_{ZI}(n) = A^n x(0) \tag{7.27}$$

Again considering the response to zero initial state and input at instant k only, we need to solve

$$x(k+1) = Bu(k), \qquad 0 < k \tag{7.28}$$

By using equation (7.27a), we obtain the zero state response to inputs at all instants from $k = 0$ to n,

$$x_{ZS}(n) = \sum_{k=0}^{n-1} A^{n-k-1}Bu(k) \tag{7.29}$$

The complete time-domain response of the state vector is, by linear superposition,

$$\begin{aligned} x(n) &= x_{ZI}(n) + x_{ZS}(n) \\ &= A^n x(0) + \sum_{k=0}^{n-1} A^{n-k-1}Bu(k) \end{aligned} \tag{7.30}$$

Finally, the output equation (7.11) enables us to obtain the complete system response

$$y(n) = CA^n x(0) + C \sum_{k=0}^{n-1} A^{n-k-1}Bu(k) + Du(n) \tag{7.31}$$

7.3.2. *Z*-Domain Solution

We first note that the Z transform of A^n is given by

$$\begin{aligned} Z[A^n] &= \sum_{n=0}^{\infty} A^n z^{-n} \\ &= z(zI - A)^{-1} \end{aligned} \tag{7.32}$$

The Z transform of equation (7.8) yields

$$zX(z) - zx(0) = AX(z) + BU(z)$$

leading to

$$X(z) = z(zI - A)^{-1}x(0) + (zI - A)^{-1}BU(z) \tag{7.33}$$

and, taking the inverse Z transform, using relation (7.32) and the convolution property, we obtain

$$x(n) = A^n x(0) + \sum_{k=0}^{n-1} A^{n-k-1}Bu(k)$$

which is the same expression as is given in equation (7.30) above.

REMARK 7.1. The impulse response of the system described by equations (7.8) and (7.11) can be obtained by replacing $u(n) = \delta(n)$ in equation (7.30) when the initial condition is zero. This yields

$$h(0) = D$$

$$h(n) = CA^{n-1}B, \qquad n > 0 \tag{7.34}$$

Thus an alternative expression for the response equation (7.31) will be

$$y(n) = CA^n x(0) + \sum_{k=0}^{n} h(n-k)u(k) \tag{7.35}$$

where the second term is nothing but the zero-state response to the input $u(n)$.

7.3.3. State Transition Matrix

We see in equation (7.32) that the matrix A^n plays a very important part in the evaluation of system performance. This matrix is called the state transition matrix of the system and it permits the determination of the state of the system at any sampling instant from the value of the state vector n sampling instants earlier and the intervening values of the input.

7.3.3.1. *Characteristic Polynomial and Cayley–Hamilton Theorem for One-Dimensional Systems*

If a system is defined as in equation (7.2), the denominator polynomial

$$B(z) = \sum_{i=0}^{N} b_i z^{-i} \tag{7.36}$$

is called the characteristic polynomial of the 1-D system. If this polynomial has roots z_j, $j = 1, 2, \ldots, N$, then

$$\sum_{i=0}^{N} b_i z_j^{-i} = 0, \qquad j = 1, 2, \ldots, N \tag{7.37a}$$

whence

$$z_j^N = \sum_{i=1}^{N} -b_i z_j^{N-i}, \qquad j = 1, 2, \ldots, N \tag{7.37b}$$

with $b_0 = 1$.

The Cayley–Hamilton Theorem. The state matrix A of a system described by equation (7.1) satisfies its own characteristic matrix equation, i.e.,

$$\sum_{i=0}^{N} b_i A^{n-i} = 0 \tag{7.38}$$

This is easily seen from the Cayley–Hamilton theorem, which expresses any power N of a matrix A as the linear combination of the first $N-1$ powers of A, namely, since $b_0 = 1$,

$$A^N = -\sum_{i=1}^{N} b_i A^{N-i} \tag{7.39}$$

This suggests a method for computing the power of a matrix.

From equation (7.14), it may be seen that the poles of the transfer function are the roots of the characteristic equation given by

$$\text{Determinant}(zI - A) = 0$$

The state matrix may thus be evaluated by first determining the coefficients b_i from the matrix A using Leverrier–Faddeeva's algorithm.[3] Then the state transition matrix is computed recursively from equation (7.39). This technique is computationally more efficient than a direct evaluation of the state transition matrix by successive use of the recursion formula

$$A^k = A \cdot A^{k-1} \tag{7.40}$$

7.3.4. Stability of One-Dimensional Systems[2]

It was shown in Chapter 3 that a 1-D system is absolutely stable if, and only if, the poles of the system transfer function lie inside the unit circle in the Z plane.

Evaluation of equation (7.14) shows that the denominator polynomial in z^{-1} of the transfer function is derived from the expansion of $(I - Az^{-1})^{-1}$, which is $\mathrm{Det}(I - Az^{-1})$, where $\mathrm{Det}(A)$ represents the determinant of the matrix A. Thus the condition for stability is that the roots of

$$\mathrm{Det}(I - Az^{-1}) = 0$$

shall not lie inside the unit circle in the z^{-1} plane. This immediately leads to the statement that a 1-D state-space system is stable if, and only if, its eigenvalues lie outside the unit circle in the z^{-1} plane.

7.3.5. Controllability and Observability in One-Dimensional Systems[2]

The concepts of controllability and observability play a very important role in the design and analysis of modern control systems. They are therefore briefly reviewed in this section.

DEFINITION 7.1 (Complete State Controllability). We consider, in general, the MIMO system in equation (7.15). The system is said to be completely state controllable if, for any initial time (say $n = 0$), there exist a set of unconstrained controls (input) $u(0), u(1), \ldots, u(M)$ that will transfer the system from the initial state $x(0)$ to any final state $x(M)$, with M finite.

THEOREM 7.1 (Complete State Controllability). *The linear time-invariant, discrete-time system in equation* (7.15) *is completely state controllable if and only if the* $N \times MP$ *matrix*

$$\hat{C}_M = [B \; AB \; \cdots \; A^{M-1}B] \tag{7.41}$$

is of rank N, *or the* $N \times N$ *matrix* $\hat{C}_M \hat{C}_M^t$ *is nonsingular.*

The proof of Theorem 7.1 has been given elsewhere.[2]

When $M = N$, matrix $\hat{C}_M$ is called the "controllability matrix" and the matrix $\hat{C}_M\hat{C}^t_M$ is known as the "controllability Gramian matrix." For single-input systems $P = 1$ and for the case when $M = N$, the controllability matrix is square and must be invertible for complete state controllability.

DEFINITION 7.2 (Complete Observability). System (7.15) is said to be completely observable if any initial state, say $x(0)$, can be determined from knowledge of the output sequence $y(n)$, $n = 0, 1, \ldots, M - 1$, with M finite. The input is assumed to be zero for simplicity.

THEOREM 7.2 (Complete Observability). *The linear, time-invariant, discrete-time system* (7.15) *is completely observable if and only if the* $N \times MQ$ *matrix*

$$\hat{O}_M = \begin{bmatrix} C \\ CA \\ \vdots \\ CA^{M-1} \end{bmatrix} \tag{7.42}$$

is of rank N.

The proof of Theorem 7.2 has been given elsewhere.[2]

For $M = N$ the matrix $\hat{O}_M$ is called the "observability matrix" and $\hat{O}_M\hat{O}^t_M$, which must be nonsingular, is known as the "observability Gramian matrix." Again for single-output systems $Q = 1$ and if $M = N$, the observability matrix is square and must be invertible for complete observability.

THEOREM 7.3 (Invariant Theorem). *We consider system* (7.15), *where the pair* $[A, B]$ *is completely controllable and/or the pair* $[A, C]$ *is completely observable. The transformation*

$$x'(n) = Tx(n) \tag{7.43}$$

where T *is nonsingular, transforms the system equations to*

$$x'(n+1) = A'x'(n) + B'u(n)$$

$$y(n) = C'x'(n) + D'u(n) \tag{7.44}$$

with A', B', C', *and* D' *defined in equations* (7.21). *Then the pair* $[A', B']$ *is also completely controllable and/or the pair* $[A', C']$ *is also completely observable.*

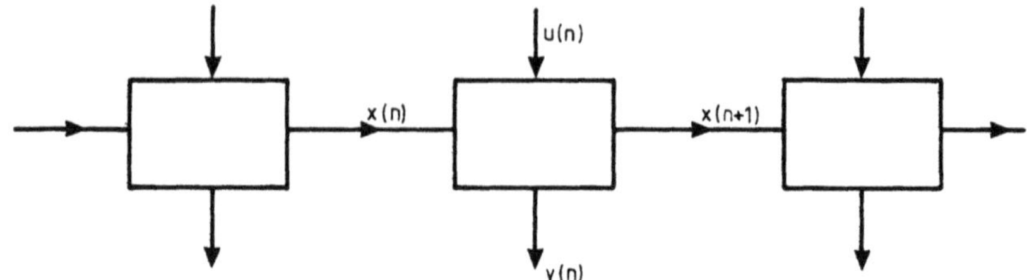

Figure 7.7. Linear-operator representation of state operations for a one-dimensional system.

7.4. REPRESENTATION OF ONE-DIMENSIONAL STATE OPERATIONS

The operation denoted by the state equations (7.8) and (7.11) is represented by the iterative graph of Figure 7.7, in which each cell performs the given linear operation at time (or space) instant n. The inputs to each cell are the present state and input vectors $x(n)$ and $u(n)$, respectively, and the outputs are the present system output vector $y(n)$ and the subsequent state vector $x(n+1)$.

EXAMPLE 7.1. Let us consider a linear, time-invariant, discrete-time system described by the difference equation

$$y(n) + 2y(n-1) + y(n-2) = u(n-1) + u(n-2)$$

We wish to obtain a state-space representation for the system and determine the complete controllability and observability of the system.

The method in Section 7.1 enables a state-space representation for the system to be obtained as follows:

$$x(n+1) = Ax(n) + Bu(n)$$

$$y(n) = Cx(n)$$

where

$$A = \begin{bmatrix} 0 & 1 \\ -1 & -2 \end{bmatrix}, \quad B = \begin{bmatrix} 0 \\ 1 \end{bmatrix}, \quad \text{and} \quad C = [1 \;\; 1]$$

Let us form the controllability and observability matrices, i.e.,

$$\hat{C}_2 = [B \;\; AB] = \begin{bmatrix} 0 & 1 \\ 1 & -2 \end{bmatrix} \quad \text{and} \quad \hat{O}_2 = \begin{bmatrix} C \\ CA \end{bmatrix} = \begin{bmatrix} 1 & 1 \\ -1 & -1 \end{bmatrix}$$

Thus the system is completely state controllable but not completely observable. In other words, not all the states $x_1(n)$ and $x_2(n)$ can be determined from knowledge of the output sequence over a finite interval.

Now consider another state-space realization for the system with new matrices

$$A' = \begin{bmatrix} 0 & 1 \\ -1 & -2 \end{bmatrix}, \qquad B' = \begin{bmatrix} 1 \\ -1 \end{bmatrix}, \qquad \text{and} \qquad C' = [1 \ \ 0]$$

The controllability and observability matrices are

$$\hat{C}'_2 = [B' \ \ A'B'] = \begin{bmatrix} 1 & -1 \\ -1 & 1 \end{bmatrix} \qquad \text{and} \qquad \hat{O}'_2 = \begin{bmatrix} C' \\ C'A' \end{bmatrix} = \begin{bmatrix} 1 & 0 \\ 0 & 1 \end{bmatrix}$$

That is, the system is not state controllable but it becomes observable. The variations in the conditions of controllability and observability depending on the state variable assignments are due to the fact that the system has a pole-zero cancellation in its transfer function, i.e.,

$$H(z) = \frac{z+1}{z^2+2z+1} = \frac{1}{z+1}$$

7.5. TWO-DIMENSIONAL STATE–SPACE MODELS[4-7]

From equation (1.2b), a 2-D infinite impulse response system may be represented by its difference equation:

$$\sum_{i=0}^{M} \sum_{j=0}^{N} b_{ij} y(m-i, n-j) = \sum_{i=0}^{M} \sum_{j=0}^{N} a_{ij} u(m-i, n-j) \qquad (7.45)$$

with $b_{00} = 1$, or by its transfer function

$$H(z_1, z_2) = \frac{\sum_{i=0}^{M} \sum_{j=0}^{N} a_{ij} z_1^{-i} z_2^{-j}}{\sum_{i=0}^{M} \sum_{j=0}^{N} b_{ij} z_1^{-i} z_2^{-j}} = \frac{A(z_1, z_2)}{B(z_1, z_2)} \qquad (7.46)$$

An alternative and probably better representation is via the 2-D state-space formulation. A similar procedure can be followed to that of Section

7.1 to yield the following general 2-D state-space model for a MIMO system:

$$r(m+1, n) = A_1 r(m, n) + A_2 s(m, n) + B_1 u(m, n) \tag{7.47a}$$

$$s(m, n+1) = A_3 r(m, n) + A_4 s(m, n) + B_2 u(m, n) \tag{7.47b}$$

$$y(m, n) = C_1 r(m, n) + C_2 s(m, n) + D u(m, n) \tag{7.47c}$$

The dimensions of the state, input and output vectors are

$$\dim(r) = N_1, \qquad \dim(s) = N_2, \qquad \dim(u) = P, \qquad \dim(y) = Q$$

Consequently the matrices have dimensions

$$\dim(A_1) = N_1 \times N_1 \qquad \dim(A_2) = N_1 \times N_2$$

$$\dim(A_3) = N_2 \times N_1 \qquad \dim(A_4) = N_2 \times N_2$$

$$\dim(B_1) = N_1 \times P \qquad \dim(B_2) = N_2 \times P$$

$$\dim(C_1) = Q \times N_1 \qquad \dim(C_2) = Q \times N_2$$

$$\dim(D) = Q \times P$$

The correspondence in two dimensions to the linear iterative graph of Figure 7.7 is shown in Figure 7.8. Each cell performs a linear transformation

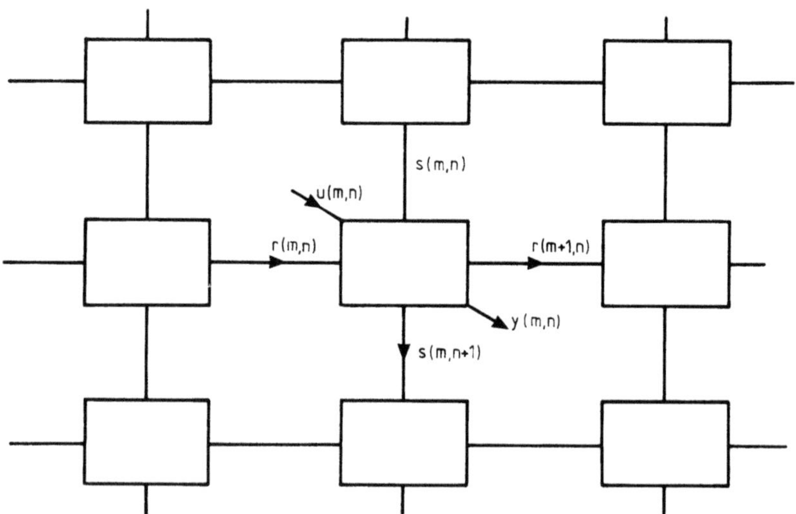

Figure 7.8. Linear-operator representation for a two-dimensional system.

on its three inputs: the horizontal and vertical state vectors $r(m, n)$ and $s(m, n)$, and the system input $u(m, n)$, to give the three cell outputs—the subsequent horizontal state vector to the right, $r(m+1, n)$, the next vertical state vector downward, $s(m, n+1)$, and the present system output vector, $y(m, n)$.

The initial conditions are given by $r(0, n)$ and $s(m, 0)$. The global state space of the 2-D system is of infinite dimensions and comprises the initial condition space necessary to propagate the state-space equation. We define two types of initial condition[7]:

1. Weak (local) initial condition $r(0,0) = \xi_1, s(0,0) = \xi_2$, and $r(0, n) = s(m, 0) = 0 \ \forall m, n \neq 0$; ξ_1 and ξ_2 are constants.
2. Strong (global) initial conditions where

$$\Pi = \Pi_r \times \Pi_s \tag{7.48}$$

$\Pi_r = \{r(0, n), \forall n\}$ and $\Pi_s = \{s(m, 0), \forall m\}$ being arbitrary.

Equations (7.47a) and (7.47b) may be written more compactly by defining a total state vector

$$x(m, n) = \begin{bmatrix} r(m, n) \\ s(m, n) \end{bmatrix} \tag{7.48a}$$

together with

$$A^{1,0} = \begin{bmatrix} A_1 & A_2 \\ 0 & 0 \end{bmatrix} \quad \text{and} \quad A^{0,1} = \begin{bmatrix} 0 & 0 \\ A_3 & A_4 \end{bmatrix} \tag{7.48b}$$

$$B^{1,0} = \begin{bmatrix} B_1 \\ 0 \end{bmatrix} \quad \text{and} \quad B^{0,1} = \begin{bmatrix} 0 \\ B_2 \end{bmatrix} \tag{7.48c}$$

The state-space equation may now be rewritten in the form

$$\begin{aligned} x(m, n) = {} & A^{1,0}x(m-1, n) + A^{0,1}x(m, n-1) \\ & + B^{1,0}u(m-1, n) + B^{0,1}u(m, n-1) \end{aligned} \tag{7.49a}$$

The output equation may similarly be written as

$$y(m, n) = Cx(m, n) + Du(m, n) \tag{7.49b}$$

with $C = [C_1 \ \ C_2]$.

The expression for the transfer function of a 2-D system in terms of the model matrices can be obtained by taking the 2-D z transform of both sides of equation (7.47). Assuming zero initial conditions, the result becomes

$$z_1R(z_1, z_2) = A_1R(z_1, z_2) + A_2S(z_1, z_2) + B_1U(z_1, z_2) \quad (7.50a)$$

$$z_2S(z_1, z_2) = A_3R(z_1, z_2) + A_4S(z_1, z_2) + B_2U(z_1, z_2) \quad (7.50b)$$

$$Y(z_1, z_2) = C_1R(z_1, z_2) + C_2S(z_1, z_2) + DU(z_1, z_2) \quad (7.50c)$$

where $R(z_1, z_2)$, $S(z_1, z_2)$, $U(z_1, z_2)$, and $Y(z_1, z_2)$ are the z transforms of sequences $\{r(m, n)\}$, $\{s(m, n)\}$, $\{u(m, n)\}$, and $\{y(m, n)\}$, respectively. From expressions (7.50a) and (7.50b) we obtain

$$\begin{bmatrix} z_1I_{N_1} - A_1 & -A_2 \\ -A_3 & z_2I_{N_2} - A_4 \end{bmatrix} \begin{bmatrix} R(z_1, z_2) \\ S(z_1, z_2) \end{bmatrix} = \begin{bmatrix} B_1 \\ B_2 \end{bmatrix} U(z_1, z_2)$$

or

$$X(z_1, z_2) = [(z_1I_{N_1} \oplus z_2I_{N_2}) - A]^{-1} \begin{bmatrix} B_1 \\ B_2 \end{bmatrix} U(z_1, z_2) \quad (7.51a)$$

where $\oplus$ denotes the direct sum operation, namely,

$$z_1I_{N_1} \oplus z_2I_{N_2} = \begin{bmatrix} z_1I_{N_1} & 0 \\ 0 & z_2I_{N_2} \end{bmatrix}$$

Substitution of equation (7.51a) into expression (7.50c) yields

$$Y(z_1, z_2) = \left\{ [C_1 \;\; C_2][(z_1I_{N_1} \oplus z_2I_{N_2}) - A]^{-1} \begin{bmatrix} B_1 \\ B_2 \end{bmatrix} + D \right\} U(z_1, z_2) \quad (7.51b)$$

Thus the transfer function for the 2-D system can be written as

$$H(z_1, z_2) = \frac{Y(z_1, z_2)}{U(z_1, z_2)} = C[(z_1I_{N_1} \oplus z_2I_{N_2}) - A]^{-1}B + D \quad (7.52)$$

where

$$A = \begin{bmatrix} A_1 & A_2 \\ A_3 & A_4 \end{bmatrix} \quad \text{and} \quad B = \begin{bmatrix} B_1 \\ B_2 \end{bmatrix}$$

7.6. TWO-DIMENSIONAL STRUCTURE TRANSFORMATION

The 2-D state vector may be scaled and rotated as in one dimension so that

$$x'(m, n) = Tx(m, n) \tag{7.53a}$$

or, in expanded form,

$$\begin{bmatrix} r'(m, n) \\ s'(m, n) \end{bmatrix} = T \begin{bmatrix} r(m, n) \\ s(m, n) \end{bmatrix} \tag{7.53b}$$

In order to maintain the separation of the horizontal states and vertical states in the composite state vector, matrix T must assume the partitioned form

$$T = \begin{bmatrix} T_1 & 0 \\ 0 & T_2 \end{bmatrix} \tag{7.54}$$

With this constraint it is now possible to set

$$(A^{0,1})' = TA^{0,1}T^{-1} \tag{7.55a}$$

$$(A^{1,0})' = TA^{1,0}T^{-1} \tag{7.55b}$$

$$(B^{1,0})' = TB^{1,0} \tag{7.55c}$$

$$(B^{0,1})' = TB^{0,1} \tag{7.55d}$$

$$C' = CT^{-1} \tag{7.55e}$$

$$D' = D \tag{7.55f}$$

7.7. REALIZATION OF STATE–SPACE EQUATIONS

The realization problem can be approached in an analagous manner to that adopted for one dimension. The transfer function (7.46) may be expressed in the form

$$H(z_1, z_2) = \frac{\sum_{j=0}^{N} a_j(z_1^{-1})z_2^{-j}}{\sum_{j=0}^{N} b_j(z_1^{-1})z_2^{-j}} \tag{7.56}$$

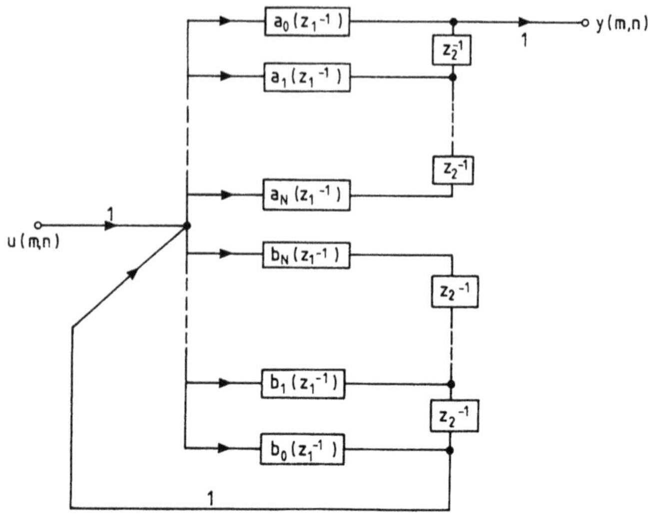

Figure 7.9. Block realization of two-dimensional state-space equations.

where the numerator and denominator have been written as polynomials in z_2^{-1} with coefficients which are functions of z_1^{-1}. By treating this as a 1-D transfer function, the direct realization takes the form of Figure 7.9.

The polynomials $a_i(z_1^{-1})$ and $b_i(z_1^{-1})$ may now be realized in a similar manner, giving rise to the implementation shown in Figure 7.10. The outputs of the delay elements in the horizontal direction, marked z_1^{-1}, represent the horizontal elements of the state vector, and the outputs of the vertical delay elements, z_2^{-1}, represent the vertical elements. These latter are in two groups, one associated with the recursive part of the system and the other with the output part. This realization can be represented by the state equations (7.47), where

$$r(m, n) = [r_1(m, n)r_2(m, n) \cdots r_M(m, n)]^t$$

$$s(m, n) = [s_1'(m, n) \cdots s_N'(m, n)s_1''(m, n) \cdots s_N''(m, n)]^t \tag{7.57a}$$

$$A_1 = \begin{bmatrix} -b_{10} & -b_{20} & \cdots & & \cdots & -b_{N0} \\ 1 & 0 & & & & \\ & 1 & 0 & & & \mathbf{0} \\ & & \ddots & \ddots & & \\ & \mathbf{0} & & \ddots & \ddots & \\ & & & & 1 & 0 \end{bmatrix}_{(N\times N)}$$

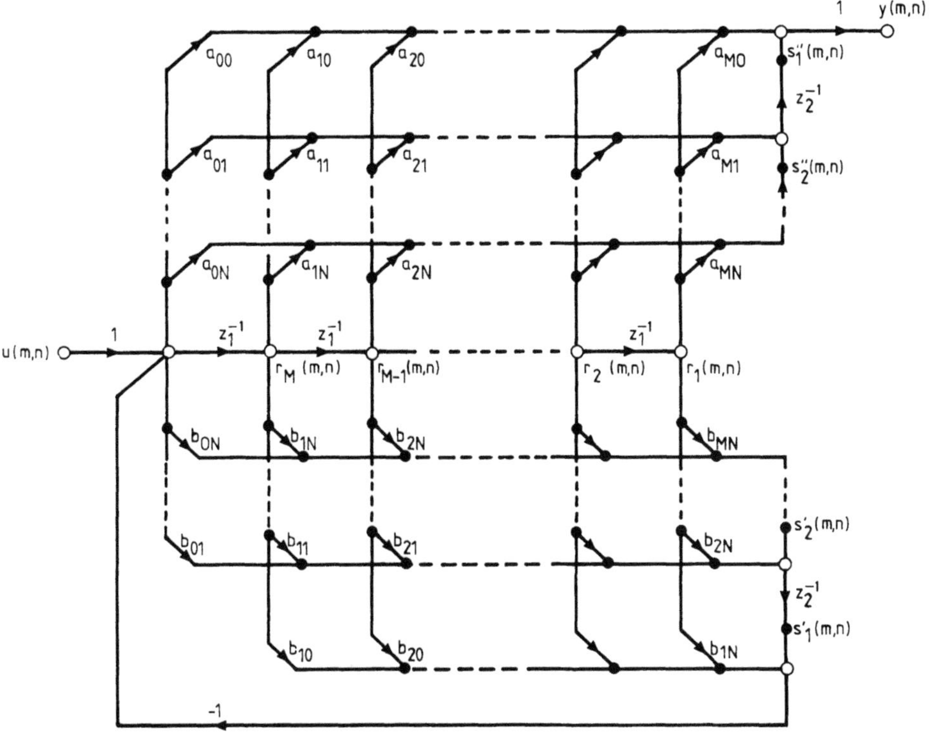

Figure 7.10. Expanded two-dimensional state-space realization.

$$A_2 = \left[\begin{array}{cccccc|c} -1 & 0 & 0 & \cdots & \cdots & & \\ 0 & & & & & & \\ 0 & & & & & & \\ \vdots & & & 0 & & & 0 \\ \vdots & & & & & & \\ 0 & & & & & & \end{array}\right]_{(N\times 2M)}$$

$$A_3 = \left[\begin{array}{ccc} b'_{11} & \cdots & b'_{N1} \\ \vdots & & \vdots \\ b'_{1M} & \cdots & b'_{NM} \\ \hline a'_{11} & \cdots & a'_{N1} \\ \vdots & & \vdots \\ a'_{1M} & \cdots & a'_{NM} \end{array}\right]_{(2M\times N)}$$

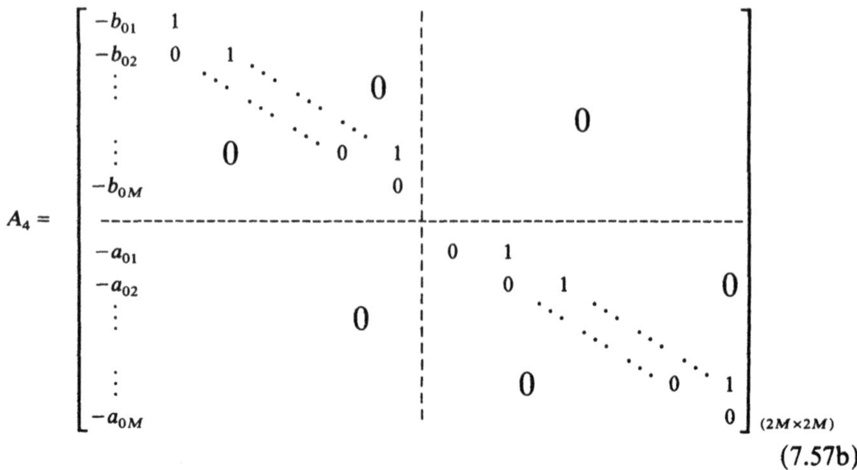

(7.57b)

$$B_1 = [1\ 0\ \cdots\ 0]^t$$

$$B_2 = [b_{01}\ b_{02}\ \cdots\ b_{0M}\ a_{01}\ a_{02}\ \cdots\ a_{0M}]^t \tag{7.57c}$$

$$C_1 = [a'_{10}\ a'_{20}\ \cdots\ a'_{N0}]$$

$$C_2 = [-a_{00}\ 0\ \cdots\ 0\ 1\ 0\ \cdots\ 0]$$

$$D = a_{00} \tag{7.57d}$$

where

$$b'_{ij} = b_{ij} - b_{i0}b_{0j}, \qquad 1 \le i \le N, 1 \le j \le M$$

$$a'_{ij} = a_{ij} - b_{i0}a_{0j}, \qquad 1 \le i \le N, 0 \le j \le M \tag{7.57e}$$

It can be seen from equation (7.49b) that the output may be derived directly from the state vector. It is thus desirable to obtain a closed-form solution of equation (7.49a). This may be addressed both in the 2-D spatial domain and in the corresponding 2-D z domain.

7.8. SOLUTION OF TWO-DIMENSIONAL STATE–SPACE EQUATIONS

7.8.1. State Transition Matrix[6]

Analogous to the 1-D case, we need to determine the state transition matrix $A^{i,j}$. This is seen to be the "two-tuple power" of the partitioned matrix A, defined in the following way:

1. $A^{i,j} = 0, \qquad i < 0 \text{ or } j < 0$
2. $A^{0,0} = I$
3. $A^{i,j} = A^{1,0}A^{i-1,j} + A^{0,1}A^{i,j-1} \qquad \text{for } (i,j) > (0,0)$ (7.58)

7.8.2. Spatial-Domain Solution[5,6]

The solution of the state equations for a 2-D system may be determined in a similar manner to that for one dimension. It may be shown that[7]

$$x(m,n) = \sum_{j=0}^{n} A^{m,n-j}\begin{bmatrix} r(0,j) \\ 0 \end{bmatrix} + \sum_{i=0}^{m} A^{m-i,n}\begin{bmatrix} 0 \\ s(i,0) \end{bmatrix} + \sum_{(0,0)\le i,j<m,n}\sum [A^{m-i-1,n-j}B^{1,0} + A^{m-i,n-j-1}B^{0,1}]u(i,j) \tag{7.59a}$$

the first two terms being the zero-input response and the last representing the zero-state response. Hence

$$y(m,n) = C\left\{\sum_{j=0}^{n} A^{m,n-j}\begin{bmatrix} r(0,j) \\ 0 \end{bmatrix} + \sum_{i=0}^{m} A^{m-i,n}\begin{bmatrix} 0 \\ s(i,0) \end{bmatrix} + \sum_{(0,0)\le i,j<m,n}\sum [A^{m-i-1,n-j}B^{1,0} + A^{m-i,n-j-1}B^{0,1}]u(i,j)\right\} + Du(m,n) \tag{7.59b}$$

which relates the state vector $x(m, n)$ to the input $u(i, j)$ and to the set of boundary conditions $r(0, j)$ along the top and $s(i, 0)$ along the left side.

In order to evaluate the time response of the system, it is seen that the state transition matrix $A^{m,n}$ must be evaluated. This may be done in a similar manner to that for one dimension, using the corresponding 2-D Cayley–Hamilton theorem. However, this is computationally inefficient, since it requires the solution of impractically large numbers of linear simultaneous equations; in two dimensions it is preferable to use the recursion formula (7.58).

In the case of systems whose transfer function has a separable denominator, a simple method is available for the determination of the state transition matrix. The necessary and sufficient condition for separability of the denominator is that either $A_2 = 0$ or $A_3 = 0$. In the former case

$$A^{1,0}A^{0,1} = 0 \tag{7.60a}$$

which results in

$$A^{i,j} = (A^{0,1})^j (A^{1,0})^i \tag{7.60b}$$

In the latter case

$$A^{0,1} A^{1,0} = 0 \tag{7.61a}$$

and we have

$$A^{i,j} = (A^{1,0})^i (A^{0,1})^j \tag{7.61b}$$

In order to determine each of these factors separately, we need only solve two simultaneous systems of order M and N.

7.8.3. Z-Domain Solution

An alternative approach to the solution may be obtained, analogously to that of the 1-D problem, by taking the 2-D Z transform of the state equation (7.47) along the two spatial dimensions as in system (7.50) with the exception that in this case we assume nonzero global initial conditions. Before we start the development consider the following results.

FACT 7.1. Having defined the state transition matrix $A^{i,j}$ in equations (7.58), it can be shown that the following relations hold:

$$\begin{aligned}(z_1, z_2)[A^{m,n}] &= \sum_{m=0}^{\infty} \sum_{n=0}^{\infty} A^{m,n} z_1^{-m} z_2^{-n} \\ &= \sum_{k=0}^{\infty} (z_1^{-1} A^{1,0} + z_2^{-1} A^{0,1})^k \end{aligned} \tag{7.62a}$$

This can also be related to

$$\sum_{k=0}^{\infty} (z_1^{-1} A^{1,0} + z_2^{-1} A^{0,1})^k = [(z_1 I_{N_1} \oplus z_2 I_{N_2}) - A]^{-1} (z_1 I_{N_1} \oplus z_2 I_{N_2}) \tag{7.62b}$$

The (z_1, z_2) transform of system (7.47) with a nonzero global initial condition gives

$$\begin{aligned} X(z_1, z_2) = {} & [(z_1 I_{N_1} \oplus z_2 I_{N_2}) - A]^{-1} \\ & \times \left\{ (z_1 I_{N_1} \oplus z_2 I_{N_2}) \begin{bmatrix} R(0, z_2) \\ S(z_1, 0) \end{bmatrix} + \begin{bmatrix} B_1 \\ B_2 \end{bmatrix} U(z_1, z_2) \right\} \end{aligned} \tag{7.63}$$

Now by considering the above result and using the convolution and shifting properties, after taking the inverse transformation the first term in equation (7.63) yields the zero-input response and the second term leads to the zero-state response. The output expression can be obtained likewise.

REMARK 7.2. The impulse response of the 2-D system can be obtained in a similar manner to that in the 1-D case by replacing $u(m, n) = \delta(m, n)$ in equation (7.59b) and considering zero global initial conditions. This gives

$$h(0,0) = D$$

$$h(m, n) = C[A^{m-1,n}B^{1,0} + A^{m,n-1}B^{0,1}], \qquad (m, n) > (0, 0) \tag{7.64}$$

Using the above result the alternative output response expression in terms of $h(m, n)$ is given by

$$y(m, n) = C\left\{\sum_{j=0}^{n} A^{m,n-j}\begin{bmatrix} r(0,j) \\ 0 \end{bmatrix} + \sum_{i=0}^{m} A^{m-i,n}\begin{bmatrix} 0 \\ s(i,0) \end{bmatrix}\right\}$$
$$+ \sum_{i=0}^{m}\sum_{j=0}^{n} h(m-i, n-j)u(i,j) \tag{7.65}$$

EXAMPLE 7.2. We consider a separable 2-D system given by the following state-space model:

$$r(m+1, n) = r(m, n) + u(m, n)$$

$$s(m, n+1) = r(m, n) + s(m, n)$$

$$y(m, n) = s(m, n)$$

Assuming that the global initial conditions are zero, find the impulse response and the expression for the output response.

SOLUTION. The matrices $A^{1,0}$, $A^{0,1}$, $B^{1,0}$, $B^{0,1}$, and C are given by

$$A^{1,0} = \begin{bmatrix} 1 & 0 \\ 0 & 0 \end{bmatrix}, \qquad A^{0,1} = \begin{bmatrix} 0 & 0 \\ 1 & 1 \end{bmatrix}, \qquad B^{1,0} = \begin{bmatrix} 1 \\ 0 \end{bmatrix}, \qquad B^{0,1} = \begin{bmatrix} 0 \\ 0 \end{bmatrix}$$

and $C = [0 \;\; 1]$. Since the system is separable and $A_2 = 0$, equation (7.60b) yields

$$A^{i,j} = (A^{0,1})^j(A^{1,0})^i = \begin{bmatrix} 0 & 0 \\ 1 & 1 \end{bmatrix}\begin{bmatrix} 1 & 0 \\ 0 & 0 \end{bmatrix} = \begin{bmatrix} 0 & 0 \\ 1 & 0 \end{bmatrix}$$

$$A^{0,j} = \begin{bmatrix} 0 & 0 \\ 1 & 1 \end{bmatrix} \quad \text{and} \quad A^{i,0} = \begin{bmatrix} 1 & 0 \\ 0 & 0 \end{bmatrix}$$

Thus we have

$$h(0, n) = CA^{-1,n}B^{1,0} = 0 \qquad \forall n \geq 0$$

$$h(m, 0) = CA^{m-1,0}B^{1,0} = [0 \ \ 1]\begin{bmatrix} 1 & 0 \\ 0 & 0 \end{bmatrix}\begin{bmatrix} 1 \\ 0 \end{bmatrix} = 0 \qquad \forall m \geq 0$$

$$h(m, n) = CA^{0,n}A^{m-1,0}B^{1,0} = [0 \ \ 1]\begin{bmatrix} 0 & 0 \\ 1 & 0 \end{bmatrix}\begin{bmatrix} 1 \\ 0 \end{bmatrix} = 1 \qquad \forall (m, n) \geq (1,1)$$

or

$$h(m, n) = u_1(m - 1, n - 1)$$

where $u_1(m, n)$ represents the 2-D unit step function.

The expression for the output is

$$y(m, n) \sum_{i=1}^{m} \sum_{j=1}^{n} u(i, j)$$

It is clear that this system is BIBO unstable. This can also be seen from the transfer function of the system which is

$$H(z_1, z_2) = \frac{1}{(z_1 - 1)(z_2 - 1)}$$

7.8.4. The Characteristic Polynomial and the Cayley–Hamilton Theorem for Two-Dimensional Systems[6]

As in the 1-D case, the denominator polynomial of the transfer function (7.46) or (7.52), namely

$$\begin{aligned} B(z_1, z_2) &= \text{Det}[I - z_1^{-1}A^{1,0} - z_2^{-1}A^{0,1}] \\ &= \sum_{i=0}^{M} \sum_{j=0}^{N} b_{ij} z_1^{-i} z_2^{-j} \end{aligned} \tag{7.66}$$

is called the characteristic polynomial of the 2-D system. One major difference between 1-D and 2-D characteristic polynomials is that the roots of the 1-D characteristic equation are isolated, while in the 2-D case they form a closed surface or manifold in four-dimensional space. Thus, it is not generally possible to determine the behavior of the 2-D system by examining the roots of the characteristic equation, as there are an infinite number of these.

We consider now the Cayley-Hamilton theorem in two dimensions.[6] Every partitioned matrix

$$A = \begin{bmatrix} A_1 & A_2 \\ A_3 & A_4 \end{bmatrix}$$

satisfies its own characteristic equation, i.e.,

$$\sum_{i=0}^{M} \sum_{j=0}^{N} b_{ij} A^{M-i,N-j} = 0 \tag{7.67a}$$

or

$$A^{M,N} = -\sum_{\substack{i=0 \\ i+j\neq 0}}^{M} \sum_{j=0}^{N} b_{ij} A^{M-i,N-j} \tag{7.67b}$$

The latter expression can be used to generate an $A^{M,N}$ matrix recursively.

7.8.5. Stability of Two-Dimensional Systems[8,9]

It has been shown in Chapter 3 that the sufficient condition for stability of a 2-D system is that the poles of the system should all lie outside the unit bidisc in the (z_1^{-1}, z_2^{-1}) plane (Shank's criterion). This test is impractical from the implementation point of view, because it requires infinite mapping. More useful criteria are those of Huang and DeCarlo-Strintzis (see Section 3.5), which are used here. Let us rewrite the transfer function (7.52) in the following form:

$$\begin{aligned} H(z_1, z_2) &= C \begin{bmatrix} z_1 I_{N_1} - A_1 & -A_2 \\ -A_3 & z_2 I_{N_2} - A_4 \end{bmatrix}^{-1} B, \qquad D = 0 \\ &= \frac{C \operatorname{Adj} \begin{bmatrix} z_1 I_{N_1} - A_1 & -A_2 \\ -A_3 & z_2 I_{N_2} - A_4 \end{bmatrix} B}{\operatorname{Det} \begin{bmatrix} z_1 I_{N_1} - A_1 & -A_2 \\ -A_3 & z_2 I_{N_2} - A_4 \end{bmatrix}} \end{aligned} \tag{7.68}$$

where Adj$[A]$ represents the adjoint of matrix A. The characteristic polynomial of the 2-D system in (z_1^{-1}, z_2^{-1}) is given by

$$B(z_1, z_2) = \operatorname{Det} \begin{bmatrix} I_{N_1} - z_1^{-1} A_1 & -z_1^{-1} A_2 \\ -z_2^{-1} A_3 & I_{N_2} - z_2^{-1} A_4 \end{bmatrix} \tag{7.69}$$

Using the property of the determinant of partitioned matrices, namely,

$$\operatorname{Det}\begin{bmatrix} A & B \\ C & D \end{bmatrix} = \operatorname{Det}[A]\operatorname{Det}[D - C\ A^{-1}\ B], \qquad \operatorname{Det}[A] \neq 0$$
$$= \operatorname{Det}[D]\operatorname{Det}[A - B\ D^{-1}\ C], \qquad \operatorname{Det}[D] \neq 0 \qquad (7.70)$$

we obtain

$$B(z_1, z_2) = \operatorname{Det}[I_{N_1} - z_1^{-1}A_1]\operatorname{Det}\{I_{N_2} - z_2^{-1}[A_4 + A_3(z_1 I_{N_1} - A_1)^{-1}A_2]\}$$
$$= \operatorname{Det}[I_{N_2} - z_2^{-1}A_4]\operatorname{Det}\{I_{N_1} - z_1^{-1}[A_1 + A_2(z_2 I_{N_2} - A_4)^{-1}A_3]\} \qquad (7.71)$$

We assume that the inverse of the relevant matrices exist.

Definition 7.3. A square matrix is said to be stable if all of its eigenvalues lie in the interior of the unit circle in the z plane.

Now, the Huang and DeCarlo–Strintzis criteria lead to the following results:

Theorem 7.4.[8] *The following statements are equivalent (sufficient condition):*

1. *System* (7.47) *is BIBO stable*
2. (i) A_1 *is stable,*
   (ii) $A_4 + A_3(z_1 I_{N_1} - A_1)^{-1}A_2$ *with* $|z_1| = 1$ *is stable.*
3. (i) A_4 *is stable,*
   (ii) $A_1 + A_2(z_2 I_{N_2} - A_4)^{-1}A_3$ *with* $|z_2| = 1$ *is stable.*
4. (i) A *is stable,*
   (ii) A_1 *has no eigenvalues on the unit circle,*
   (iii) $A_4 + A_3(z_1 I_{N_1} - A_1)^{-1}A_2$ *with* $|z_1| = 1$ *has no eigenvalues on the unit circle.*
5. (i) A *is stable,*
   (ii) A_4 *has no eigenvalues on the unit circle,*
   (iii) $A_1 + A_2(z_2 I_{N_2} - A_4)^{-1}A_3$ *with* $|z_2| = 1$ *has no eigenvalues on the unit circle.*

Corollary 7.1.[8] *The following three conditions are necessary for BIBO stability of system* (7.47): (1) A *is stable*; (2) A_1 *is stable*; (3) A_4 *is stable.*

Corollary 7.2. *For a separable two-dimensional system when* $A_2 = 0$ *or* $A_3 = 0$, *the system is BIBO stable if and only if* A_1 *and* A_4 *are stable.*

COROLLARY 7.3.[8] *For the case when* $N_1 = N_2 = 1$ *and* $a_2a_3 \neq 0$, *system* (7.47) *is BIBO stable if and only if*

(i) $|a_1| < 1$,
(ii) $\max\{|a_4 + a_2a_3/(1 - a_1)|, |a_4 - a_2a_3/(1 + a_1)|\} < 1$.

EXAMPLE 7.3. We consider two systems with state matrices

$$A = \begin{bmatrix} 0.5 & 0.25 \\ -0.25 & 1 \end{bmatrix} \quad \text{and} \quad A' = \begin{bmatrix} -0.5 & 0.75 \\ 1 & 0.5 \end{bmatrix}$$

Determine the stability of these systems.

The first system is unstable because a_4 is unstable, even though A and a_1 are stable. The second system is also unstable because A' is unstable (eigenvalues of A' are ± 1).

EXAMPLE 7.4. We consider

$$A = \begin{bmatrix} -0.4 & -0.6 \\ 0.6 & 0.8 \end{bmatrix}$$

Determine the stability of the system.

The eigenvalues of A are $\lambda_1 = \lambda_2 = 0.2$ so that A is stable. Quantities a_1 and a_4 are also stable. However, we have

$$\text{Max}\left\{\left|a_4 + \frac{a_2a_3}{1 - a_1}\right|, \left|a_4 - \frac{a_2a_3}{1 + a_1}\right|\right\} = 0.8 + \frac{0.36}{1 - 0.4}$$
$$= 1.4$$

Hence the system is unstable.

7.8.6. Controllability and Observability in Two-Dimensional Systems[7,10,11]

In 1-D systems the controllability and observability properties are very important in finding a minimal realization for the system. As a matter of fact, the minimal realization[1,2] exists if and only if the system is both controllable and observable. It is interesting to see that the notions of controllability and observability introduced by Roesser[6] are not related to the minimal realization of 2-D systems. These notions will be referred to as "local controllability and observability"[7] in this book. A state-space model can be locally controllable and observable without being minimal, and conversely, a system can be minimal without being locally controllable or locally observable.

DEFINITION 7.4.[7] The 2-D system (7.47) is said to be locally controllable if for any initial state (global condition) there exist a set of inputs $u(m, n)$, $(0,0) \leq m, n < P, Q$ that will transfer the system from the initial state to any final state $X(P, Q)$, with P, Q finite.

THEOREM 7.5.[7] *The two-dimensional system* (7.47) *is locally controllable if and only if the controllability matrix*

$$\hat{C}_{N_1,N_2} = [B^{0,1} B^{0,2} \cdots B^{0,N_2} B^{1,0} B^{1,1} \cdots B^{1,N_2} \cdots B^{N_1,0} B^{N_1,1} \cdots B^{N_1,N_2}] \tag{7.72}$$

where

$$B^{i,j} = A^{i-1,j} B^{1,0} + A^{i,j-1} B^{0,1}, \qquad i, j > (0,0)$$

is of full rank.

DEFINITION 7.5. The 2-D system (7.47) is said to be locally observable if any weak nonzero initial condition can be determined from knowledge of the output sequence $\{y(m, n)\}$, $(0,0) \leq m, n \leq (P, Q)$, with P, Q being finite. Input is assumed to be zero for simplicity.

THEOREM 7.6.[7] *The two-dimensional system* (7.47) *is locally observable if and only if the observability matrix*

$$\hat{O}_{N_2,N_1} = \begin{bmatrix} C \\ CA^{0,1} \\ \vdots \\ CA^{0,N_2} \\ CA^{1,0} \\ \vdots \\ CA^{1,N_2} \\ \vdots \\ CA^{N_{1,0}} \\ \vdots \\ CA^{N_1,N_2-1} \end{bmatrix} \tag{7.73}$$

is of full rank.

The following examples serve to illustrate the fact that the local controllability and observability are not adequate as far as minimality is concerned.

EXAMPLE 7.5.[7] We consider the state–space model

$$\begin{bmatrix} r_1(m+1,n) \\ s_1(m,n+1) \\ s_2(m,n+1) \end{bmatrix} = \begin{bmatrix} 1 & 1 & 1 \\ 1 & 1 & 1 \\ 0 & 0 & 1 \end{bmatrix} \begin{bmatrix} r_1(m,n) \\ s_1(m,n) \\ s_2(m,n) \end{bmatrix} + \begin{bmatrix} 1 \\ 0 \\ 1 \end{bmatrix} u(m,n)$$

$$y(m,n) = [1 \;\; 1 \;\; 1] \begin{bmatrix} r_1(m,n) \\ s_1(m,n) \\ s_2(m,n) \end{bmatrix}$$

The controllability matrix is

$$\hat{C}_{1,2} = [B^{0,1} \;\; B^{1,0} \;\; B^{1,1}] = \begin{bmatrix} 0 & 1 & 1 \\ 0 & 0 & 1 \\ 1 & 0 & 0 \end{bmatrix}$$

which is of rank 3, i.e., the system is locally controllable. The observability matrix is

$$\hat{O}_{2,1} = \begin{bmatrix} C \\ CA^{0,1} \\ CA^{0,2} \\ CA^{1,0} \\ CA^{1,1} \end{bmatrix} = \begin{bmatrix} 1 & 1 & 1 \\ 1 & 1 & 2 \\ 1 & 1 & 3 \\ 1 & 1 & 1 \\ 2 & 2 & 3 \end{bmatrix}$$

which is of rank 2, i.e., the system is not locally observable. Despite the fact that the system is not locally observable, the minimal realization nevertheless exists. This can be seen by forming the transfer function

$$H(z_1, z_2) = [1 \;\; 1 \;\; 1] \begin{bmatrix} z_1 - 1 & -1 & -1 \\ -1 & z_2 - 1 & -1 \\ 0 & 0 & z_2 - 1 \end{bmatrix}^{-1} \begin{bmatrix} 1 \\ 0 \\ 1 \end{bmatrix}$$

$$= \frac{z_2(z_1 + z_2 - 1)}{(z_2 - 1)(z_1 z_2 - z_1 - z_2)} = \frac{A(z_1, z_2)}{B(z_1, z_2)}$$

The realization of the above transfer function requires at least one horizontal state and two vertical states; thus the structure originally given is indeed minimal.

EXAMPLE 7.6.[7] We consider the transfer function

$$H(z_1, z_2) = \frac{z_1 - z_2}{z_1 z_2 - 1}$$

Then, using Nerode equivalence,[7] the controller realization is obtained in the form

$$\begin{bmatrix} r_1(m+1, n) \\ s_1(m, n+1) \\ s_2(m, n+1) \end{bmatrix} = \begin{bmatrix} 0 & 0 & 1 \\ 0 & 0 & 1 \\ 1 & 0 & 0 \end{bmatrix} \begin{bmatrix} r_1(m, n) \\ s_1(m, n) \\ s_2(m, n) \end{bmatrix} + \begin{bmatrix} 1 \\ 1 \\ 0 \end{bmatrix} u(m, n)$$

$$y(m, n) = [-1 \;\; 1 \;\; 0] \begin{bmatrix} r_1(m, n) \\ s_1(m, n) \\ s_2(m, n) \end{bmatrix}$$

The controllability matrix is

$$\hat{C}_{1,2} = [B^{0,1} \;\; B^{1,0} \;\; B^{1,1}] = I_3$$

i.e., an identity matrix of order 3. Thus the system is locally controllable. The observability matrix is

$$\hat{O}_{2,1} = \begin{bmatrix} C \\ CA^{0,1} \\ CA^{1,0} \end{bmatrix} = \begin{bmatrix} -1 & 1 & 0 \\ -1 & 0 & 0 \\ 0 & 0 & -1 \end{bmatrix}$$

which is also of full rank. Thus the system is both locally controllable and observable. However, it can easily be shown that

$$\begin{bmatrix} r_1(m+1, n) \\ s_1(m, n+1) \end{bmatrix} = \begin{bmatrix} 0 & -1 \\ -1 & 0 \end{bmatrix} \begin{bmatrix} r_1(m, n) \\ s_1(m, n) \end{bmatrix} + \begin{bmatrix} -1 \\ 1 \end{bmatrix} u(m, n)$$

$$y(m, n) = [1 \;\; 1] \begin{bmatrix} r_1(m, n) \\ s_1(m, n) \end{bmatrix}$$

is also a possible realization for $H(z_1, z_2)$ which is clearly minimal, since we need at least one horizontal and one vertical delay to realize $H(z_1, z_2)$.

As a result, the original controller realization is not minimal, even though it is locally controllable and observable.

Kung *et al.*[7] have shown that, by reformulating the local controllability and observability notions, one can guarantee that a minimal realization would indeed lead to a locally controllable and observable structure, but the converse will not still be valid. Let us consider the partitioned $\hat{C}_{N_1,N_2}$ and $\hat{O}_{N_2,N_1}$ matrices, i.e.,

$$\hat{C}_{N_1,N_2} = \begin{bmatrix} \hat{C}^r_{N_1,N_2} \\ \hline \hat{C}^s_{N_1,N_2} \end{bmatrix} \begin{matrix} N_1 \text{ rows} \\ N_2 \text{ rows} \end{matrix} \qquad \hat{O}_{N_2,N_1} = \begin{matrix} N_1 \text{ columns} & N_2 \text{ columns} \\ [\hat{O}^r_{N_2,N_1} & \vdots \; \hat{O}^s_{N_2,N_1}] \end{matrix} \tag{7.74}$$

where the superscripts r and s refer to the parts associated with the horizontal and vertical state vector components r and s in system (7.47).

PROPOSITION 7.1.[7] *The states r and s are separately locally controllable (observable) if and only if* $\hat{C}^r_{N_1,N_2}$ *and* $\hat{C}^s_{N_1,N_2}$ $(\hat{O}^r_{N_2,N_1}, \hat{O}^s_{N_2,N_1})$ *in matrices* (7.74) *are separately of full rank.*

As a consequence of the above proposition, if a system is minimal, then the horizontal and vertical states are separately locally controllable and observable.

EXAMPLE 7.7. Let us consider the system in Example 7.5 and form $\hat{O}^r_{2,1}$ and $\hat{O}^s_{2,1}$, i.e.,

$$\hat{O}^r_{2,1} = \begin{bmatrix} 1 \\ 1 \\ 1 \\ 1 \\ 2 \end{bmatrix} \quad \text{and} \quad \hat{O}^s_{2,1} = \begin{bmatrix} 1 & 1 \\ 1 & 2 \\ 1 & 3 \\ 1 & 1 \\ 2 & 3 \end{bmatrix}$$

It can easily be seen that these matrices are of full rank. Thus the minimal realization in Example 7.5 is separately locally controllable and observable.

Below, alternative definitions of controllability and observability are given which relate better to the notion of minimality.

DEFINITION 7.6.[7,10,11] The state-space model (7.47) with transfer function (7.52) is said to be modally controllable if

$$[(z_1 I_{N_1} \oplus z_2 I_{N_2}) - A], B \tag{7.75a}$$

are left coprime with respect to (w.r.t.) $\mathbf{R}(z_1, z_2)$, and is said to be modally observable if

$$C, [(z_1 I_{N_1} \oplus z_2 I_{N_2}) - A] \tag{7.75b}$$

are right coprime w.r.t. $\mathbf{R}(z_1, z_2)$, where $\mathbf{R}(z_1, z_2)$ denotes the set of polynomials in z_1 and z_2 with real coefficients. For definitions of coprimers, the reader is referred elsewhere.[12,13]

The following results were developed by Eising.[10]

DEFINITION 7.7.[10] Suppose $A(z_2) \in \mathbf{R}^{n\times n}(z_2)$ and $B(z_2) \in \mathbf{R}^{n\times p}(z_2)$, where $\mathbf{R}^{m\times n}(z_2)$ denotes the set of $m \times n$ matrices with entries in $\mathbf{R}(z_2)$. The pair $(A(z_2), B(z_2))$ is said to be controllable w.r.t. $\mathbf{R}(z_2)$ if

$$\operatorname{rank}[B(z_2) \;\; A(z_2)B(z_2) \;\; \cdots \;\; A^{n-1}(z_2)B(z_2)] = n \tag{7.76}$$

where the rank is considered over the field $\mathbf{R}(z_2)$.

THEOREM 7.7.[10] *The following statements are equivalent*:

1. $(A(z_2), B(z_2))$ *is a controllable pair w.r.t.* $\mathbf{R}(z_2)$.
2. $(z_1 I - A(z_2))$ *and* $B(z_2)$ *are left coprime w.r.t.* $\mathbf{R}(z_2)[z_1]$ *where* $\mathbf{R}(z_2)[z_1]$ *represents the set of polynomials in* z_1 *with coefficients in* $\mathbf{R}(z_2)$.

We now state the main result.

THEOREM 7.8.[10] *The state-space model* (7.47) *is modally controllable if and only if*:

1. *The pair* $(A_1 + A_2(z_2 I_{N_2} - A_4)^{-1}A_3,\ B_1 + A_2(z_2 I_{N_2} - A_4)^{-1}B_2)$ *is controllable w.r.t.* $\mathbf{R}(z_2)$

and

2. *The pair* $(A_4 + A_3(z_1 I_{N_1} - A_1)^{-1}A_2,\ B_2 + A_3(z_1 I_{N_1} - A_1)^{-1}B_1)$ *is controllable w.r.t.* $\mathbf{R}(z_1)$.

REMARK 7.3. By duality, all the results given in the above are also valid for modal observability. As an example, the observability counterpart of Theorem 7.8 is that given below.

THEOREM 7.9.[10] *The state-space model* (7.47) *is modally observable if and only if*

1. *The pair* $(C_1 + C_2(z_2 I_{N_2} - A_4)^{-1}A_3,\ A_1 + A_2(z_2 I_{N_2} - A_4)^{-1}A_3)$ *is observable w.r.t.* $\mathbf{R}(z_2)$

and

2. *The pair* $(C_2 + C_1(z_1 I_{N_1} - A_1)^{-1}A_2,\ A_4 + A_3(z_1 I_{N_1} - A_1)^{-1}A_2)$ *is observable w.r.t.* $\mathbf{R}(z_1)$.

THEOREM 7.10. *The state-space model* (7.47) *is minimal if and only if it is modally controllable and observable.*

It must be mentioned that the obtained minimal realization of order $M + N$ may have complex gains which are not desirable from the practical implementation point of view. Therefore, the real-gain structure with orders higher than $N_1 + N_2$ may be more appropriate. Kung *et al.*[7] provided several examples which indicate that the $(M + 2N)$-order real-gain realization in Section 7.7 may in fact be more desirable than the complex-gain minimal realization.

7.9. ROUNDOFF ERROR IN STATE-SPACE DIGITAL FILTERS[14-19]

One advantage of the state-space design of digital filters is that, by modification of the state transformation, we may vary the roundoff error performance of the system. The source of roundoff accumulation in digital systems can be attributed to

1. Input quantization,
2. Coefficient quantization,
3. Product quantization.

In this section the problem of roundoff-error minimization for 1-D and 2-D digital filters represented by state-space formulations is considered.

7.9.1. Roundoff Errors in One-Dimensional Systems[14,16]

Let us consider a system with input, state, and output $[u(n), x(n)$, and $y(n)]$ which are quantized by a finite word length machine to $\bar{u}(n)$, $\bar{x}(n)$, and $\bar{y}(n)$ such that

$$\bar{u}(n) = u(n) + \Delta u(n)$$

$$\bar{x}(n) = x(n) + \Delta x(n)$$

$$\bar{y}(n) = y(n) + \Delta y(n)$$

The multipliers of this machine are quantized such that the quantized matrices $\bar{A}$, $\bar{B}$, $\bar{C}$, $\bar{D}$, are related to the design values by $\bar{A} = A + \Delta A$, etc. The state equations may then be written as

$$\bar{x}(n+1) = [\bar{A}\bar{x}(n)]_E + [\bar{B}\bar{u}(n)]_E \tag{7.77a}$$

and

$$\bar{y}(n) = [\bar{C}\bar{x}(n)]_E + [\bar{D}\bar{u}(n)]_E \tag{7.77b}$$

where $[\cdot]_E$ represents quantization by rounding off the enclosed product to a step size of value E. An alternative form is

$$\bar{x}(n+1) = \bar{A}\bar{x}(n) + \bar{B}\bar{u}(n) + \alpha(n) + \beta(n) \tag{7.78a}$$

and

$$\bar{y}(n) = \bar{C}\bar{x}(n) + \bar{D}\bar{u}(n) + \gamma(n) + \delta(n) \tag{7.78b}$$

By subtracting the state equations (7.15) from system (7.78) and neglecting second-order quantities, we obtain

$$\begin{aligned}\Delta x(n+1) = \bar{A}\Delta x(n) + \bar{B}\Delta u(n) + \Delta A x(n)\\ + \Delta B u(n) + \alpha(n) + \beta(n)\end{aligned} \tag{7.79a}$$

and

$$\begin{aligned}\Delta y(n) = \bar{C}\Delta x(n) + \bar{D}\Delta u(n) + \Delta C x(n)\\ + \Delta D u(n) + \gamma(n) + \delta(n)\end{aligned} \tag{7.79b}$$

These two equations may be regarded as a set of equations relating the noise vector $\Delta x(n)$ and output $\Delta y(n)$ to the multiple inputs consisting of the last five elements of equation (7.79a).

If the initial state error vector is assumed to be zero and we apply equation (7.31), then equations (7.79) may be solved to give the output

$$\Delta y(n) = \Delta y_1(n) + \Delta y_2(n) + \Delta y_3(n) \tag{7.80}$$

where

$$\Delta y_1(n) = \bar{C}\sum_{k=0}^{n-1} \bar{A}^{n-k-1}\bar{B}\Delta u(k) + \bar{D}\Delta u(n) \tag{7.81a}$$

$$\begin{aligned}\Delta y_2(n) = \bar{C}\sum_{k=0}^{n-1} \bar{A}^{n-k-1}[\Delta A x(k) + \Delta B u(k)]\\ + \Delta C x(n) + \Delta D u(n)\end{aligned} \tag{7.81b}$$

and

$$\Delta y_3(n) = \bar{C}\sum_{k=0}^{n-1} \bar{A}^{n-k-1}[\alpha(k) + \beta(k)] + \gamma(n) + \delta(n) \tag{7.81c}$$

The first of these components of output error $\Delta y_1(n)$ gives the error due to the input quantization, the second component $\Delta y_2(n)$ is caused by the quantization of the coefficients of the filter, and the third $\Delta y_3(n)$ is related to the errors caused by product quantization.

From equation (7.81b), since $x(n)$ is related to $u(n)$, the noise due to coefficient quantization is proportional to $u(n)$, and hence the corresponding signal-to-noise ratio is fixed for a given network. The other two components of error are, however, independent of input, and so improvement in signal-to-noise ratio may be achieved by increasing the input level.

Again, considering the effect of scaling and rotation on the noise components, since from equations (7.21) CA^kB and D are invariant to the transformation matrix T, the noise due to input quantization is constant for all transformations while the errors caused by coefficient and product quantization are dependent on the transformation. This suggests that a search may be made for an appropriate transformation which minimizes the noise subject to some other constraints.

7.9.2. Dynamic-Range Constraint in One-Dimensional Systems

In the previous section, we considered the possibility of reducing noise by variation of the transformation matrix and hence the structure for implementation. We shall now determine the effect of structure transformation on the dynamic range of the filter. We shall then be in a position to minimize the noise without concomitantly causing overflow in the computation.

If L is the maximum number that can be represented in the computer, then

$$u(n) \leq L \qquad \text{for all } n \tag{7.82}$$

Assuming zero initial conditions, we obtain from equation (7.30)

$$\begin{aligned} x(n) &= \sum_{k=0}^{n} A^k B u(n-k) \\ &= \sum_{k=0}^{n} f(k) u(n-k) \end{aligned} \tag{7.83}$$

where $f(k) \triangleq A^k B$ is the state for impulse input.

On applying a structural transformation T and denoting by a prime the transformed variables, we have

$$f'(k) = T^{-1} f(k) = T^{-1} A^k B \tag{7.84}$$

and

$$x'(n) = \sum_{k=0}^{n} f'(k)u(n-k) \tag{7.85}$$

Now the l_p norm for the sequence $u(k)$; $k = 0, 1, \ldots, n$ is defined as

$$\|u(k)\|_p = \left\{ \sum_{k=0}^{n} |u(k)|^p \right\}^{1/p}$$

Application of Holders inequality to equation (7.85) yields

$$|x'(n)| \leq \|f'(k)\|_p \|u(k)\|_q \tag{7.86}$$

for $1/p + 1/q = 1$, $p \geq 1$.

In order that no overflow occurs in the state vector,

$$|x'(n)| \leq L[1 \;\; 1 \;\; \cdots \;\; 1]^t \tag{7.87a}$$

and this will hold for the input constrained by relation (7.82) if, and only if,

$$\|f'(k)\|_p \leq [1 \;\; 1 \;\; \cdots \;\; 1]^t \tag{7.87b}$$

In practice, $p = 2$ is usually considered sufficiently stringent and this leads to the dynamic-range constraint given by

$$\sum_{k=0}^{\infty} (T^{-1}A^k B) \cdot (T^{-1}A^k B) = [1 \;\; 1 \;\; \cdots \;\; 1]^t \tag{7.88}$$

where the dot product of two vectors represents the multiplication of corresponding elements.

In an alternative form, this may be written as

$$K' = T^{-1}K(T^{-1})^t = \sum_{k=0}^{\infty} (T^{-1}A^k B)(T^{-1}A^k B)^t = \begin{bmatrix} 1 & & & \times \\ & 1 & & \\ & & \ddots & \\ & & & \ddots \\ \times & & & 1 \end{bmatrix} \tag{7.89}$$

where the product is the normal matrix multiplication.

The off-diagonal elements in equation (7.89) are immaterial, and

$$K = \sum_{k=0}^{\infty} A^k B (A^k B)^t \tag{7.90}$$

is a symmetric, positive-definite matrix which satisfies the Lyapunov matrix equation

$$K = AKA^t + BB^t \tag{7.91}$$

7.9.3. Output Noise Power in One-Dimensional Systems

If we assume that the result of multiplication is approximated by rounding to the nearest least-significant bit, the variance of each such error is $E_0^2/12$ where E_0 is the quantization step. The expected square error in the product quantization is then

$$E[\Delta y_3^2] = \frac{E_0^2}{12} \sum_{k=0}^{\infty} CA^k Q (CA^k)^t + (\mu + \nu) E_0^2/12 \tag{7.92}$$

where Q is a diagonal matrix whose element q_i is the number of elements in the ith rows of A and B that are neither zero nor one, while μ and ν are the number of similar elements in C and D, respectively. We note that it is assumed that the product quantization error is a white noise process.

As a result of a structure transformation T, the expected square error becomes

$$E[\Delta y_3'^2] = \frac{E_0^2}{12} \sum_{k=0}^{\infty} CA^k T Q T^t (CA^k)^t + (\mu' + \nu) E_0^2/12 \tag{7.93}$$

The major element in this expression is the first term, which may be expressed by the noise power gain

$$G' = \sum_{k=0}^{\infty} CA^k T Q T^t (CA^k)^t \tag{7.94a}$$

$$= \mathrm{tr}[QW'] \tag{7.94b}$$

where $\mathrm{tr}[A]$ denotes the trace of the matrix A and $W' = T^t W T$ where

$$W = \sum_{k=0}^{\infty} (A^k)^t C^t C A^k \tag{7.95}$$

Matrix W also satisfies the Lyapunov matrix equation

$$W = A^{t}WA + C^{t}C \tag{7.96}$$

The solution to the problem of determining the appropriate transformation T to minimize the noise subject to the dynamic-range constraint has been solved by Hwang.[15,16]

7.9.4. Solution of the Minimization Problem in the One-Dimensional Case

Hwang[16] has proposed a technique for minimizing G' in equation (7.94) subject to the dynamic-range constraint of equation (7.89). He has further shown that the minimum noise gain attainable by a state-space realization subject to the dynamic-range constraint is

$$G_{\min} = \left(\sum_{i=1}^{N} \theta_i \right)^2 \Big/ N \tag{7.97}$$

where θ_i are the positive square roots of the eigenvalues of the matrix KW and N is the order of the filter. However, this minimum is achieved if, and only if, all the eigenvalues are identical.

The transformation to achieve the minimum unit noise filter is given by

$$T_{\min} = T_0 R_1 \Lambda^* R_0^{t} \tag{7.98}$$

where

$$\Lambda^* = \begin{bmatrix} \lambda_1^* & & & & 0 \\ & \lambda_2^* & & & \\ & & \ddots & & \\ & & & \ddots & \\ 0 & & & & \lambda_N^* \end{bmatrix} \tag{7.99}$$

with

$$\lambda_i^* = \left(\sum_{m=1}^{N} \theta_m / N\theta_i \right)^{1/2}$$

R_0 is an orthogonal matrix which satisfies the constraint

$$R_0(\Lambda^*)^{-2}R_0^t = \begin{bmatrix} 1 & & & & \times \\ & 1 & & & \\ & & \ddots & & \\ & & & \ddots & \\ \times & & & & 1 \end{bmatrix} \tag{7.100}$$

The orthogonal matrix R_1 satisfies

$$R_1^t T_0^t W T_0 R_1 = \begin{bmatrix} \theta_1^2 & & & & 0 \\ & \theta_2^2 & & & \\ & & \ddots & & \\ & & & \ddots & \\ 0 & & & & \theta_N^2 \end{bmatrix} \tag{7.101}$$

where T_0 is a solution of

$$T_0 T_0^t = K \tag{7.102}$$

7.9.5. Roundoff Errors in Two-Dimensional Systems[17-19]

In a manner similar to the development of equations (7.77), we may obtain from equation (7.49) for a 2-D system

$$\begin{aligned} \bar{x}(m, n) = {} & [\bar{A}^{1,0}\bar{x}(m-1, n)]_E + [\bar{A}^{0,1}\bar{x}(m, n-1)]_E \\ & + [\bar{B}^{1,0}\bar{u}(m-1, n)]_E + [\bar{B}^{0,1}\bar{u}(m, n-1)]_E \end{aligned} \tag{7.103a}$$

and

$$\bar{y}(m, n) = [\bar{C}\bar{x}(m, n)]_E + [\bar{D}\bar{u}(m, n)]_E \tag{7.103b}$$

or alternatively

$$\begin{aligned} \bar{x}(m, n) = {} & \bar{A}^{1,0}\bar{x}(m-1, n) + \bar{A}^{0,1}\bar{x}(m, n-1) + \bar{B}^{1,0}\bar{u}(m-1, n) \\ & + \bar{B}^{0,1}\bar{u}(m, n-1) + \alpha_{10}(m-1, n) + \alpha_{01}(m, n-1) \\ & + \beta_{10}(m-1, n) + \beta_{01}(m, n-1) \end{aligned} \tag{7.104a}$$

and

$$\bar{y}(m, n) = \bar{C}\bar{x}(m, n) + \bar{D}\bar{u}(m, n) + \gamma(m, n) + \delta(m, n) \tag{7.104b}$$

Assuming zero-boundary-error states, this leads to a total error in the output

$$\Delta y(m, n) = \Delta y_1(m, n) + \Delta y_2(m, n) + \Delta y_3(m, n)$$

where, from equation (7.59b),

$$\Delta y_1(m, n) = \bar{C} \sum\sum_{(0,0)\le i,j<m,n} [\bar{A}^{m-i-1,n-j}\bar{B}^{1,0} + \bar{A}^{m-i,n-j-1}\bar{B}^{0,1}]\Delta u(i, j) + \bar{D}\Delta u(m, n) \tag{7.105a}$$

$$\begin{aligned}\Delta \bar{y}_2(m, n) = \bar{C} \sum\sum_{(0,0)\le i,j<m,n} \{&[\bar{A}^{m-i-1,n-j}\Delta A^{1,0} + \bar{A}^{m-i,n-j-1}\Delta A^{0,1}]x(i, j) \\ &+ [\bar{A}^{m-i-1,n-j}\Delta B^{1,0} + \bar{A}^{m-i,n-j-1}\Delta B^{0,1}]u(i, j)\} \\ &+ \Delta Cx(m, n) + \Delta Du(m, n)\end{aligned} \tag{7.105b}$$

and

$$\begin{aligned}\Delta \bar{y}_3(m, n) = \bar{C} \sum\sum_{(0,0)\le i,j<m,n} \{&\bar{A}^{m-i-1,n-j}[\alpha_{10}(i, j) + \beta_{10}(i, j)] \\ &+ \bar{A}^{m-i,n-j-1}[\alpha_{01}(i, j) + \beta_{01}(i, j)]\} \\ &+ \gamma(m, n) + \delta(m, n)\end{aligned} \tag{7.105c}$$

7.9.6. Dynamic-Range Constraint in Two-Dimensional Systems

If zero initial conditions are assumed, then equation (7.59a) may be written in the form

$$\begin{aligned}x(m, n) &= \sum\sum_{(0,0)\le i,j<m,n} (A^{i-1,j}B^{1,0} + A^{i,j-1}B^{0,1})u(m - i, n - j) \\ &= \sum\sum_{(0,0)\le i,j<m,n} f(i, j)u(m - i, n - j)\end{aligned} \tag{7.106a}$$

where $f(i, j)$ is the state for an impulse input, i.e.,

$$f(i, j) \triangleq A^{i-1,j}B^{1,0} + A^{i,j-1}B^{0,1} \tag{7.106b}$$

Using a similar approach to that used in the 1-D case, the dynamic-range constraint for a 2-D system is given by

$$K' = T^{-1}K(T^{-1})^t = \begin{bmatrix} 1 & & & \times \\ & 1 & & \\ & & \ddots & \\ & & & \ddots \\ \times & & & 1 \end{bmatrix} \tag{7.107a}$$

where

$$K = \sum_{i=0}^{\infty} \sum_{j=0}^{\infty} f(i,j)f^t(i,j) \tag{7.107b}$$

and T is a structure transformation matrix. Methods for computing matrix K are discussed elsewhere.[19]

7.9.7. Output Noise Power in Two-Dimensional Systems

If quantization of products is carried out by rounding, the expected square error is[19]

$$\begin{aligned} E[\Delta y_3^2] &= \frac{E_0^2}{12} \sum_{i=0}^{\infty} \sum_{j=0}^{\infty} CA^{i,j}Q(CA^{i,j})^t + \frac{E_0^2}{12}(\mu + \nu) \\ &= \frac{E_0^2}{12} \operatorname{tr}[QW] + \frac{E_0^2}{12}(\mu + \nu) \end{aligned} \tag{7.108a}$$

where $Q = Q_1 \oplus Q_2$ and

$$W = \sum_{i=0}^{\infty} \sum_{j=0}^{\infty} (A^{i,j})^t C^t C A^{i,j} \tag{7.108b}$$

Quantity Q_1 is a diagonal matrix whose ith diagonal elements q_{1i} are the number of coefficients in the ith rows of $A^{1,0}$ and $B^{1,0}$ that are neither zero nor one, and Q_2 is a similar matrix with respect to $A^{0,1}$ and $B^{0,1}$.

Under a structural transformation T, the output noise power is given by

$$E[\Delta y_3'^2] = \frac{E_0^2}{12} \operatorname{tr}[QW'] + \frac{E_0^2}{12}(\mu' + \nu) \tag{7.109}$$

where $W' = T^t W T$.

Having formulated the output noise power and the dynamic-range constraint, the objective of the optimization problem is that, given a controllable, observable, and stable realization $\{A, B, C, D\}$, we must find a 2-D similarity transformation $T = T_1 \oplus T_2$ such that the transformed realization $\{A', B', C', D'\}$ minimizes the output noise power $E[\Delta y_3'^2]$ subject to constraint (7.107).

7.9.8. Solution of the Minimization Problem in the Two-Dimensional Case

The procedure for finding the transformation matrix $T_{\min}$ in the 2-D case is similar in nature to that of the 1-D case with perhaps some minor differences. The following procedure has been proposed by Lu and Antoniou.[19]

1. Compute positive-definite matrices K and W using expressions (7.107b) and (7.108b) as suggested in their paper.[19]
2. Find a matrix $P = P_1 \oplus P_2$ such that

$$\bar{K} \equiv P^{-1}K(P^{-1})^{t} = \begin{bmatrix} I_{N_1} & \Gamma \\ \Gamma^{t} & I_{N_2} \end{bmatrix} \tag{7.110}$$

   holds, in which Γ is a certain $N_1 \times N_2$ real matrix.
3. Compute matrix $\bar{W}$ using

$$\bar{W} = P^{t}WP = \begin{bmatrix} \bar{W}_{11} & \bar{W}_{12} \\ \bar{W}_{21} & \bar{W}_{22} \end{bmatrix} \tag{7.111}$$

4. Find a block-diagonal matrix $R = R_1 \oplus R_2$ such that

$$R_1 \bar{W}_{11} R_1^{t} = \text{diag}\{\theta_1, \ldots, \theta_{N_1}\} \tag{7.112a}$$

   and

$$R_2 \bar{W}_{22} R_2^{t} = \text{diag}\{\theta_{N_1+1}, \ldots, \theta_{N_1+N_2}\} \tag{7.112b}$$

5. Compute the block-diagonal matrix $\Lambda^* = \Lambda_1^* \oplus \Lambda_2^*$ where

$$\Lambda_1^* = \text{diag}\{\lambda_{11}^*, \ldots, \lambda_{1N_1}^*\} \tag{7.113a}$$

   and

$$\Lambda_2^* = \text{diag}\{\lambda_{21}^*, \ldots, \lambda_{2N_2}^*\} \tag{7.113b}$$

   while λ_{1j}^* and λ_{2j}^* are obtained using

$$\lambda_{1j}^* = \left[\frac{\dfrac{1}{N_1}\sum_{i=1}^{N_1}\theta_i}{\theta_j}\right]^{1/2}, \qquad i \le j \le N_1 \tag{7.114a}$$

and

$$\lambda_{2j}^* = \left[\frac{\frac{1}{N_2}\sum_{i=1}^{N_2}\theta_{N_1+i}}{\theta_{N_1+j}}\right]^{1/2}, \quad 1 \le j \le N_2 \tag{7.114b}$$

6. Find a block-diagonal matrix $S = S_1 \oplus S_2$ such that

$$S_1(\Lambda_1^*)^{-2}S_1^t = \begin{bmatrix} 1 & & \times \\ & \ddots & \\ & & \ddots \\ \times & & 1 \end{bmatrix}_{N_1\times N_1} \tag{7.115a}$$

and

$$S_2(\Lambda_2^*)^{-2}S_2^t = \begin{bmatrix} 1 & & \times \\ & \ddots & \\ & & \ddots \\ \times & & 1 \end{bmatrix}_{N_2\times N_2} \tag{7.115b}$$

using the algorithm given in the appendix of Hwang's paper.[16]
7. Form

$$T_{\min} = PR\Lambda^* S^t \tag{7.116}$$

7.9.9. Example of Minimization of the Noise Figure for a One-Dimensional Filter

As an illustration of the minimization procedure we shall consider a 1-D filter realized, in the first instance, by the canonical form and subsequently optimized for minimum noise figure constrained by limitations on the dynamic range. A computer program has been written to effect this. An example of the application of the 2-D procedure has been given by Lu and Antoniou.[19]

The filter considered is

$$H(z) = \frac{0.015941494(z+1)^3}{z^3 - 1.97486115z^2 + 1.55616123z - 0.45376813}$$

In canonical-form state-space realization, equations (7.9a), (7.9b), and

(7.12a) yield

$$A = \begin{bmatrix} 0 & 1.0 & 0 \\ 0 & 0 & 1.0 \\ 0.45376813 & -1.55616123 & 1.97486115 \end{bmatrix}$$

$$B = [0 \;\; 0 \;\; 1.0]^t \qquad \text{and} \qquad C = [0.0231752 \;\; 0.0330169 \;\; 0.0893067]$$

In this representation the noise gain is 0.794853 and the dynamic range is 17.061988. If the dynamic range is restricted to unity, the transformation to realize the minimum noise filter is

$$T = \begin{bmatrix} 0.6419587 & -4.5485092 & 0.9652688 \\ 4.0332241 & -4.9278728 & 3.3482255 \\ 5.9732865 & -3.9924622 & 5.0902377 \end{bmatrix}$$

and the minimum unit noise gain is 2.98569. Although this is greater than that of the original filter, it must be appreciated that the dynamic range has been constrained. The resulting state-space matrices for this filter are

$$A' = \begin{bmatrix} 0.9232052 & 0.5879984 & 0.4782213 \\ -0.8926182 & 1.1096006 & -0.6809177 \\ -0.6418073 & -0.2676111 & -0.0579444 \end{bmatrix}$$

$$B' = [-1.349572 \;\; 0.224706 \;\; 1.956393]^t \qquad \text{and}$$

$$C' = [0.681497 \;\; -0.624670 \;\; 0.587511]$$

It should also be noted that this filter involves a total of 15 multipliers compared with the six of the prototype.

REFERENCES

1. P. K. Sinha, *Multivariable Control—An Introduction*, Dekker, New York (1984).
2. B. C. Kuo, *Digital Control Systems*, Holt, Reinhart and Winston, New York (1980).
3. P. K. Faddeev and V. N. Faddeeva, *Computational Methods of Linear Algebra*, Freeman, San Francisco (1963).
4. E. Fornasini and G. Marchesini, State-space realisation theory of two-dimensional filters, *IEEE Trans. Autom. Control* **AC-21**, 484–492 (1976).
5. D. D. Givone and R. P. Roesser, Multidimensional linear iterative circuits—General properties, *IEEE Trans. Comput.* **C-21**, 1067–1073 (1972).
6. R. P. Roesser, A discrete state-space model for linear image processing, *IEEE Trans. Autom. Control* **AC-20**, 1–10 (1975).

7. S-Y. Kung, B. C. Levy, M. Morf, and T. Kailath, New results in 2-D systems theory, Part II: 2-D state-space models—Realization and the notions of controllability, observability and minimality, *Proc. IEEE* **65**, 945-961 (1977).
8. W. S. Lu and E. B. Lee, Stability analysis for two-dimensional systems, *IEEE Trans. Circuits Syst.* **CAS-30,** 455-461 (1983).
9. W. S. Lu and E. B. Lee, Stability analysis for two-dimensional systems via a Lyapunov approach, *IEEE Trans. Circuits Syst.* **CAS-32,** 61-68 (1985).
10. R. Eising, Controllability and observability of 2-D systems, *IEEE Trans. Autom. Control* **AC-24,** 132-133 (1979).
11. C. Ciftcibasi and O. Yuksel, Sufficient or necessary conditions for modal controllability and observability of Roesser's 2-D system model, *IEEE Trans. Autom. Control* **AC-28,** 527-529 (1983).
12. M. Morf, B. C. Levy, and S. Y. Kung, New results in 2-D systems theory, Part I: 2-D polynomial matrices, factorization and coprimeness, *Proc. IEEE* **65,** 861-872 (1977).
13. N. K. Bose, *Applied Multidimensional System Theory*, Van Nostrand Reinhold, New York (1982).
14. S. Y. Hwang, Roundoff noise in state-space digital filtering: a general analysis, *IEEE Trans. Acoust., Speech, Signal Process.* **ASSP-24,** 256-262 (1976).
15. S. Y. Hwang, Dynamic range constraint in state-space digital filtering, *IEEE Trans. Acoust., Speech, Signal Process.* **ASSP-23,** 591-593 (1975).
16. S. Y. Hwang, Minimum uncorrelated unit noise in state-space digital filtering, *IEEE Trans. Acoust., Speech, Signal Process.* **ASSP-25,** 273-281 (1977).
17. M. R. Azimi-Sadjadi, *The New Approach towards Minimum Roundoff Noise in 2-D State-Space Digital Filtering*, MSc Thesis, Imperial College (1978).
18. T. Lin, M. Kawamata, and T. Higuchi, A unified study on the roundoff noise in 2-D state-space digital filters, *IEEE Trans. Circuits Syst.* **CAS-33,** 724-730 (1986).
19. W. Lu and A. Antoniou, Synthesis of 2-D state-space fixed point digital filter structures with minimum roundoff noise, *IEEE Trans. Circuits Syst.* **CAS-33,** 965-973 (1986).

8

Block Implementation of Digital Filters

A crucial problem occurs when attempting to filter an image of large size, caused by computational limitations in processing time and storage capacity. In particular, it is possible to obtain computational gain when operating on a large image by an FIR filter whose impulse response size is much smaller than the region of support of the input image. (The region of support is the set of values on which the image is defined.) To achieve this, a technique known as sectioning may be used, whereby the image is broken up into overlapping sections of size of the same order as the impulse response of the filter, and the filtering operation performed on the sections in turn. Breaking the image into blocks in this manner facilitates the implementation of the algorithm on a parallel processor. There are two sectioning procedures, known as the "select-save" and "overlap-add" methods,[1-3] that may be used to eliminate any wrap-around error caused by the circular convolution inherent in the use of FFT algorithms.

IIR filters, on the other hand, cannot directly be implemented using FFT algorithms since the impulse response sequence is of infinite extent. However, if the input and output sequences are arranged in nonoverlapping blocks of size greater than or equal to the order of the filter, the difference equation describing the system becomes a block recursive equation.[4-10] The matrices involved in this latter equation are in lower- or upper-triangular Toeplitz (or convolutional) forms. This permits the use of FFT or other fast transform algorithms for performing the matrix/vector operations and thus significantly increases the efficiency of the recursive filtering operation. This scheme will be referred to as "block processing."

In this chapter sectioning and block processing methods for implementation of one-dimensional (1-D) and two-dimensional (2-D) FIR and IIR filters are considered.

8.1. IMPLEMENTATION OF FINITE-IMPULSE-RESPONSE DIGITAL FILTERS

The output response $y(n)$ of a causal FIR filter having an impulse response $h(n)$ to an input $u(n)$ is given by the linear convolution

$$y(n) = \sum_{k=0}^{n} h(n-k)u(k) \tag{8.1}$$

In the 2-D case, where the impulse response $h(m, n)$ of the system has a quarter-plane region of support, the linear convolution becomes

$$y(m, n) = \sum_{i=0}^{m} \sum_{j=0}^{n} h(m-i, n-j)u(i, j) \tag{8.2}$$

If both input and impulse response sequences are of finite extent, then it is possible to utilize the 2-D FFT to evaluate their linear convolution and to obtain the output sequence. However, due to the periodicity property of the kernel function of the DFT, the IDFT of the product of the DFTs of $\{u(m, n)\}$ and $\{h(m, n)\}$ will not generate the desired $\{y(m, n)\}$. Instead, the circular convolution provides an aliased version of the desired output sequence. It has been shown[1] that the circular convolution of two finite-extent sequences $\{u(m, n)\}$ and $\{h(m, n)\}$, both of size $K \times L$, generates another sequence $\{\bar{y}(m, n)\}$ of the same size, which is given by

$$\begin{aligned} \bar{y}(m, n) = & \sum_{i=0}^{m} \sum_{j=0}^{n} h(m-i, n-j)u(i, j) \\ & + \sum_{i=m+1}^{K-1} \sum_{j=n+1}^{L-1} h(m+K-i, n+L-j)u(i, j) \\ & + \sum_{i=0}^{m} \sum_{j=n+1}^{L-1} h(m-i, n+L-j)u(i, j) \\ & + \sum_{i=m+1}^{K-1} \sum_{j=0}^{n} h(m+K-i, n-j)u(i, j) \end{aligned} \tag{8.3}$$

The first term on the right of equation (8.3) is just the desired convolution, while the second, third, and fourth terms represent the effect of "wrap-around" error caused by the circular nature of the convolution. It is possible to eliminate this error by zero-padding both sequences to make them of the same size as the result of their linear convolution, namely $(2K-1) \times (2L-1)$, and then performing the circular convolution using 2-D DFT. The following example in one dimension serves to illustrate the differences between linear and circular convolution.

EXAMPLE 8.1. The convolution of two sequences $u(n) = \{1, 2\}$ and $h(n) = \{-1, 1\}$, both being zero outside the interval, is desired to be determined.

The direct use of equation (8.1) yields the sought result $y(n) = \{-1, -1, 2\}$. Note that the size of this convolution is $M + N - 1 = 3$, where $M = N = 2$ are the size of the input and impulse response sequences.

However, if we use the DFT method we first compute the DFT of the two sequences giving

$$U(k) = \{3, -1\} \qquad \text{and} \qquad H(k) = \{0, -2\}$$

The product of these two sequences is

$$\begin{aligned} \bar{Y}(k) &= H(k)U(k) \\ &= \{0, 2\} \end{aligned}$$

and taking the IDFT we obtain

$$\bar{y}(n) = \{1, -1\}$$

The discrepancies may be easily seen; by wrapping the sequence $y(n)$ around on itself and adding the overlapping first and last terms, we obtain $\bar{y}(n)$.

To overcome this difficulty the following method is used. Choose the smallest P to satisfy

$$\text{Min}[P = 2^k] \geq M + N - 1 = 3$$

which gives $P = 4$. Then zero-pad both sequences to this size. Thus

$$u'(n) = \{1, 2, 0, 0\} \qquad \text{and} \qquad h'(n) = \{-1, 1, 0, 0\}$$

The DFTs of these sequences are

$$U'(k) = \{3, 1 - 2j, -1, 1 + 2j\} \qquad \text{and} \qquad H'(k) = \{0, -1 - j, -2, -1 + j\}$$

and, multiplying these as before, we obtain

$$Y'(k) = \{0, -3 + j, 2, -3 - j\}$$

while taking the IDFT gives

$$y'(n) = \{-1, -1, 2, X\}$$

where X indicates that the element may be discarded. The above result thus gives the desired linear convolution output.

Although the foregoing procedure is useful in eliminating the wrap-around error in the circular convolution, in most practical situations the image to be filtered is considerably larger in size (such as 512×512) than the support of the filter impulse response (e.g., 20×20). As a result the evaluation of the output using the zero-padding method will be computationally laborious. The sectioning schemes may be used to circumvent this problem.

8.1.1. Select-Save Method[1,2]

When filtering large arrays of data by means of an FIR filter whose impulse response is short compared with the region of support of the data, considerable economy can be achieved by sectioning the data and performing the convolution between data and filter response by means of the FFT. This necessitates the use of some technique to eliminate the wrap-around error which would thereby be introduced.

In this approach, the size of the section $K \times L$ is chosen greater than the size $M \times N$ of the support of the impulse response. The impulse response is then made up to a $K \times L$ array by padding it with zeros. The FFT is taken of both the section of the input array and the augmented impulse response and, finally, the inverse FFT of their product. The resulting array will contain wrap-around error as shown by equation (8.3). However, the error terms in this equation will be zero for all those terms corresponding to the zeros in the augmented impulse response, and these will correspond to the correct linear convolution, the other terms being discarded. This may be outlined in the following computer algorithm for an image of size $P \times Q$.

1. Choose the sections $K \times L$ so that $K > M$ and $L > N$.
2. Extend the impulse response sequence to size $K \times L$ by adding rows and columns of zeros. Compute the DFT of the extended impulse response.
3. Read in the first complete row of input sections $K \times Q$. Form the first input section $u(m, n)$ from the first L columns.
4. Compute the DFT of $u(m, n)$. Form the product of these two DFTs and then compute its IDFT.
5. Discard the first subsection defined by $m = 0$ to $M - 2$, and $n = 0$ to $N - 2$. The remainder represents a valid part of the output $y(m, n)$.

6. Form a new section $u(m, n)$ whose first $N-1$ columns are the same as the last $N-1$ columns of the previous section and repeat steps (4) and (5). Note that the last section may need to be padded by zeros to fill up to size $K \times L$.
7. Repeat along the next row of sections, the first $M-1$ rows being identical with the last $M-1$ rows of the preceding row of sections.
8. Continue in this way until the whole image has been processed.

It should be noted that there will be no entries in the output image corresponding to a band, $M-1$ pixels wide along the top of the picture and $N-1$ pixels wide along the left side. In order to fill these, it is necessary to augment the picture with $M-1$ rows at the top and $N-1$ rows to the left. Hunt[1] suggested that a constant value should be used, but better results are obtained if the global mean of the image is used.

8.1.2. Overlap-Add Method[1,2]

This method requires the overlapping of sections and adding the results. However, the rejection of the wrap-around error is here achieved by supplementing each section of the image by columns of zeros along the right-hand boundary and rows of zeros along the bottom equal to the number of columns and rows in the impulse response.

The relevant steps of the algorithm are:

1. Identical to the select-save algorithm.
2. Identical to the select-save algorithm.
3. Read a complete row of input sections of size $K-M+1$. Form the first section from the first $L-N+1$ columns of this and augment it by rows and columns of zeros up to a $K \times L$ array.
4. Proceed as in (4) above.
5. This forms the first partial section of the output convolutional product.
6. The next input section is taken starting from column $L-N+2$ to include the next $L-N+1$ columns similarly padded with zeros.
7. This is processed as above and the results of the previous section supplemented with these values. Where the two augmented sections overlap, the results are added to give the true output sequence. This is continued to the end of the first row of sections.
8. The technique is repeated in subsequent rows to form the complete output sequence.

8.1.3. Minimization of Computation Time[1,3]

In order to determine the optimum size of the partitioned blocks of the input data, one may minimize the time for computation of the convolution. Since computation time is very dependent on the type of computer and operating system, we shall restrict this analysis to optimizing the number of multiplications involved. This completely ignores the not inconsiderable time occupied in input/output requests, but as this is very dependent on computer architecture and the environment, it cannot be generalized.

If we use either the select-save or overlap-add method, the size of each acceptable part of the blocks processed in each iteration of the sectioning algorithm is of size $(K - M + 1) \times (L - N + 1)$.

The total number of blocks to be processed will be

$$N_{\text{blocks}} = \left\lceil \frac{P}{K - M + 1} \right\rceil \left\lceil \frac{Q}{L - N + 1} \right\rceil \tag{8.4}$$

where $\lceil \cdot \rceil$ indicates rounding up to the next higher integer.

The number of complex multiplications required for each block is

$$2N_{\text{FFT}} + KL \tag{8.5}$$

where N_{FFT} is the number of complex multiplications required to compute an FFT of a $K \times L$ array. This depends on the radix of the FFT algorithm used. For a radix-2 transform, as shown in equation (1.61), this number is approximately

$$N_{\text{FFT}} = \frac{KL}{2} \log_2(KL) \tag{8.6}$$

If a vector radix (2×2) transform[6] is employed, with $K = L = 2^p$, this number can be reduced to approximately

$$N_{\text{FFT}} = \frac{3K^2}{4} \log_2 K \tag{8.7}$$

Optimization with respect to block size may now be carried out numerically.

8.2. IMPLEMENTATION OF INFINITE-IMPULSE-RESPONSE DIGITAL FILTERS

The essence of the sectioning technique is that the impulse response of the filter will be smaller than the size of the sections into which the input array is divided.

For an IIR filter, this is not possible. However, if the impulse response dies away sufficiently fast so that all the significant components are contained within a region $M \times N$, we may use either the select-save or overlap-add method and introduce only small errors, which are caused by aliasing or truncation.[3] An alternative and more accurate approach would be to use block recursive processing. In the following section, the 1-D block processing method is first discussed to provide a firm ground for understanding the 2-D block processing, which is the topic of Section 8.2.3.

8.2.1. One-Dimensional Block Processing[7-9]

We consider a 1-D causal linear time-invariant recursive (IIR) digital filter described by the difference equation of order N,

$$\sum_{i=0}^{N} b_i y(n-i) = \sum_{i=0}^{M} a_i u(n-i) \tag{8.8}$$

In matrix/vector form this difference equation may be written in a manner similar to that used in Section 1.8.1. For $M = 2$ and $N = 3$, this takes the form

$$\begin{bmatrix} 1 & & & & \\ b_1 & 1 & & & \\ b_2 & b_1 & 1 & 0 & \\ b_3 & b_2 & b_1 & 1 & \\ 0 & b_3 & b_2 & b_1 & \ddots \\ 0 & 0 & & & \\ \vdots & & & & \\ \vdots & 0 & & & \end{bmatrix} \begin{bmatrix} y(0) \\ y(1) \\ y(2) \\ \vdots \\ \vdots \\ \vdots \end{bmatrix} = \begin{bmatrix} a_0 & & & & \\ a_1 & a_0 & & & \\ a_2 & a_1 & a_0 & 0 & \\ 0 & a_2 & a_1 & a_0 & \\ 0 & 0 & a_2 & a_1 & \ddots \\ \vdots & & & & \\ \vdots & & 0 & & \end{bmatrix} \begin{bmatrix} u(0) \\ u(1) \\ u(2) \\ \vdots \\ \vdots \\ \vdots \end{bmatrix} \tag{8.9}$$

(Note that $b_0 = 1$.)

The input vector may be partitioned into blocks of length L:

$$U_0 = [u(0)\;\; u(1)\;\; \cdots\;\; u(L-1)]^t \quad \text{and}$$

$$U_1 = [u(L)\;\; u(L+1)\;\; \cdots\;\; u(2L-1)]^t$$

and in general

$$U_j = [u(jL)\;\; u(jL+1)\;\; \cdots\;\; u(jL+L-1)]^t$$

Similarly

$$Y_j = [y(jL)\;\; y(jL+1)\;\; \cdots\;\; y(jL+L-1)]^t \tag{8.10}$$

The A and B matrices may be similarly partitioned into $L \times L$ submatrices. If $L \geq \text{Max}(M, N)$ then we may write A and B as block Toeplitz matrices, since there will be only two nonzero submatrices:

$$A = \begin{bmatrix} A_0 & & & 0 & \\ A_1 & A_0 & & & \\ 0 & A_1 & A_0 & & \\ & & A_1 & \ddots & \\ & 0 & & \cdots & \cdots \end{bmatrix} \tag{8.11}$$

and B similarly, where (for $L = 5$)

$$A_0 = \begin{bmatrix} a_0 & & & 0 & \\ a_1 & a_0 & & & \\ a_2 & a_1 & a_0 & & \\ 0 & a_2 & a_1 & a_0 & \\ 0 & 0 & a_2 & a_1 & a_0 \end{bmatrix} \tag{8.12a}$$

$$A_1 = \begin{bmatrix} 0 & 0 & 0 & a_2 & a_1 \\ & & & & a_2 \\ & & & & 0 \\ & & & & 0 \\ & 0 & & & 0 \end{bmatrix} \tag{8.12b}$$

and

$$B_0 = \begin{bmatrix} 1 & & & 0 & \\ b_1 & 1 & & & \\ b_2 & b_1 & 1 & & \\ b_3 & b_2 & b_1 & 1 & \\ 0 & b_3 & b_2 & b_1 & 1 \end{bmatrix} \tag{8.12c}$$

$$B_1 = \begin{bmatrix} 0 & 0 & b_3 & b_2 & b_1 \\ & & & b_3 & b_2 \\ & & & & b_3 \\ & & & & 0 \\ & 0 & & & 0 \end{bmatrix} \tag{8.12d}$$

The recursion equation can now be written in the form

$$BY = AU \tag{8.13a}$$

where A and B are defined by expression (8.11) and

$$U = [U_0 \; U_1 \; \cdots]^t \qquad \text{and} \qquad Y = [Y_0 \; Y_1 \; \cdots]^t \tag{8.13b}$$

By considering the kth row of blocks in equation (8.13a), we may write

$$B_0 Y_k + B_1 Y_{k-1} = A_0 U_k + A_1 U_{k-1} \tag{8.14}$$

which may be rearranged to give the block recursion equation

$$Y_k = -B_0^{-1} B_1 Y_{k-1} + B_0^{-1} A_0 U_k + B_0^{-1} A_1 U_{k-1} \tag{8.15}$$

This equation may be represented by the block flow diagram of Figure 8.1a or, in canonical form, by Figure 8.1b; Δ represents the block delay.

The convolution operation

$$a_n = b_n * h_n \tag{8.16a}$$

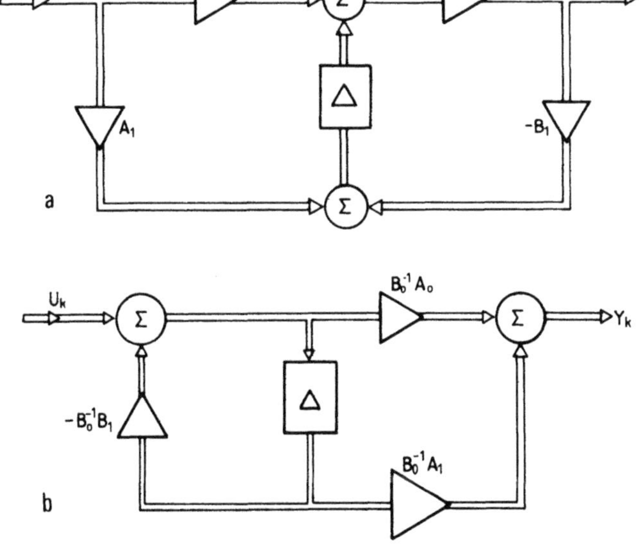

Figure 8.1. One-dimensional block recursion flow graph: (a) direct form; (b) canonic form.

can also be written in matrix/vector form to give

$$A = BH \tag{8.16b}$$

where

$$H = \begin{bmatrix} H_0 & & & 0 & \\ H_1 & H_0 & & & \\ H_2 & H_1 & H_0 & & \\ \vdots & & & \ddots & \\ \vdots & & & & \ddots \end{bmatrix} \tag{8.16c}$$

The constituent block matrices of H are defined by

$$H_i = \begin{bmatrix} h_{iL} & h_{iL-1} & \cdots & h_{(i-1)L-1} \\ h_{iL+1} & h_{iL} & & \\ \vdots & & & \vdots \\ h_{(i+1)L-1} & & \cdots & h_{iL} \end{bmatrix} \tag{8.17}$$

and $h_j = 0$ for $j < 0$.

Equations (8.11) and (8.16) yield

$$\begin{aligned} A_0 &= B_0 H_0 \\ A_1 &= B_0 H_1 + B_1 H_0 \\ 0 &= B_0 H_i + B_1 H_{i-1}, \qquad i = 2, 3, \ldots \end{aligned} \tag{8.18}$$

Therefore

$$B_0^{-1} A_0 = H_0 \qquad \text{and} \qquad \tilde{H}_1 \triangleq B_0^{-1} A_1 = H_1 + B_0^{-1} B_1 H_0 \tag{8.19}$$

and, defining

$$K \triangleq -B_0^{-1} B_1 \tag{8.20}$$

the block recursive equation (8.15) becomes

$$Y_k = K Y_{k-1} + H_0 U_k + \tilde{H}_1 U_{k-1} \tag{8.21}$$

It can be shown[9] that $B_0^{-1} = H_0'$, where H_0' is a convolution matrix having a form similar to H_0 and containing, as its elements, the impulse response samples of the associated all-pole filter. As a result, all the matrices involved in equation (8.21) are convolution operators. This suggests that

FFT algorithms may be applied profitably for performing the matrix/vector operations in this equation. Each output block in equation (8.21) is obtained recursively using the past output block and present and past input blocks.

REMARK 1. The third equation in system (8.18) can be used recursively to obtain the other constituent block matrices, namely H_i, $i = 2, 3, \ldots$, where

$$\begin{aligned} H_i &= -B_0^{-1} B_1 H_{i-1} \\ &= K H_{i-1}, \qquad i = 1, 2, \ldots \end{aligned} \tag{8.22}$$

REMARK 2. For FIR filters where no recursion is involved, $b_i = 0$ for all $i \neq 0$ and $b_0 = 1$. In this case $B_0 = \mathrm{I}$ and $B_1 = 0$, so $H_i = 0$ for $i = 2, 3, \ldots$. The block equation becomes

$$Y_k = H_0 U_k + H_1 U_{k-1} \tag{8.23}$$

Thus each output block in this equation is evaluated using only the present and past input blocks. The operations in this equation are equivalent to those in the overlap-add method.

REMARK 3. Yet another block equation can be obtained by writing

$$y_n = h_n * u_n \tag{8.24}$$

in matrix form and then partitioning the resultant matrix to yield

$$Y = HU \tag{8.25}$$

By using equations (8.13b) and (8.16c), we obtain the block convolution equation for the kth output block in the form

$$Y_k = \sum_{j=0}^{k} H_{k-j} U_j \tag{8.26}$$

Each of the matrix products on the right side of this equation is itself a convolution. This block convolution may be represented by the flow diagram of Figure 8.2. Each of the section convolution operators in equation (8.26) may be implemented either directly, or using FFT. In general, the computational efficiency is increased if the block length is large and also if the block length is considerably greater than the length of the impulse response of the FIR filter. In this latter case the only filter blocks which are nonzero are H_0 and H_1.

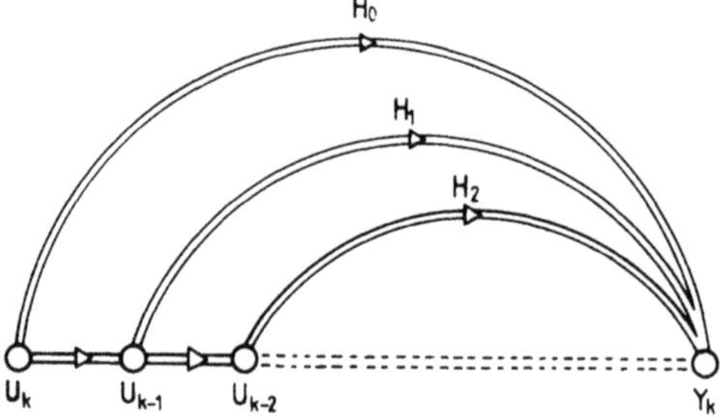

Figure 8.2. Flow graph of one-dimensional block convolution filter.

8.2.2. Computational Methods for One-Dimensional Block Recursive Implementation

The 1-D block recursive equation (8.15) may be implemented by carrying out the various convolutions and matrix multiplications involved in a variety of ways. It has been shown by Burrus[8] that in the case of 1-D implementation, the speed of computation depends on the algorithms used.

Assuming that the order of the numerator of the transfer function is equal to the denominator ($N = M$), we may compute the output from equation (8.21). The product involving H_0 may be achieved using the FFT, but those involving K and $\tilde{H}_1$, since they contain only $L \times M$ nonzero entries, may be performed by direct multiplication. This leads to the average number of real multiplications for each output point,

$$N_1 = \frac{4L \log_2 L + 6L + 2ML}{L} = 4 \log_2 L + 6 + 2M \qquad (8.27a)$$

Alternatively, if we write equation (8.15) in the form

$$Y_k = B_0^{-1}[-B_1 Y_{k-1} + A_0 U_k + A_1 U_{k-1}]$$

we note that A_1 and B_1 each have $(M + 1)M/2$ nonzero terms and A_0 has $(M + 1)(L - M/2)$ such terms, and so the computation within brackets may efficiently be carried out by direct multiplication. The product involving B_0^{-1} is an $L \times L$ convolution performed by FFT. The total number of real multiplications per output point is thus

$$N_2 = \frac{4L \log_2 L + 6L + (M + 1)(L + M/2)}{L}$$

$$= 4 \log_2 L + M + 7 + M(M + 1)/2L \qquad (8.27b)$$

Table 8.1. Number of Computations for Convolution Algorithms[a]

M	N_1	N_2	N_3	N_3 (mod)
2	14 (2)	14 (2)	30 (2)	30 (2)
8	34 (8)	31 (8)	54 (8)	53 (16)
40	110 (64)	81 (128)	90 (64)	83 (128)
100	234 (128)	152 (1024)	102 (128)	92 (512)

[a] The values of N_1 were obtained using equation (8.21), the values of N_2 were obtained using equation (8.15), and the values of N_3 were computed using FFT.

Again, all the matrix multiplications B_0^{-1}, B_1, and $[A_1 A_0]$ may be computed by FFT; the number of real multiplications per output point needed in this case is

$$N_3 = \frac{12L\log_2 L + 18L}{L} = 12\log_2 L + 18 \qquad (8.27c)$$

It should be noted that these factors related to the computational time are dependent on the choice of block size. For convenience, it is usual to choose the block size as a power of two, in order to utilize the radix-two FFT algorithms.

Table 8.1 shows how the number of computations varies with the three methods of computation considered, for different values of filter order M; they are all computed for the optimum value of block length. The optimum block length is given in parentheses. Table 8.2 shows the effect of variation of block length on number of multiplications for N_1 and N_2 for a filter of size $M = 100$; the block length is always assumed to be not less than the filter size.

It can be seen that the number of multiplications per output point is not critically dependent on the block size. For the method indicated by N_3, further reduction in number of multiplications is possible when $L > 2M$, although this necessitates an increase in block size.

Table 8.2. Effect of Varying Block Size on Number of Multiplications for a Filter of Size $M = 100$[a]

L	2048	1024	512	256	128
N_1	250	246	242	238	234
N_2	153	152	153	158	174

[a] The values of N_1 were obtained using equation (8.21) and the values of N_2 were obtained using equation (8.15).

As a generalization, it may be seen that, when compared with the direct realization in which the number of multiplications equals $2M$, the implementation using N_1 is always greater, but the others become advantageous for filter orders greater than about 40.

8.2.3. Two-Dimensional Block Processing[4,5]

We consider a 2-D causal quarter-plane IIR digital filter described by the difference equation

$$\sum_{i=0}^{M_2} \sum_{j=0}^{N_2} b_{i,j} y(m-i, n-j) = \sum_{i=0}^{M_1} \sum_{j=0}^{N_1} a_{i,j} u(m-i, n-j) \tag{8.28}$$

In two dimensions, the same procedure as previously may be followed. We may subdivide the input and output arrays into nonoverlapping blocks of size $K \times L$ where

$$K \geq \max(M_1, M_2) \qquad \text{and} \qquad L \geq \max(N_1, N_2) \tag{8.29}$$

If these blocks are now arranged as vectors with the element subscripts ordered lexicographically, equation (8.28) may be written as

$$\begin{aligned} C_{00} Y_{i,j} + C_{01} Y_{i,j-1} + C_{10} Y_{i-1,j} + C_{11} Y_{i-1,j-1} \\ = D_{00} U_{i,j} + D_{01} U_{i,j-1} + D_{10} U_{i-1,j} + D_{11} U_{i-1,j-1} \end{aligned} \tag{8.30}$$

Quantity $U_{i,j}$ represents the (i, j)th input block defined as a column vector of length KL such that

$$\begin{aligned} U(i,j) &= U_{i,j} \\ &= [U_{iK}^{(j)} \; U_{iK+1}^{(j)} \; \cdots \; U_{(i+1)K-1}^{(j)}]^{\mathrm{t}} \end{aligned} \tag{8.31a}$$

and

$$U_m^{(j)} = [u_{m,jL} \; u_{m,jL+1} \; \cdots \; u_{m,(j+1)L-1}] \tag{8.31b}$$

Similarly $Y(i, j) = Y_{i,j}$, also of length KL, is defined as

$$Y(i,j) = [Y_{iK}^{(j)} \; Y_{iK+1}^{(j)} \; \cdots \; Y_{(i+1)K-1}^{(j)}]^{\mathrm{t}} \tag{8.32a}$$

and

$$Y_m^{(j)} = [y_{m,jL} \; y_{m,jL+1} \; \cdots \; y_{m,(j+1)L-1}] \tag{8.32b}$$

The quantities C_{00} and C_{10} are lower- and upper-triangular $(KL \times KL)$ block Toeplitz matrices whose (i, j)th elements are

$$C_{00}(i, j) = \begin{cases} B_m, & m = i - j, \quad 0 \leq m \leq M_2 \\ 0, & \text{otherwise}, \quad i, j = 1, \ldots, K \end{cases} \tag{8.33a}$$

and

$$C_{10}(i, j) = \begin{cases} B_m & m = K - (j - i), \quad 1 \leq m \leq M_2 \\ 0 & \text{otherwise}, \quad i, j = 1, \ldots, K \end{cases} \tag{8.33b}$$

Quantities C_{01} and C_{11} are defined similarly in terms of matrices B'_m. The block matrices B_m and B'_m are lower- and upper-triangular $L \times L$ Toeplitz having (i, j)th elements given by

$$B_m(i, j) = \begin{cases} b_{m,n} & n = i - j, \quad 0 \leq n \leq N_2 \\ 0 & \text{otherwise}, \quad i, j = 1, \ldots, L \end{cases} \tag{8.34a}$$

and

$$B'_m(i, j) = \begin{cases} b_{m,n} & n = L - (j - i), \quad 0 \leq n \leq N_2 \\ 0 & \text{otherwise}, \quad i, j = 1, \ldots, L \end{cases} \tag{8.34b}$$

The matrices D_{pq} may be defined similarly to equation (8.33) in terms of block matrices A_m and A'_m, which themselves are defined in terms of $a_{m,n}$, as in equations (8.34).

Equation (8.30) indicates that the output block is dependent not only on the data in the input block, but also on the immediate preceding output blocks. If the dimensions of the block are less than the order of the filter, the block recursive equation will contain additional terms which depend on more remote past output blocks. However, it is desirable to keep the block size less than the order of the filter in order that the above efficiency of implementation may be achieved. If the block size is reduced to 1×1, then equation (8.30) becomes identical with the ordinary scalar finite-difference equation.

Equation (8.30) may be written explicitly in the form

$$\begin{aligned} Y_{i,j} = C_{00}^{-1}(&-C_{01} Y_{i,j-1} - C_{10} Y_{i-1,j} - C_{11} Y_{i-1,j-1} + D_{00} U_{i,j} + D_{01} U_{i,j-1} \\ &+ D_{10} U_{i-1,j} + D_{11} U_{i-1,j-1}) \end{aligned} \tag{8.35a}$$

or

$$\begin{aligned} Y_{i,j} = {} & E_{10} Y_{i-1,j} + E_{01} Y_{i,j-1} + E_{11} Y_{i-1,j-1} + F_{00} U_{i,j} + F_{10} U_{i-1,j} \\ &+ F_{01} U_{i,j-1} + F_{11} U_{i-1,j-1} \end{aligned} \tag{8.35b}$$

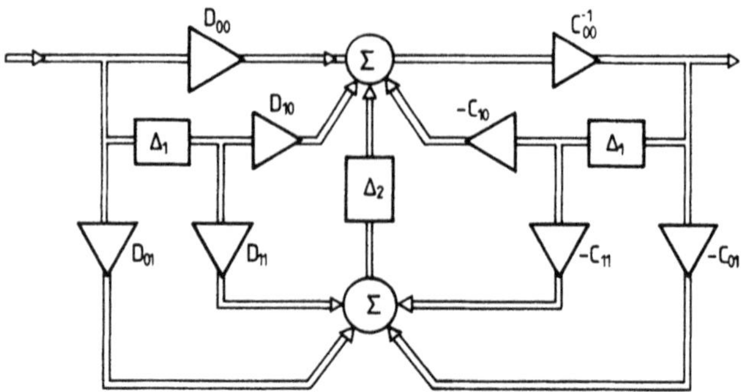

Figure 8.3. Two-dimensional block recursion flow graph.

Matrices E_{pq} and F_{pq} are defined by

$$E_{pq} = -C_{00}^{-1} C_{pq} \qquad \text{for } (p, q) = (0, 1), (1, 0), \text{ and } (1, 1) \tag{8.36a}$$

and

$$F_{pq} = C_{00}^{-1} D_{pq} \qquad \text{for } (p, q) = (0, 0), (0, 1), (1, 0), \text{ and } (1, 1) \tag{8.36b}$$

The 2-D block recursive equation (8.30) may be realized in a variety of ways. One canonical block realization is shown in Figure 8.3. In this, Δ_1 and Δ_2 represent the block delay operators associated with the first and second subscripts, respectively.

8.2.3.1. Special Block Structures(4)

8.2.3.1a. All-Pole Filter. The block recursive equation (8.35) may be simplified in a number of cases. In the all-pole filter, we have $a_{00} = 1$, $a_{ij} = 0$ for all $(i, j) \neq (0, 0)$. In this case

$$D_{00} = I_{KL} \qquad \text{and} \qquad D_{01} = D_{10} = D_{11} = 0$$

where I_n is an identity matrix of order n.

The block recursive equation (8.35) becomes

$$Y_{i,j} = E_{10} Y_{i-1,j} + E_{01} Y_{i,j-1} + E_{11} Y_{i-1,j-1} + C_{00}^{-1} U_{i,j} \tag{8.37}$$

8.2.3.1b. Separable-Product Filter. The transfer function $H(z_1, z_2)$ of a separable-product filter may be written

$$H(z_1, z_2) = H_1(z_1) H_2(z_2) \tag{8.38a}$$

where

$$H_1(z_1) = \sum_{i=0}^{M_2} \alpha_i z_1^{-i} \Big/ \sum_{i=0}^{M_1} \beta_i z_1^{-i} \tag{8.38b}$$

and

$$H_2(z_2) = \sum_{i=0}^{N_2} \alpha'_i z_2^{-i} \Big/ \sum_{i=0}^{N_1} \beta'_i z_2^{-i} \tag{8.38c}$$

In this case, the matrices C_{pq} and D_{pq} for the complete filter will also be separable product, hence

$$C_{pq} = \Lambda_p \otimes \Lambda'_q, \qquad p, q = 0, 1 \tag{8.39a}$$

and

$$D_{pq} = \Pi_p \otimes \Pi'_q, \qquad p, q = 0, 1 \tag{8.39b}$$

where Λ_p, Λ'_q, Π_p, and Π'_q are Toeplitz matrices associated with the corresponding 1-D block structures. The sign $\otimes$ represents the Kronecker product operator.

We define

$$\Psi_p = (\Lambda_p \otimes I_L), \qquad \Theta_p = (\Pi_p \otimes I_L), \qquad p, q = 0, 1 \tag{8.40a}$$

$$\Psi'_q = (I_K \otimes \Lambda'_q), \qquad \Theta'_q = (I_K \otimes \Pi'_q), \qquad p, q = 0, 1 \tag{8.40b}$$

and introduce

$$Z_i = \Psi'_0 Y_{i,j} + \Psi'_1 Y_{i,j-1} \tag{8.41a}$$

and

$$W_i = \Theta'_0 U_{i,j} + \Theta'_1 U_{i,j-1} \tag{8.41b}$$

Hence Z_{i-1} can be determined by 1-D convolution from the past and current output blocks $[Y_{i-1,j} \;\; Y_{i-1,j-1}]$ in the j direction; also W_i and W_{i-1} may be determined from the input blocks $[U_{i,j} \;\; U_{i,j-1}]$ and $[U_{i-1,j} \;\; U_{i-1,j-1}]$. One may then determine Z_i from Z_{i-1} by the 1-D block recursive equation in the i direction,

$$Z_i = \Psi_0^{-1}(-\Psi_1 Z_{i-1} + \Theta_0 W_i + \Theta_1 W_{i-1}) \tag{8.42}$$

Finally, the output may be determined by the 1-D block recursive equation in the j direction,

$$Y_{i,j} = (\Psi'_0)^{-1}(-\Psi'_1 Y_{i,j-1} + Z_i) \tag{8.43}$$

8.2.3.1c. Separable-Denominator Filter. In this case, the operations on the output blocks can be separated into two 1-D convolutions, while the input blocks will be of nonseparable form.

8.2.3.1d. Non-Recursive Filter. Here we find that $C_{00} = I_{KL}$ and $C_{01} = C_{10} = C_{11} = 0$, and the block equation is of nonrecursive form as

$$Y_{i,j} = F_{00}U_{i,j} + F_{10}U_{i-1,j} + F_{01}U_{i,j-1} + F_{11}U_{i-1,j-1} \tag{8.44}$$

The operations in this equation are analogous to those of the 2-D overlap-add sectioning method.

REMARK 8.4. The convolution operation

$$a_{m,n} = b_{m,n} * h_{m,n}$$

may be expressed in block matrix form in a similar manner to its 1-D counterpart in equations (8.16)-(8.18). In this case the following results would be obtained[10]:

$$H_{00} = C_{00}^{-1}D_{00} = F_{00} \tag{8.45a}$$

$$H_{10} = -C_{00}^{-1}C_{10}H_{00} + C_{00}^{-1}D_{10} = E_{10}H_{00} + F_{10} \tag{8.45b}$$

$$H_{01} = -C_{00}^{-1}C_{01}H_{00} + C_{00}^{-1}D_{01} = E_{01}H_{00} + F_{01} \tag{8.45c}$$

$$\begin{aligned} H_{11} &= -C_{00}^{-1}C_{10}H_{01} - C_{00}^{-1}C_{01}H_{10} - C_{00}^{-1}C_{11}H_{00} + C_{00}^{-1}D_{11} \\ &= E_{10}H_{01} + E_{01}H_{10} + E_{11}H_{00} + F_{11} \end{aligned} \tag{8.45d}$$

and

$$H_{i,j} = E_{10}H_{i-1,j} + E_{01}H_{i,j-1} + E_{11}H_{i-1,j-1} \tag{8.45e}$$

$$\text{for all } (i,j) \neq (0,0),\ (1,0),\ (0,1), \text{ and } (1,1)$$

In the latter equation, $H_{i,j}$ is a $KL \times KL$ block Toeplitz matrix whose blocks are themselves Toeplitz and which is defined by

$$H_{i,j} = \begin{bmatrix} H_{iK}^{(j)} & H_{iK-1}^{(j)} & \cdots & H_{(i-1)K+1}^{(j)} \\ H_{iK+1}^{(j)} & H_{iK}^{(j)} & \cdots & H_{(i-1)K+2}^{(j)} \\ \vdots & \vdots & & \vdots \\ H_{(i+1)K-1}^{(j)} & H_{(i+1)K-2}^{(j)} & \cdots & H_{iK}^{(j)} \end{bmatrix} \tag{8.46a}$$

where

$$H_m^{(j)} = \begin{bmatrix} h_{m,jL} & h_{m,jL-1} & \cdots & h_{m,(j-1)L+1} \\ h_{m,jL+1} & h_{m,jL} & \cdots & h_{m,(j-1)L+2} \\ \vdots & \vdots & & \vdots \\ h_{m,(j+1)L-1} & h_{m,(j+1)L-2} & \cdots & h_{m,jL} \end{bmatrix} \tag{8.46b}$$

Consequently all the matrices involved in system (8.35) are convolutional operators and the corresponding matrix/vector operations can be carried out using 2-D FFT algorithms. We note that C_{00}^{-1} is equivalent to the convolution operator of the associated all-pole filter.

REMARK 8.5. The input-output convolution operation

$$y_{m,n} = h_{m,n} * u_{m,n} \tag{8.47}$$

may also be written in block form. This leads to the 2-D block convolution sum[10]

$$Y_{i,j} = \sum_{m=0}^{i} \sum_{n=0}^{j} H_{m,n} U_{i-m,j-n} \tag{8.48}$$

For an FIR filter

$$H_{i,j} = 0 \qquad \text{for all } (i,j) \neq (0,0),\ (0,1),\ (1,0), \text{ and } (1,1) \tag{8.49}$$

Thus for these filters the block convolution sum becomes

$$Y_{i,j} = H_{00} U_{i,j} + H_{10} U_{i-1,j} + H_{01} U_{i,j-1} + H_{11} U_{i-1,j-i} \tag{8.50}$$

8.2.4. Computational Methods for Two-Dimensional Block Recursive Implementation

A study of 2-D computational techniques has been carried out by Azimi-Sadjadi.[4] He compares methods of implementation based on direct multiplication, fast Fourier transform, number theoretic transform, polynomial transform, and the short convolution algorithm, and assesses their efficiency on the basis of computational complexity.

8.2.4.1. Direct Multiplication

A consideration of equation (8.30a) shows that the terms within the brackets may be computed by direct matrix-vector multiplication, and the number of multiplications required for each term is T_i, where

$$T_1 = (M_1 + 1)(N_1 + 1)(K - M_1/2)(N_1/2)$$

$$T_2 = (M_1 + 1)(N_1 + 1)(M_1/2)(L - N_1/2)$$

$$T_3 = (M_1 + 1)(N_1 + 1)(M_1/2)(N_1/2)$$

$$T_4 = (M_2 + 1)(N_2 + 1)(K - M_2/2)(L - N_2/2)$$

$$T_5 = (M_2 + 1)(N_2 + 1)(K - M_2/2)(N_2/2)$$

$$T_6 = (M_2 + 1)(N_2 + 1)(M_2/2)(L - N_2/2)$$

$$T_7 = (M_2 + 1)(N_2 + 1)(M_2/2)(N_2/2) \tag{8.51}$$

Finally, assuming that the inversion of C_{00} has been performed in advance, the number of multiplications required for C_{00}^{-1} in equation (8.35a) is

$$T_8 = (K/2)(K + 1)(L/2)(L + 1)$$

Thus the total number of multiplications is

$$\begin{aligned}\bar{N}_{\text{DMVP}} = &(M_1 + 1)(N_1 + 1)(M_1L/2 + N_1K/2 - M_1N_1/4)\\ &+ (M_2 + 1)(N_2 + 1)KL + K(K + 1)L(L + 1)/4\end{aligned} \tag{8.52a}$$

Condition (8.24a) may be employed to obtain an upper bound on the number of operations:

$$\bar{N}_{\text{DMVP}} \leq 2KL(K + 1)(L + 1) \tag{8.52b}$$

As in the 1-D case, this may be expressed as the average number of multiplications per output pixel,

$$\bar{N}_{\text{DMVP}}/KL = N_{\text{DMVP}} \leq 2(K + 1)(L + 1) \tag{8.52c}$$

The number of multiplications required is given in Table 8.3, for various block sizes and for a filter transfer function with $(M_1, N_1) = (M_2, N_2) = (M, N)$ having several values.

8.2.4.2. Using the Fast Fourier Transform

The matrix operator C_{00}^{-1} representing (K, L) linear convolutions may be performed by appending the appropriate number of zeros and performing

Table 8.3. Number of Multiplications for Two-Dimensional Computation using Direct Multiplication for Various Block Sizes (K, L) and Filter Sizes (M, N)

	N_{DMVP}			
(K, L)	$(M, N) = (2, 2)$	$(M, N) = (3, 3)$	$(M, N) = (4, 4)$	$(M, N) = (8, 8)$
(2, 2)	18	—	—	—
(4, 4)	19	32	50	—
(8, 8)	31	42	56	162
(16, 16)	82	91	103	189
(32, 32)	282	290	300	372
(64, 64)	1065	1073	1083	1147
(128, 128)	4169	4177	4186	4246

a $(2K, 2L)$ circular convolution via FFT. Similarly, the rectangular operator $[D_{11}\ D_{01}\ D_{10}\ D_{00}]$ can be performed by $(K + M_2, L + N_2)$ circular convolutions evaluated using FFT. Finally, the operator $[C_{01}, C_{10}, C_{11}]$ can be applied by a 2-D FFT, representing a circular convolution of size $(K + M_1, L + N_1)$. Taking the upper limit of the constraint imposed by condition (8.29), all the Fourier transforms have maximum dimensions of $2K \times 2L$.

If the dimensions of the block are integral powers of two, the ordinary radix-2 FFT may be used. For other block sizes, various combined radices may be more economical.

If all these operations are carried out by FFT using an efficient algorithm for real data, the number of multiplications using radix-2 for blocks whose dimensions are integral powers of two is

$$\bar{N}_{FFT} = 20KL \log_2 KL + 40(K + L) - 68KL \tag{8.53a}$$

and the number of multiplications per output pixel using FFT is

$$N_{FFT} = \bar{N}_{FFT}/KL = 20 \log_2 KL + 40(K + L)/KL - 68 \tag{8.53b}$$

The values for this are shown in Table 8.4. It should be noted that the number of multiplications is essentially independent of the order of the filter so long as conditions (8.29) are satisfied.

Comparison with Table 8.3 shows that, for block sizes greater than (16, 16), computation using the FFT requires fewer multiplications than the direct vector multiplication for the second-order filter considered. As was stated for 1-D implementation, the FFT reduces the number of computations considerably for large block size and also permits the processing of blocks

Table 8.4. Number of Multiplications for Two-Dimensional Computations using the Fast Fourier Transform

(K, L)	N_{FFT}
2, 2	20
4, 4	32
8, 8	62
16, 16	97
32, 32	134.5
64, 64	173.25
128, 128	212.63

having large dimension. However, it suffers the disadvantage that the computations involve trigonometric functions and also complex arithmetic, even for arrays involving real convolutions; there is consequently considerably greater roundoff error in the computations. Furthermore, the FFT requires considerable storage for all the complex functions, which may prove a constraint on the use of a small computer.

8.2.4.3. Short Convolution Algorithm

Since the block Toeplitz matrices in equation (8.30) are relatively sparse, we may use the technique of Section 1.8.2.1 for computing the convolutions by partitioning the matrices into 2×2 submatrices. Since many of these will be null submatrices, a considerable saving in computation time may be achieved. For different block sizes, restricted to integral powers of two, the number of multiplications per output pixel, N_{SC}, is given in Table 8.5 for various values of the order of the filter (M, N).

8.2.4.4. Number Theoretic Transforms (NTT) and Polynomial Transforms (PT)

The use of number theoretic transforms and polynomial transforms will be discussed in Section 9.4. They are capable of providing error-free computation and, in addition, all computations are carried out in the real domain. The main drawback is that all calculations are performed in finite rings of real integers and polynomials; this imposes a limit on the maximum possible transform length constrained by the word length of the machine. The net effect of these constraints is that 2-D block recursive processing is computationally very inefficient using NTT. Polynomial transforms are considerably more efficient, as will be seen from Table 9.2 in the next

Table 8.5. Number of Multiplications for Two-Dimensional Computation using the Short Convolution Algorithm for Various Block Sizes (K, L) and Filter Sizes (M, N)

	N_{SC}		
(K, L)	$(M, N) = (2, 2)$	$(M, N) = (4, 4)$	$(M, N) = (8, 8)$
(2, 2)	19.75	not efficient	not efficient
(4, 4)	24.94	not efficient	not efficient
(8, 8)	35.11	56.36	not efficient
(16, 16)	48.1	65.77	not efficient
(32, 32)	62.55	78.22	not efficient
(64, 64)	77.78	92.38	144.17
(128, 128)	93.4	107.45	155.9

chapter. A detailed study of the applications of these transforms for 2-D block processing is given elsewhere.[4,10]

8.3. BLOCK STATE–SPACE IMPLEMENTATION

An alternative approach to block implementation is via the state-space approach of Chapter 7. This opens up the possibility of using linear transformation techniques to generate new state-space structures.

8.3.1. One-Dimensional Block State–Space Implementation[9]

Equations (7.8) and (7.11) give the state equations for a 1-D linear discrete-time system in which the input and output are scalar sequences. These equations, reproduced here, are

$$x(n+1) = Ax(n) + Bu(n) \qquad \text{and} \qquad y(n) = Cx(n) + Du(n) \tag{8.54}$$

where the state vector $x(n)$ has dimensions equal to the order of the filter function, N; A is an $N \times N$ square matrix, B an N-length column vector, C an N-length row vector, and D is a scalar.

The input and output data sequences may be partitioned into blocks of length L such that the kth input and output block vectors are

$$\tilde{U}(k) = \begin{bmatrix} u(kL) \\ u(kL+1) \\ \vdots \\ u(kL+L-1) \end{bmatrix} \qquad \text{and} \qquad \tilde{Y}(k) = \begin{bmatrix} y(kL) \\ y(kL+1) \\ \vdots \\ y(kL+L-1) \end{bmatrix} \tag{8.55}$$

Equation (8.54) may now be rewritten in terms of these block vectors as

$$\tilde{X}(k+1) = \tilde{A}\tilde{X}(k) + \tilde{B}\tilde{U}(k) \qquad \text{and} \qquad \tilde{Y}(k) = \tilde{C}\tilde{X}(k) + \tilde{D}\tilde{U}(k) \tag{8.56}$$

where $\tilde{X}(k)$ is an $N \times 1$ vector given by

$$\tilde{X}(k) = x(kL) \tag{8.57}$$

The state matrix $\tilde{A}$ possesses dimensions $N \times N$ and is defined by

$$\tilde{A} = A^L \tag{8.58a}$$

$\tilde{B}$ is an $N \times L$ matrix given by

$$\tilde{B} = [A^{L-1}B \; A^{L-2}B \; \cdots \; AB \; B] \tag{8.58b}$$

$\tilde{C}$ is an $L \times N$ matrix where

$$\tilde{C} = \begin{bmatrix} C \\ CA \\ CA^2 \\ \vdots \\ CA^{L-1} \end{bmatrix} \tag{8.58c}$$

and $\tilde{D}$ is an $L \times L$ lower-triangular Toeplitz matrix having (i, j)th element

$$\tilde{D}_{(i,j)} = \begin{cases} 0, & i < j \\ D, & i = j \\ CA^{i-j-1}B, & i > j \end{cases} \tag{8.58d}$$

This block state-space representation possesses properties similar to those of the scalar input form. It is amenable to state transformations which permit the generation of a wide range of different realizations, some of which may have desirable properties such as reduced roundoff noise.

This state-space representation may be implemented by the structure of Figure 8.4a or, in expanded form, by Figure 8.4b.

It may be observed that in equation (8.56) the state vector $X(k)$ is evaluated only once at the beginning of each block. This may be unacceptable in control applications, where it is generally assumed that the state variables must be available at every instant n. This problem may be solved by designing interpolated state observers. The block state-space structure in system (8.56) has been found to offer several advantages such as improved

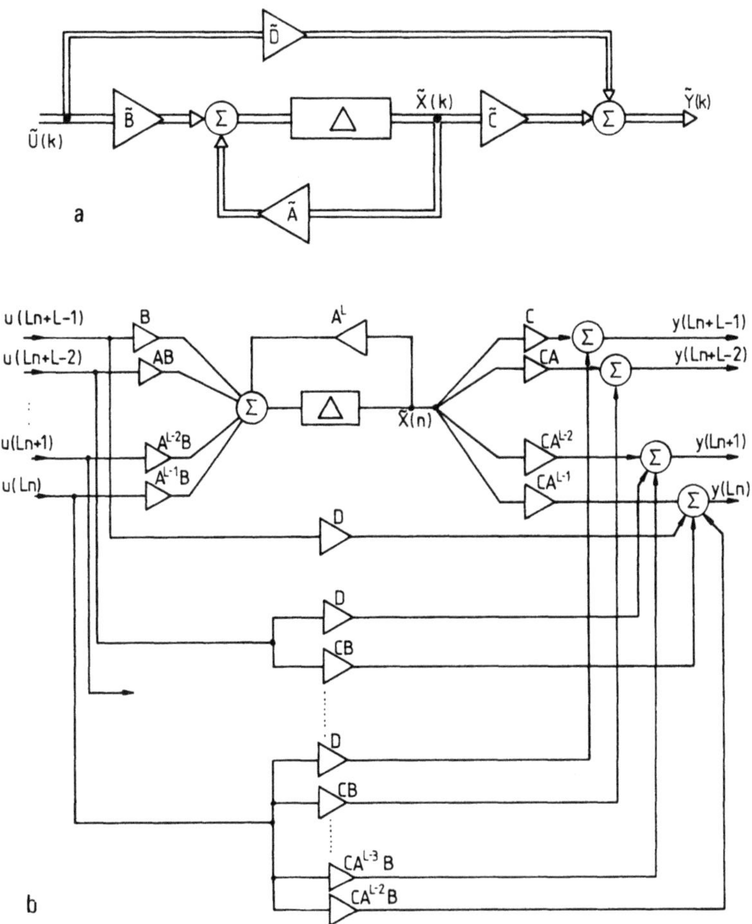

Figure 8.4. One-dimensional block state-space recursion filter: (a) block flow graph; (b) expanded flow graph.

stability, roundoff error, and sensitivity performance. Lu *et al.*[11] employed this model to design parallel and pipeline architectures using systolic arrays which operate at a block rate of $1/L$.

8.3.2. Stability, Observability, and Controllability of Block State–Space Systems

8.3.2.1. Stability

In Section 7.3.4, it was noted that the stability of a SISO system depended on the location of the poles of the transfer function. For a 1-D system, the poles are identical with the eigenvalues of the state matrix A. In the block state realization we note that, owing to the state matrix $\tilde{A} = A^L$ property, the eigenvalues of the block state matrix are the Lth power of the eigenvalues of the scalar state matrix.

The condition for stability is that the poles of the transfer function or the eigenvalues of the state matrix must lie within the unit circle on the z plane. The closer the poles are to the boundary of the unit circle, the more likely is the system to become unstable for small variations in the parameters. It is thus apparent that the poles of the block state-space realization are moved from their position for the original state-space model toward the origin and away from the boundary of the unit circle, thus implying an improvement in stability.

8.3.2.2. Observability

It has been shown in Section 7.3.5 that the state vector $x(n)$ in equation (8.54) can be determined at any instant n_0 from knowledge of the output $y(n)$ and input $u(n)$ if, and only if, the system is completely observable. The necessary and sufficient condition for this is that the rank of the observability matrix be equal to N; that is, the matrix must be of full rank.

THEOREM 8.1.[9] *The rank of the observability matrix is invariant under the transformation from scalar state-space realization to block state-space realization.*

COROLLARY 8.1. *The system given by the block state-space model of equation* (8.56) *is completely observable if, and only if, the scalar state-space counterpart is completely observable.*

8.3.2.3. Controllability

In Section 7.3.5, the complete state controllability has been defined to be an important property of a linear system which allows the transfer of any initial state $x(0)$ to any final state $x(m)$, given a set of unconstrained control signals $u(n)$, $n = 0, 1, \ldots, m$. The necessary and sufficient condition for complete state controllability is that the rank of the controllability matrix be equal to N; i.e., the matrix is of full rank.

THEOREM 8.2.[9] *The rank of the controllability matrix is invariant under the mapping from the scalar state-space realization to the block state-space realization.*

COROLLARY 8.2.[9] *The system given by the block state-space model of equation* (8.56) *is completely state controllable if, and only if, the scalar state-space counterpart is completely state controllable.*

8.3.3. Two-Dimensional Block State–Space Implementation[4]

From equations (7.49) the scalar 2-D finite-difference equation written in state-space form is

$$\begin{aligned} x(m, n) &= A^{1,0}x(m-1, n) + A^{0,1}x(m, n-1) \\ &\quad + B^{1,0}u(m-1, n) + B^{0,1}u(m, n-1) \\ y(m, n) &= Cx(m, n) + Du(m, n) \end{aligned} \tag{8.59}$$

where

$$x(m, n) = \begin{bmatrix} r(m, n) \\ s(m, n) \end{bmatrix}$$

$r(m, n)$ is an $N_1 \times 1$ vector and $s(m, n)$ an $N_2 \times 1$ vector, with $x(m, n)$ having dimensions $N \times 1$ and $N = N_1 + N_2$.

The data in $u(m, n)$ may be reordered in blocks $U(m, n)$ of size $K \times L$, such that the following block state vectors in the vertical and horizontal directions are defined:

$$\tilde{R}(m, n) = \begin{bmatrix} r'(mK, nL) \\ r'(mK, nL+1) \\ \vdots \\ r'(mK, nL+L-1) \end{bmatrix} \tag{8.60a}$$

and

$$\tilde{S}(m, n) = \begin{bmatrix} s'(mK, nL) \\ s'(mK+1, nL) \\ \vdots \\ s'(mK+K-1, nL) \end{bmatrix} \tag{8.60b}$$

where

$$r'(m, n) = \begin{bmatrix} r(m, n) \\ 0 \end{bmatrix} \tag{8.61a}$$

and

$$s'(m, n) = \begin{bmatrix} 0 \\ s(m, n) \end{bmatrix} \tag{8.61b}$$

The elements of the block state vectors represent the states associated with the boundary elements of the (m, n)th block, $\tilde{R}(m, n)$ along the vertical direction and $\tilde{S}(m, n)$ along the horizontal direction. However, $\tilde{R}(m, n)$ propagates horizontally to generate $\tilde{R}(m+1, n)$ and $\tilde{S}(m, n)$ propagates vertically to produce $\tilde{S}(m, n+1)$.

Using these two vectors we may express the elements of $\tilde{R}(m+1, n)$ and $\tilde{S}(m, n+1)$ in terms of the elements of the block state vectors $\tilde{R}(m, n)$ and $\tilde{S}(m, n)$. By successive application of the scalar equations (7.47) to each element of the block state vectors and rearrangement of terms, we may express the 2-D block state realization equations as

$$\tilde{R}(m+1, n) = \tilde{A}_1\tilde{R}(m, n) + \tilde{A}_2\tilde{S}(m, n) + \tilde{B}_1U(m, n)$$

$$\tilde{S}(m, n+1) = \tilde{A}_3\tilde{R}(m, n) + \tilde{A}_4\tilde{S}(m, n) + \tilde{B}_2U(m, n)$$

$$Y(m, n) = \tilde{C}_1\tilde{R}(m, n) + \tilde{C}_2\tilde{S}(m, n) + \tilde{D}U(m, n) \tag{8.62}$$

Vector $Y(m, n)$ represents the corresponding output block of size $K \times L$ with elements ordered appropriately.

In the above, $\tilde{A}_1$ and $\tilde{A}_4$ are lower-triangular block Toeplitz matrices of size $LN \times LN$ and $KN \times KN$ respectively; and they may be defined by their (i, j)th elements,

$$\tilde{A}_{1(i,j)} = \begin{cases} 0, & i < j \\ A^{1,0}A^{K-1,i-j}, & i \geq j, \end{cases} \qquad i, j = 1, \ldots, L \tag{8.63a}$$

and

$$\tilde{A}_{4(i,j)} = \begin{cases} 0, & i < j \\ A^{0,1}A^{i-j,L-1}, & i \geq j, \end{cases} \qquad i, j = 1, \ldots, K \tag{8.63b}$$

Similarly, $\tilde{A}_2$ and $\tilde{A}_3$ of dimension $LN \times KN$ and $KN \times LN$, respectively, are defined by their (i, j)th elements

$$\tilde{A}_{2(i,j)} = A^{1,0}A^{K-j,i-1}, \qquad i = 1, \ldots, L; j = 1, \ldots, K \tag{8.63c}$$

and

$$\tilde{A}_{3(i,j)} = A^{0,1}A^{i-1,L-j}, \qquad i = 1, \ldots, K; j = 1, \ldots, L \tag{8.63d}$$

Again, $\tilde{B}_1$ and $\tilde{B}_2$ are lower-triangular block Toeplitz matrices of size $LN \times LK$ and $KN \times KL$, respectively, whose (i, j)th elements are given by

$$\tilde{B}_{1(i,j)} = \begin{cases} 0, & i < j \\ B'_{i-j}, & i \geq j, \end{cases} \qquad i, j = 1, \ldots, L \tag{8.64a}$$

and

$$\tilde{B}_{2(i,j)} = \begin{cases} 0, & i < j \\ B''_{i-j}, & i \geq j, \end{cases} \qquad i, j = 1, \ldots, K \tag{8.64b}$$

where the block matrices B'_m and B''_m are defined by

$$B'_m = [A^{1,0}B_{K-1,m} \;\; A^{1,0}B_{K-2,m} \;\; \cdots \;\; A^{1,0}B_{0,m}]$$

$$B''_m = [A^{0,1}B_{m,L-1} \;\; A^{0,1}B_{m,L-2} \;\; \cdots \;\; A^{0,1}B_{m,0}] \tag{8.65a}$$

for $m \geq 1$ and

$$B'_0 = [B_{K,0} \;\; B_{K-1,0} \;\; \cdots \;\; B_{1,0}]$$

$$B''_0 = [B_{0,L} \;\; B_{0,L-1} \;\; \cdots \;\; B_{0,1}] \tag{8.65b}$$

In the above $B_{m,n}$ is defined by

$$B_{m,n} = A^{m-1,n}B_{1,0} + A^{m,n-1}B_{0,1} \tag{8.65c}$$

The matrix $\tilde{C}_1$, of size $KL \times LN$, is given by

$$\tilde{C}_1 = \begin{bmatrix} C_0 \\ C_1 \\ \vdots \\ C_{K-1} \end{bmatrix} \tag{8.66a}$$

where C_m are lower-triangular block Toeplitz matrices of size $L \times LN$ whose (i, j)th elements are

$$C_{m(i,j)} = \begin{cases} 0, & i < j \\ CA^{m,i-j}, & i \geq j, \end{cases} \qquad i, j = 1, \ldots, L \tag{8.66b}$$

Also, $\tilde{C}_2$ is a lower-triangular Toeplitz matrix of size $KL \times KN$, having (i, j)th element defined by

$$\tilde{C}_{2(i,j)} = \begin{cases} 0, & i < j \\ C'_{i-j}, & i \geq j, \end{cases} \qquad i, j = 1, \ldots, K \tag{8.66c}$$

where

$$C'_m = \begin{bmatrix} CA^{m,0} \\ CA^{m,1} \\ \vdots \\ CA^{m,L-1} \end{bmatrix} \tag{8.66d}$$

Finally, $\tilde{D}$ is a $KL \times KL$ lower-triangular block Toeplitz matrix whose (i, j)th element is

$$\tilde{D}_{(i,j)} = \begin{cases} 0, & i < j \\ D_{i-j}, & i \geq j, \end{cases} \qquad i, j = 1, \ldots, K \tag{8.67a}$$

and D_m is identified by its (p, q)th element

$$D_{m(p,q)} = \begin{cases} 0, & p < q \\ CB_{m,p-q} + D\delta(m, p-q), & p \geq q, \end{cases} \qquad p, q = 1, \ldots, L \tag{8.67b}$$

and $\delta(\cdot)$ denotes the Kronecker delta.

It is thus seen that equations (8.62), having a form similar to the 2-D scalar state-space equations developed by Roesser, give the desired block state-space formulation. The realization for this equation is shown in block diagram form in Figure 8.5.

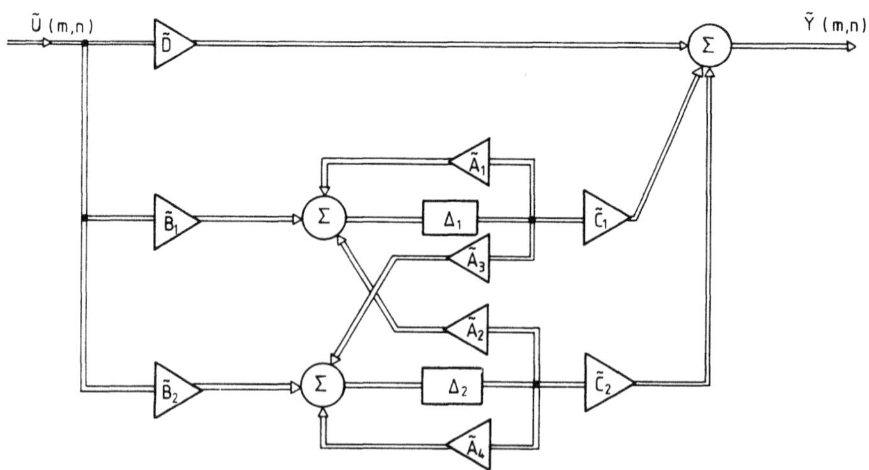

Figure 8.5. Two-dimensional block state-space filter.

The structure of the block state-space realization lends itself to linear matrix transformations in the same way as the SISO state-space system and thus permits generation of additional new 2-D block state-space structures, which may have advantageous properties. A new version of the block state-space equation (8.62) has recently been developed[12] and used to model the image generation and degradation processes in 2-D block Kalman filtering for image restoration.

8.3.4. Stability of the Two-Dimensional Block State–Space System

A necessary (but not sufficient) condition for a SISO scalar system to be stable is that the eigenvalues λ_{1i} and λ_{4i} of A_1 and A_4 lie within a unit polydisk. From the definitions of $\tilde{A}_1$ and $\tilde{A}_4$ for a MIMO system from

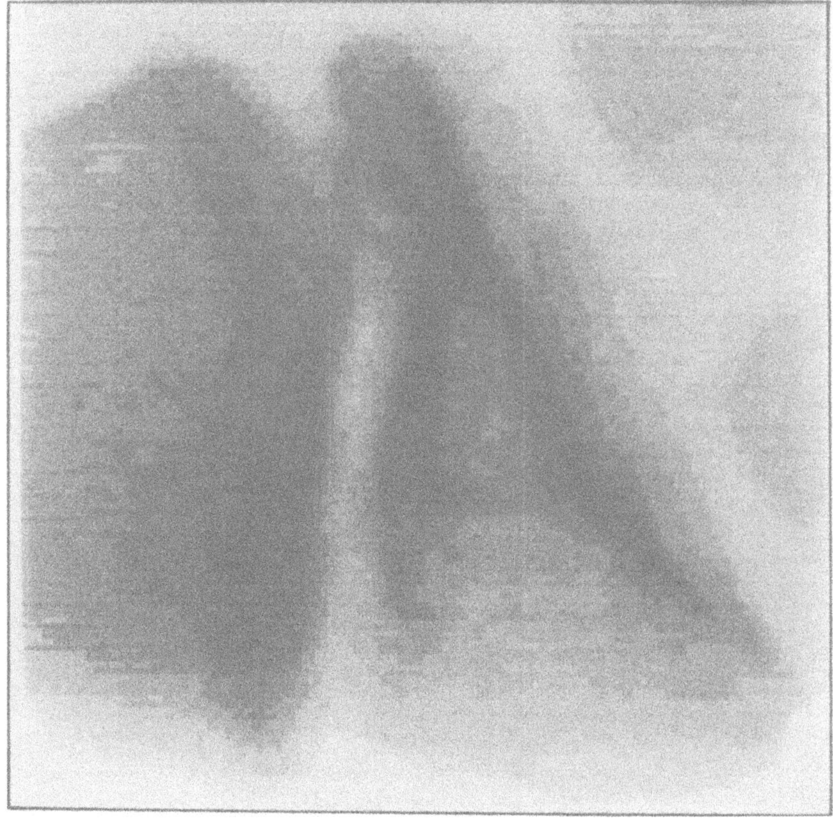

Figure 8.6. Original X-ray image.

equations (8.62), it may be seen that the eigenvalues $\tilde{\lambda}_{1i}$ and $\tilde{\lambda}_{4i}$ are equal to the KLth power of λ_{1i} and λ_{4i}. Thus, if the scalar state-space system is stable, the eigenvalues of the block state-space system will move closer to the origin of the unit polydisk, and hence it becomes even more stable.

8.4. APPLICATION OF BLOCK-PROCESSING TECHNIQUES

A filter, implemented using the block-processing technique, has been used to remove the effects of quantization noise from an X-ray image of part of a boy's hand, digitized to an array of 128 × 145 pixels having 32 gray levels. This is shown in Figure 8.6.

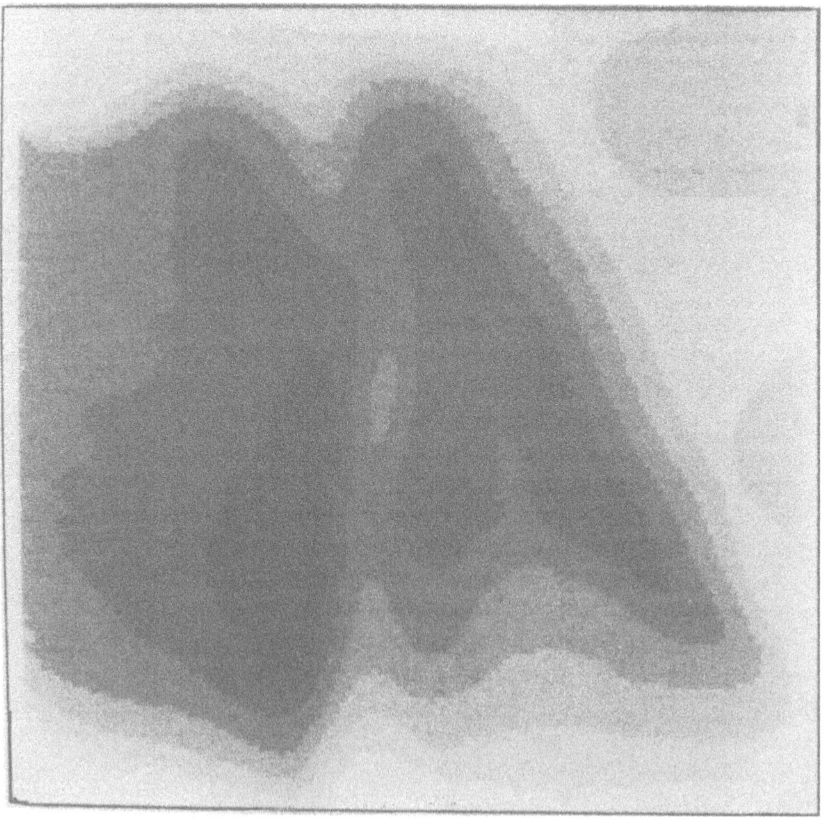

Figure 8.7. Image filtered using block recursion techniques with blocks of size 8 × 8.

The filter designed for this procedure has a 2-D low-pass circularly symmetric transfer function given by

$$H(z_1^{-1}, z_2^{-1}) = 0.0122 \times \frac{[1\ z_1^{-1}\ z_1^{-2}]\begin{bmatrix} 1.0 & 0.410191 & 0.594957 \\ 0.240013 & -0.887865 & 0.423221 \\ 0.560841 & 0.453500 & 0.360962 \end{bmatrix}\begin{bmatrix} 1 \\ z_2^{-1} \\ z_2^{-2} \end{bmatrix}}{[1\ z_1^{-1}\ z_1^{-2}]\begin{bmatrix} 1.0 & -0.500549 & -0.138282 \\ -0.690435 & -0.195020 & 0.346731 \\ -0.043308 & 0.342758 & -0.093572 \end{bmatrix}\begin{bmatrix} 1 \\ z_2^{-1} \\ z_2^{-2} \end{bmatrix}}$$

The block-processing algorithm using the above 2-D filter has been applied to the X-ray image of Figure 8.6. The resultant image for an 8×8 block is shown in Figure 8.7. The response when working with a 4×4 block is indistinguishable. It is usually found that small-size blocks are preferable in terms of reduced computation time, and this consequently results in reduced storage time.

REFERENCES

1. B. R. Hunt, Minimization of the computation time by using the technique of sectioning for digital filtering of pictures, *IEEE Trans. Comput.* **C-21,** 1219–1222 (1972).
2. R. E. Twogood, M. P. Ekstrom, and S. K. Mitra, Optimal sectioning procedure for the implementation of 2-D digital filters, *IEEE Trans. Circuits Syst.* **CAS-25,** 260–269 (1978).
3. M. R. Azimi-Sadjadi, R. A. King, and S. K. Pal, Optimisation technique for the implementation of two-dimensional recursive digital filters by sectioning, *IEE Proc.* **129,** Part F, No. 5, 373–380 (1982).
4. M. R. Azimi-Sadjadi and R. A. King, Two-dimensional block processors—Structures and implementations, *IEEE Trans. Circuits Syst.* **CAS-33,** 42–50 (1986).
5. S. K. Mitra and R. Gnanasekaran, Block implementation of two-dimensional digital filters, *J. Franklin Inst.* **316,** 299–316 (1983).
6. D. E. Dudgeon and R. M. Mersereau, *Multidimensional Digital Signal Processing,* Prentice-Hall, Englewood Cliffs, NJ (1984).
7. C. S. Burrus, Block implementation of digital filters, *IEEE Trans. Circuit Theory* **CT-18,** 697–701 (1971).
8. C. S. Burrus, Block realization of digital filters, *IEEE Trans. Audio Electroacoust.* **AU-20,** 230–235 (1972).
9. C. W. Barnes and S. Shinnaka, Block shift invariance and block implementation of discrete-time filters, *IEEE Trans. Circuits Syst.* **CAS-27,** 667–672 (1980).
10. M. R. Azimi-Sadjadi, Block Implementation of Two-Dimensional Digital Filters, PhD Thesis, University of London (1981).
11. H. H. Lu, E. A. Lee, and D. G. Messerschmitt, Fast recursive filtering with multiple slow processing elements, *IEEE Trans. Circuits Syst.* **CAS-32,** 1119–1129 (1985).
12. M. R. Azimi-Sadjadi and P. W. Wong, Two-dimensional block Kalman filtering for image restoration, *IEEE Trans. Acoust., Speech, Signal Process.* **ASSP-35,** 1736–1749 (1987).

9

Number Theoretic Transformation Techniques

9.1. ORTHOGONAL TRANSFORMS AND ELEMENTS OF MODULAR ARITHMETIC

Processing signals with a digital computer or with dedicated digital hardware involves the implementation of computational schemes on sequences of numbers. Practically, it is not possible to process an infinitely long sequence, although it is common practice to analyze many processing systems as though this were the case. Thus, we deal with finite-length sequences which have finite values in the range $0 \leq n \leq N-1$ and whose values are zero outside this range.

Typically, the input-output relationship for a signal processing system, as given in equation (1.1a), is that of convolution

$$y(n) = \sum_{k=0}^{N-1} u(n)h(n-k), \qquad n = 0, 1, \ldots, N-1 \tag{9.1}$$

or

$$y(n) = u(n) * h(n)$$

where $u(n)$ is the input sequence, $y(n)$ the output sequence, and $h(n)$ the impulse response of the system or filter.

In order to implement equation (9.1) directly, N^2 multiplications are required. However, if the two sequences are transformed by a linear transform which has digital convolution properties, the convolution process will involve N-point multiplications as given by

$$Y(k) = U(k)H(k) \tag{9.2}$$

where $Y(k)$, $U(k)$, and $H(k)$ are the corresponding transformed data.

Discrete Fourier transform (DFT) and number theoretic transform (NTT) are two examples of orthogonal transforms which can implement expression (9.1) very efficiently. The bulk of the computation is the amount of arithmetic required to convert the input sequences into the transform domain, and back.

As for the DFT, fast algorithms exist for the NTT. These transforms are defined over finite fields and rings of integers with all arithmetic performed modulo an integer and, recently, some hardware structures have been devised to implement NTTs in a fast and efficient way.[1,2,60]

The family of NTTs is wide and includes, among others, Mersenne, Fermat, pseudo-Mersenne, pseudo-Fermat, complex Mersenne, and complex Fermat transforms.[3-14] An exhaustive coverage of all these transforms is thus beyond the scope of this book. However, we shall treat some of these transforms in detail, along with other transforms in the number theoretic family, such as polynomial transforms,[10,13,14] and the more recent p-adic number transforms.[14-20]

Since all these transforms are based on number theoretic concepts in general and on modular arithmetic in particular, some results from the properties of integers will be presented below.[21-26]

9.1.1. Divisibility of Integers

We consider two integers a and b, with b positive. Division of a by b is defined by

$$a = bq + r, \qquad 0 \leq r \leq b \tag{9.3}$$

where q is the quotient and r the remainder. If the latter is equal to zero, then b is said to be a divisor of a, and this operation is denoted by $b \mid a$. The integer a is prime if it has no divisors other than one and itself; otherwise a is composite. The fundamental theorem of arithmetic states that any composite number a can be uniquely factorized as

$$a = \prod_i p_i^{m_i} \tag{9.4}$$

where p_i is a prime number and m_i is a positive integer.

9.1.2. Congruences and Residues

If two integers, a_1 and a_2, give the same remainder when divided by an integer b (the modulus), i.e.,

$$a_1 = bq_1 + r \qquad \text{and} \qquad a_2 = bq_2 + r \tag{9.5}$$

then a_1 and a_2 are two elements of the residual class r modulo b (also said to be "congruent modulo b"), and this is denoted as

$$a_1 \equiv a_2 \pmod{b} \tag{9.6}$$

Alternatively, the latter relationship can be expressed as

$$b \mid (a_1 - a_2) \tag{9.7}$$

The remainder r is called the residue and is of particular interest. It can be written in the form

$$r = \langle a_1 \rangle_b = \langle a_2 \rangle_b \tag{9.8}$$

Equation (9.5) shows that addition, subtraction, and multiplication can be performed directly on residues. In other words

$$\langle a_1 + a_2 \rangle_b = \langle \langle a_1 \rangle_b + \langle a_2 \rangle_b \rangle_b \tag{9.9a}$$

$$\langle a_1 - a_2 \rangle_b = \langle \langle a_1 \rangle_b - \langle a_2 \rangle_b \rangle_b \tag{9.9b}$$

$$\langle a_1 \cdot a_2 \rangle_b = \langle \langle a_1 \rangle_b \cdot \langle a_2 \rangle_b \rangle_b \tag{9.9c}$$

9.1.3. Groups and Abelian Groups

A group G is a nonempty set of elements which, with a binary operation $\odot$, satisfies the following postulates:

1. *Closure*: for every a and b in G, $a \odot b$ is also in G.
2. *Associativity*: for every a, b, c in G, $(a \odot b) \odot c = a \odot (b \odot c)$.
3. *Identity*: there is a unique element e in G, called the identity element, such that $a \odot e = e \odot a = a$ for all a in G.
4. *Inverse*: for every a in G there is a unique inverse b in G, such that $a \odot b = b \odot a = e$.

A group G is called an Abelian group if every pair of elements commute, i.e., $a \odot b = b \odot a$ for all a and b in G.

9.1.4. Rings and Fields

Rings and fields are sets of integers whose elements obey certain properties. A ring Z is defined if its elements adhere to the following conditions:

1. $a \cdot b$ is a legitimate element of the set.

2. $(Z, +)$ is an Abelian group, where $(Z, +)$ denotes additive elements of the ring Z.
3. The associativity property is satisfied.
4. The two distributive laws are satisfied, i.e.,

$$a \cdot (b + c) = a \cdot b + a \cdot c$$

and

$$(b + c) \cdot a = b \cdot a + c \cdot a$$

If multiplication is commutative, then Z is a commutative ring. In a ring $Z_M = \{0, 1, 2, \ldots, M - 1\}$, all integers are congruent modulo M to some integer in Z_M. If, in Z_M, each integer has a unique multiplicative inverse, then the ring becomes a field. It can be shown[27] that a ring is a field if, and only if, M is a prime number.

9.1.5. Modular Arithmetic

The following arithmetic operations can be performed with modular arithmetic.

1. *Addition*: addition modulo some arbitrary modulus can be performed as ordinary p-ary arithmetic followed by a division of the result by the modulus. The remainder is the required sum.
2. *Negation*: to negate a number a, it is subtracted from the modulus M as follows: $-a = -a + M \pmod{M}$.
3. *Subtraction*: this consists in negating a number as in (2) and then adding it according to (1).
4. *Multiplication*: modular multiplication is, again, similar to p-ary multiplication followed by a division of the result by the modulus. The remainder is the required quotient.
5. *Multiplicative Inverse*: a multiplicative inverse of an integer a exists in Z_M if, and only if, a and M are relatively prime. The inverse, a^{-1}, is then given by $a \cdot a^{-1} \equiv 1 \pmod{M}$.

9.1.6. The Chinese Remainder Theorem (CRT)

It is assumed that $m_1, \ldots, m_k$ are positive integers, relatively prime in pairs:

$$(m_i, m_j) = 1 \qquad \text{if } i \neq j$$

We let $b_1, \ldots, b_j$ be arbitrary integers. Then the system of congruences

$$x \equiv b_1 \pmod{M_1}$$
$$\vdots$$
$$x \equiv b_j \pmod{M_j}$$

has exactly one solution modulo M where $M = \sum_{i=1}^{k} m_i$.

To construct this solution, we define

$$M_i = M/m_i \tag{9.10}$$

We note that $(m_i, M_i) = 1$. Thus, there are solutions N_i of

$$N_i M_i \equiv 1 \pmod{m_i} \tag{9.11}$$

With these N_i, the solution x to the simultaneous congruences is

$$x = b_1 N_1 M_1 + \cdots + b_k N_k M_k \pmod{M} \tag{9.12}$$

According to the notation introduced in equation (9.8), the above congruence can be expressed as

$$\langle x \rangle_{m_i} \equiv b_i N_i M_i \equiv b_i \pmod{m_i} \tag{9.13}$$

since all other terms in the sum (9.12) making up x contain the factor m_i, and therefore do not contribute to the residue modulo m_i. Furthermore, because $N_i M_i \equiv 1 \pmod{m_i}$, the solution is unique modulo M.

9.1.7. Euler's Totient Function, Euler's Theorem, and Fermat's Little Theorem

If $M \geq 1$, Euler's totient function $\phi(M)$ is defined to be the number of positive integers not exceeding M that are relatively prime to M. Hence

$$\phi(M) = \sum_{i=1}^{M} * \, i \tag{9.14}$$

where $*$ indicates that the sum is extended over those i relatively prime to M.

Euler's theorem states that, for every α relatively prime to M,

$$\alpha^{\phi(M)} \equiv 1 \pmod{M} \tag{9.15}$$

On the other hand, if $M = p$ is a prime, it can be observed that all integers less than p are relatively prime to it. Consequently

$$\phi(p) = p - 1 \tag{9.16}$$

and expression (9.15) reduces to

$$\alpha^{p-1} \equiv 1 \pmod{p} \tag{9.17}$$

This is Fermat's little theorem and α is said to be a primitive root of order $p - 1$. By raising α to the powers n such that $n = 1, \ldots, p - 1$, it will generate all the elements in $I_p = \{0, 1, 2, \ldots, p - 1\}$.

9.1.8. Quadratic Residues and the Legendre Symbol

In a quadratic congruence of the form $x^2 \equiv a \pmod{M}$, a is said to be a quadratic residue of M if the congruence has a solution. If no solution can be found, then a is a quadratic nonresidue of M.

If $M = p$ denotes an odd prime and $(a, p) = 1$, the Legendre symbol $[\frac{a}{p}]$ is defined such that

$$\left[\frac{a}{p}\right] = \begin{cases} 1 \text{ if } a \text{ is a quadratic residue of } p \\ -1 \text{ if } a \text{ is a quadratic nonresidue of } p \end{cases} \tag{9.18}$$

9.1.9. Extension Fields of Integer Fields, I_p

We let I_p represent a Galois field GF(p), where $I_p = \{0, 1, 2, \ldots, p - 1\}$. Then, extension fields of I_p can be represented by GF(p^2) or by $I_p(a^{1/2})$, where a is a quadratic nonresidue in I_p. So $x^2 \equiv a \pmod{p}$ has no solution, and the elements of the extension field are represented by

$$A = \alpha + a^{1/2}\beta \tag{9.19}$$

where A, $a^{1/2} \in I_p(a^{1/2})$ and $\alpha, \beta \in I_p$.

If a is an integer value, then quadratic extension fields are considered. If $a = -1$, then complex extension fields are considered.

9.1.10. Residue Polynomials

The theory of residue polynomials is closely related to the theory of integer residue classes. A polynomial $P(Z)$ divides a polynomial $H(Z)$ if a third polynomial $D(Z)$ can be found such that $H(Z) = P(Z) \cdot D(Z)$. Polynomial $H(Z)$ is said to be irreducible if its only divisors are of degree equal to zero. If, on the other hand, $P(Z)$ is not a divisor of $H(Z)$, the division of $H(Z)$ by $P(Z)$ will produce a residue $R(Z)$:

$$H(Z) = P(Z)D(Z) + R(Z) \tag{9.20}$$

where the degree of $R(Z)$ is less than the degree of $P(Z)$. All polynomials having the same residue when divided by $P(Z)$ are said to be congruent modulo $P(Z)$ and hence

$$R(Z) \equiv H(Z) \pmod{P(Z)} \tag{9.21}$$

All the concepts in integer fields and rings can be applied to polynomial residues.

9.2. NUMBER THEORETIC TRANSFORMS

The forward and inverse number theoretic transforms (NTT) of an n-point sequence $x(n)$ are defined respectively as[3,27]

$$X(k) = \sum_{n=0}^{N-1} x(n)\alpha^{nk} \pmod{M}, \qquad k = 0, 1, \ldots, N-1 \tag{9.22}$$

and

$$x(n) = N^{-1} \sum_{k=0}^{N-1} X(k)\alpha^{-nk} \pmod{M}, \qquad n = 0, 1, \ldots, N-1 \tag{9.23}$$

where $X(k)$ is the transformed sequence, M an integer, and α^{nk} represents the basis functions.

It is noted that the transform has a DFT structure and, if $\alpha = e^{-j2\pi/N}$, then equation (9.22) reduces to the DFT.

In equation (9.23), N^{-1} is the multiplicative inverse in the field in which the arithmetic is carried out, i.e., $N \cdot N^{-1} \equiv 1 \pmod{M}$. Number theoretic transforms may be defined over a field or a ring of integers if arithmetic is performed modulo a prime or a composite integer, respectively. It should be noted, however, that over a ring N^{-1} may not exist.

9.2.1. Number Theoretic Transform over a Field, I_p

If the NTT is defined over a prime integer, the conditions for existence and for possessing a circular convolution property are given by

$$\alpha^N \equiv 1 \pmod{p} \tag{9.24}$$

and

$$N \cdot N^{-1} \equiv 1 \pmod{p} \tag{9.25}$$

where α is a root of unity of order N and p is a prime integer. The parameters α, N, and p are all interrelated by expressions (9.24) and (9.25). The maximum transform length is given by Euler's totient function $\phi(p)$. Thus

$$\phi(p) = p - 1$$

In order for the transform to exist, it must satisfy

$$\alpha_{\phi}^{p-1} \equiv 1 \pmod{p} \tag{9.26}$$

where α_{ϕ} is known as the primitive root giving the maximum transform length. Quantity α_{ϕ} is not the only primitive root; there are $\phi(\phi(p))$ other primitive roots computed by Euler's function.[28] From the above, it is seen that an integer α_{ϕ} can be found in the field of a prime integer.

By Fermat's little theorem, roots of unity of order N where $N \mid \phi(p)$ can be obtained from the expression

$$\alpha = \alpha_{\phi}^{(p-1)/N} \tag{9.27}$$

Thus equation (9.24) can be solved for a transform length N, where $p - 1$ is divisible by N [denoted $N \mid (p-1)$].

9.2.2. Number Theoretic Transform over a Ring, Z_M

If the transform pair defined by equations (9.22) and (9.23) is defined over a rational integer ring Z_M, the ring is decomposed into several fields and the conditions for the existence of a transform are derived for each field. The unique decomposition of Z_M is defined by

$$Z_M = I_{p_1} \otimes I_{p_2} \otimes \cdots \otimes I_{p_j} \tag{9.28}$$

where $\otimes$ denotes the element-by-element multiplication of the corresponding fields and

$$M = p_1 p_2 \cdots p_j \tag{9.29}$$

The conditions for the existence of an NTT in the respective fields are given by

$$\left.\begin{array}{ll}1. & \alpha^N \equiv 1 \pmod{p_i} \\ 2. & NN^{-1} \equiv 1 \pmod{p_i} \\ 3. & N|(p_i - 1)\end{array}\right\} \quad \text{for } i = 1, 2, \ldots, j \tag{9.30}$$

If $O(M)$ is defined as the greatest common division (gcd) of $(p_i - 1)$ for $i = 1, 2, \ldots, j$, then the maximum transform length in Z_M is

$$N_{\max} = O(M) \tag{9.31}$$

The advantage of defining NTTs over a composite modulus is that a larger dynamic range is obtained and, by using the Chinese Remainder Theorem, the NTT can be decomposed and implemented modulo each prime p_i, in parallel, and combined together finally by the CRT. This also has the added advantage of using small word length processors, where the arithmetic is evaluated modulo relatively small primes p_i; we still need a large enough p_i to get the required transform length.

The NTT in Z_M is given by

$$[X_1(k), X_2(k), \ldots, X_j(k)] = \sum_{n=0}^{N-1} [x_1(n), x_2(n), \ldots, x_j(n)][\alpha_1, \alpha_2, \ldots, \alpha_j]^{nk} \tag{9.32}$$

It can also be written as

$$[X_1(k), X_2(k), \ldots, X_j(k)] = \left[\sum_{n=0}^{N-1} x_1(n)\alpha_1^{nk} \sum_{n=0}^{N-1} x_2(n)\alpha_2^{nk}, \ldots, \sum_{n=0}^{N-1} x_j(n)\alpha_j^{nk}\right] \tag{9.33}$$

where each summation is evaluated modulo the corresponding prime p_i and with $X_i(k)$, $x_i(n)$, $\alpha_i \in I_{p_i}$ for $1 \le i \le j$.

9.2.3. Mersenne and Fermat Number Transforms

In the previous sections, the NTT was defined over a field and ring. Here, we consider the hardware implementation of NTTs. Since arithmetic is performed modulo a prime integer, some of these prime integers can be implemented very efficiently if they have a particular structure. Prime integers having a power-of-two structure enable modular arithmetic to be performed by shift and add operations only. Such attractive primes are known as Mersenne primes, defined by $M_p = 2^p - 1$ where p is a prime,[29] and Fermat numbers which are defined by $F_t = 2^{2^t} + 1$ for $t = 1, 2, 3, \ldots$. Of all the Fermat numbers, only those up to and including F_4 are known to be prime.[30]

In the following subsections, transforms using such prime integer structures are considered.

9.2.3.1. Mersenne Number Transform

We consider arithmetic modulo primes of the form $M_p = 2^p - 1$; then transforms in the field I_{M_p} have a maximum transform length given by

$$N_{\max} = \phi(M_p) = 2^p - 2 \tag{9.34}$$

and a primitive root of order $N_{\max}$ can be found. However, transforms of length p are found to have roots of $\alpha = 2$. These transforms are known as Mersenne number transforms (MNT) and are defined as

$$X(k) \equiv \sum_{n=0}^{p-1} x(n) 2^{nk} \pmod{2^p - 1} \qquad \text{for } k = 0, 1, \ldots, p-1 \tag{9.35}$$

These transforms suffer from the fact that although p, which determines the processor word length, could be long for a very large Mersenne prime, it may still be quite short when considering a small prime. Consequently, MNTs with large transform lengths require large word length processors. Also, the transform length is not highly composite and so a simple FFT algorithm cannot be used. However, MNTs can be implemented using more complex methods, such as Winograd and prime factor algorithms.[31,32] These transforms can be used to implement circular convolutions of p points.

The inverse MNT is defined by

$$x(n) \equiv p^{-1} \sum_{k=0}^{p-1} X(k) 2^{-nk} \pmod{2^p - 1} \qquad \text{for } n = 0, 1, \ldots, p-1 \tag{9.36}$$

where

$$2^{-nk} \equiv 2^{p-nk} \pmod{2^p - 1} \tag{9.37}$$

Mersenne number transforms can also be defined over a composite modulus, which can be factorized into several Mersenne prime factors. MNTs with $\alpha = 2$ can be implemented by $p(p-1)$ additions and $(p-1)^2$ shifts.

9.2.3.2. Fermat Number Transform

If the prime integer is chosen such that $F_t = 2^{2^t} + 1$ with $t = 1, 2, 3, 4$, then the maximum transform length is given by

$$N_{\max} = 2^{2^t} \tag{9.38}$$

which is highly composite and can then be implemented by a radix-2 FFT algorithm. For a length $N = 2^t$, the primitive root $\alpha = 2$. The corresponding Fermat number transform (FNT) is defined by

$$X(k) \equiv \sum_{n=0}^{N-1} x(n)2^{nk} \pmod{F_t} \qquad \text{for } k = 0, 1, \ldots, F_t - 1 \tag{9.39}$$

having an inverse of the form

$$x(n) \equiv N^{-1} \sum_{k=0}^{N-1} X(k)2^{-nk} \pmod{F_t} \qquad \text{for } n = 0, 1, \ldots, F_t - 1 \tag{9.40}$$

FNTs can implement circular convolution through shift and add operations, as discussed in Section 8.3. However their disadvantage, as for MNTs, is that the transform length is limited by the word length of the processor. Thus large data sequences cannot be transformed directly. It is possible, though, to convert a long sequence into a multidimensional sequence and then use FNT.(32)

9.2.4. Complex Number Theoretic Transform

In some applications, like synthetic aperture radar image formation, complex convolution is required. Thus, to compute complex convolution by real number theoretic transforms, several NTTs must be implemented.(8) So it becomes advantageous to define a complex number theoretic transform (CNTT), where a complex sequence can be transformed directly. A CNTT pair is defined by

$$X(k) = \sum_{n=0}^{N-1} x(n)\alpha^{nk}, \qquad k = 0, 1, \ldots, N - 1 \tag{9.41}$$

and

$$x(n) = \frac{1}{N} \sum_{k=0}^{N-1} X(k)\alpha^{-nk}, \qquad n = 0, 1, \ldots, N - 1 \tag{9.42}$$

where $x(n)$, $X(k)$, $\alpha \in Q_p$; p is a prime, where p is of the form $4k + 3$, $k \in I$, the field Q_p being the second-order extension of the integer field I_p.

An element A of the extension field Q_p can be represented as $A = a + jb$ where $a, b \in I_p$ and $j^2 = -1$ in Q_p. The number of elements in a finite field Q_p is p^2. Quantity Q_p is also known as a Galois field, in this case denoted by GF(p^2). The complex extension field Q_p, which can also be denoted as $Q_p(\sqrt{-1})$, only exists if the expression

$$X^2 + 1 = 0$$

is irreducible over I_p.

The above argument can be generalized to all quadratic extension fields $Q_p(\sqrt{-m})$ of I_p if the equation $X^2 + m = 0$ is irreducible over I_p. By Euler's theorem[28]

$$m^{(p-1)/2} \equiv \left[\frac{m}{p}\right] (\text{mod } p) \tag{9.43}$$

The quadratic congruence $X^2 + m \equiv 0 \,(\text{mod } p)$ is reducible over I_p if $[\frac{m}{p}] = 1$. On the other hand, if $[\frac{m}{p}] = -1$, it is irreducible over I_p and $Q_p(\sqrt{-m})$ is a field of p^2 elements isomorphic to GF(p^2).[28]

In the case of a Fermat prime, $F_t = 2^{2^t} + 1$, it can be shown[10] that the congruence $X^2 + 1 \equiv 0 \,(\text{mod } F_t)$ is reducible in I_{F_t} since

$$\begin{aligned} (-1)^{(F_t-1)/2} &\equiv [-1/F_t] \,(\text{mod } F_t) \\ &= 1 \end{aligned} \tag{9.44}$$

Thus there is no complex extension field with a Fermat prime. However there exist quadratic extension fields, which are examined in the next section.

If Mersenne primes $M_p = 2^p - 1$ are considered, it can be shown that $X^2 + 1 \equiv 0 \,(\text{mod } M_p)$ is irreducible in I_{M_p} and hence complex Mersenne transforms can be defined. In general, the conditions for the existence of a CNTT are

1. α is a root of unity of order N in $Q_p(\sqrt{-1})$.
2. $N \cdot N^{-1} \equiv 1 \,(\text{mod } p)$ in $Q_p(\sqrt{-1})$.
3. $N \mid (p^2 - 1)$ since there are $p^2 - 1$ distinct elements in $Q_p(\sqrt{-1})$.

9.2.5. Quadratic Number Theoretic Transform

NTTs can be defined in quadratic extension fields as[33]

$$X(k) = \sum_{n=0}^{N-1} x(n)\alpha^{nk}, \qquad k = 0, 1, \ldots, N-1 \tag{9.45}$$

and

$$x(n) = N^{-1} \sum_{k=0}^{N-1} X(k)\alpha^{-nk}, \qquad n = 0, 1, \ldots, N-1 \tag{9.46}$$

in $Q_p(\sqrt{m})$, where $Q_p(\sqrt{m})$ is a quadratic extension field of I_p and of $p^2 - 1$ distinct elements with p a prime integer. Quantities $x(n)$ and $X(k)$ are elements of $Q_p(\sqrt{m})$ where each element is represented in the form

$$Q_p(\sqrt{m}) = a + m^{1/2}b \qquad \text{with } a, b \in I_p$$

Reed and Truong[33] have shown that quadratic extension fields exist with Fermat and Mersenne primes. For example, with Fermat primes $F_t = 2^{2^t} + 1$, the maximum transform length is

$$\begin{aligned} N_{\max} &= F_t^2 - 1 \\ &= 2^{2^{(t+1)}} + 2^{(2^t+1)} \end{aligned} \tag{9.47}$$

Transforms of order 2^{2^t+1} can be defined and implemented by radix-2 FFT algorithms. Similarly, quadratic number theoretic transforms can be defined with Mersenne primes. These can be implemented by the Winograd algorithm.[31]

9.2.6. Number Theoretic Transforms in Galois Rings GF(q^2)

In the previous two sections, transforms were defined over Galois fields GF(p^2) where p is a prime integer. Here, we define NTTs in GR(q^2) where q is a composite integer which can be factorized into its prime factors given by

$$q = p_1 p_2 \cdots p_j$$

and $X^2 + 1 \equiv 0 \pmod{p_i}$ is irreducible for $i = 1, 2, \ldots, j$.

Consequently, GR(q^2) can be decomposed into its corresponding Galois fields represented as

$$\mathrm{GR}(q^2) = \mathrm{GF}(p_1^2) \otimes \mathrm{GF}(p_2^2) \otimes \cdots \otimes \mathrm{GF}(p_j^2) \tag{9.48}$$

where every element A in GR(q^2) can be represented in its factorized form $A = (A_1, A_2, \ldots, A_i, \ldots, A_j)$ in which $A_i \in \mathrm{GF}(p_i^2)$. Thus a NTT defined in GR(q^2) must be satisfied in every GF(p_i^2) and the same conditions for the existence of such a transform apply here, namely,

$$\left.\begin{array}{ll} 1. & \alpha^N \equiv 1 \pmod{p_i} \\ 2. & N \cdot N^{-1} \equiv 1 \pmod{p_i} \\ 3. & N \mid (p_i^2 - 1) \end{array}\right\} \quad \text{for } i = 1, 2, \ldots, j \tag{9.49}$$

where α is the root of unity of order N in $\mathrm{GR}(q^2)$.

9.3. POLYNOMIAL TRANSFORM

Recently, several new techniques for implementing two-dimensional (2-D) DFT have been introduced using polynomial transform (PT) algorithms.[34,35] These algorithms are not explained here, but the concept of polynomial transform is discussed.

Polynomial transforms can be viewed as discrete Fourier-like transforms defined in a residue-class polynomial ring $R(Z)/P(Z)$, where R is either a ring or a field. The polynomial algebra is performed modulo $P(Z)$, in which the coefficients of the polynomials are taken to lie in R. A general definition of polynomial transforms can be obtained by considering a polynomial convolution $Y(Z)$ of length N defined modulo a polynomial $P(Z)$:

$$Y_n(Z) \equiv \sum_{m=0}^{N-1} H_m(Z) X_{n-m}(Z) \pmod{P(Z)} \tag{9.50}$$

and

$$H_m(Z) \equiv \sum_{i=0}^{b-1} h_{i,m} Z^i \pmod{P(Z)} \tag{9.51}$$

$$X_r(Z) \equiv \sum_{j=0}^{b-1} x_{j,r} Z^j \pmod{P(Z)} \tag{9.52}$$

where $Y_i(Z), H_i(Z), X_i(Z) \in R(Z)/P(Z)$ with R being the field of rational numbers and b the degree of $P(Z)$.

The one-dimensional (1-D) polynomial convolution can be transformed into N element-by-element multiplications in $R(Z)/P(Z)$ by a polynomial transform defined as[36]

$$\bar{H}_k(Z) \equiv \sum_{m=0}^{N-1} H_m(Z)[G(Z)]^{mk} \pmod{P(Z)}, \qquad k = 0, 1, \ldots, N-1 \tag{9.53}$$

and similarly for

$$\bar{X}_k(Z) \equiv \sum_{r=0}^{N-1} X_r(Z)[G(Z)]^{rk} \pmod{P(Z)}, \qquad k = 0, 1, \ldots, N-1 \tag{9.54}$$

where $\{G(Z)\}$ is an nth primitive root of unity in $R(Z)/P(Z)$. Then

$$\bar{Y}_k(Z) \equiv \bar{H}_k(Z)\bar{X}_k(Z) \pmod{P(Z)} \qquad \text{for } k = 0, 1, \ldots, N-1 \tag{9.55}$$

and $\{Y_i(Z)\}$ can be recovered from $\bar{Y}_k(Z)$ using the inverse polynomial transform given by

$$Y_i(Z) \equiv N^{-1} \sum_{k=0}^{N-1} \bar{Y}_k(Z)[G(Z)]^{-ik} \pmod{P(Z)}, \qquad i = 0, 1, \ldots, N-1 \tag{9.56}$$

In order to establish that the polynomial transforms support circular convolution, the transforms $\bar{H}_k(Z)$ and $\bar{X}_k(Z)$ of $H_m(Z)$ and $X_r(Z)$ are calculated. Then, element-by-element multiplication of $\bar{H}_k(Z)$ by $\bar{X}_k(Z)$ modulo $P(Z)$ is evaluated and the inverse transform of $\bar{Y}_k(Z)$ is computed. This can be represented as

$$Y_n(Z) \equiv N^{-1} \sum_{m=0}^{N-1} \sum_{r=0}^{N-1} H_m(Z)X_r(Z) \sum_{k=0}^{N-1} [G(Z)]^{(m+r-n)k} \pmod{P(Z)}$$
$$n = 0, 1, \ldots, N-1 \tag{9.57}$$

The above expression is valid provided the following three conditions are met[37]:

1. $G(Z)$ is an Nth primitive root of unity modulo $P(Z)$,

$$[G(Z)]^N \equiv 1 \pmod{P(Z)} \tag{9.58}$$

2. N and $G(Z)$ have inverses modulo $P(Z)$,

$$NN^{-1} \equiv 1 \pmod{P(Z)} \tag{9.59}$$

$$G(Z)[G(Z)]^{-1} \equiv 1 \pmod{P(Z)} \tag{9.60}$$

3. $S \equiv \sum_{k=0}^{N-1} [G(Z)]^{(m+r-n)k} \pmod{P(Z)}$

$$\equiv \begin{cases} 0 \text{ for } (m+r-n) \not\equiv 0 \pmod{N} \\ N \text{ for } (m+r-n) \equiv 0 \pmod{N} \end{cases} \tag{9.61}$$

Polynomial transforms have the same structure as DFTs, but with the complex exponential roots of unity replaced by the polynomials $G(Z)$ and with all operations defined modulo $P(Z)$. By investigating the classes of polynomial rings that possess a simple kernel $G(Z)$, it was found[37] that there always exists a PT with simple primitive roots $G(Z) = Z$ [on Z^t where $(t, r) = 1$] in the ring $R(Z)/C_r(Z)$, which satisfies the above three conditions, and $C_r(Z)$ is a cyclotomic polynomial of order N defined as

$$C_r(Z) = \prod_{Y_i \in S} (Z - Y_i) \tag{9.62}$$

where S is the set containing all primitive rth roots of unity.

The principal application of PTs concerns the computation of 2-D circular convolutions. A circular convolution of size $N \times N$ can be represented[38] as a polynomial convolution of length N, all polynomials being defined modulo $(Z^N - 1)$ as given by

$$y_{p,q} = \sum_{m=0}^{N-1} \sum_{n=0}^{N-1} h_{n,m} x_{p-n,q-m}, \qquad p, q = 0, 1, \ldots, N-1 \tag{9.63}$$

and

$$Y_j(Z) \equiv \sum_{m=0}^{N-1} H_m(Z) X_{j-m}(Z) \pmod{(Z^N - 1)}, \qquad j = 0, 1, \ldots, N-1 \tag{9.64}$$

where $H_i(Z)$ and $X_i(Z)$ are defined by equations (9.51) and (9.52), respectively.

For the coefficients in the field of rationals $R[Z]$, $(Z^N - 1)$ is the product of d cyclotomic polynomials $C_{r_i}(Z)$, where d is the number of divisors r_i of N, including 1 and N, with

$$Z^N - 1 = \prod_{i=1}^{d} C_{r_i}(Z) \tag{9.65}$$

The degree of each cyclotomic polynomial $C_{r_i}(Z)$ is $\phi(r_i)$.

Since the various polynomials $C_{r_i}(Z)$ are irreducible, the polynomial convolution defined modulo $(Z^N - 1)$ can be computed separately modulo each cyclotomic polynomial $C_{r_i}(Z)$ and by reconstructing the final result by the CRT defined for polynomials.

We now consider expression (9.65). If $N = p$ where p is an odd prime, then $Z^p - 1$ is the product of two cyclotomic polynomials

$$Z^p - 1 = (Z - 1)P(Z) \qquad \text{where } P(Z) = Z^{p-1} + Z^{p-2} + \cdots + 1 \quad (9.66)$$

and hence the polynomial convolution is defined modulo $(Z - 1)$ and $P(Z)$. Similarly, if $N = 2^t$, then

$$\begin{aligned} Z^{2^t} - 1 &= (Z - 1) \prod_{i=1}^{t} Z^{2^{t-i}} + 1 \\ &= (Z^{2^{t-1}} + 1)(Z^{2^{t-1}} - 1) \end{aligned} \quad (9.67)$$

and, again, the polynomial convolution is evaluated for each modulus for $i = 1, 2, \ldots, t$ and the final result is obtained by the CRT.[39]

A mapping which translates a circular convolution into a skew circular one, and vice versa, has also been defined.[40,41] This mapping results in a polynomial convolution modulo a cyclotomic polynomial, which removes the need for a decomposition through the CRT.

Finally, inverse polynomial transform algorithms have also been recently presented,[42,43] together with their applications to convolution and DFT.

9.4. p-ADIC TRANSFORM

p-adic transforms (PAT) have only recently been introduced[14-20] in the field of digital signal processing, due to their attractive features in relation to error-free computations and larger dynamic range. Error-free computation can be performed in the p-adic field even with rational numbers. In the following sections, we discuss the two forms of p-adic fields and, after an introduction to p-adic numbers and to arithmetic in the segmented p-adic field, we define PATs in their various forms. Complex and quadratic PATs are also investigated in relation to Fermat and Mersenne primes.

9.4.1. The p-adic Field, Q_p, and the Segmented p-adic Field, $\hat{Q}_p$

The concept of a p-adic field Q_p was originated by Hensel.[44,45] Hensel states that every p-adic number can be represented uniquely by an infinite series of the form

$$\alpha = \sum_{n=-m}^{\infty} a_n p^n \tag{9.68}$$

where $a_n \in I_p$ and $m \in Z$, the set of integers, and this infinite series converges to α with respect to the p-adic norm.[6]

However, in order to perform arithmetic using p-adic numbers, a finite segment p-adic number has to be defined. A finite segment p-adic field $\hat{Q}_p$ is thus introduced,[48-51] by truncating the infinite p-adic expansion series of a p-adic number to a fixed number of digits, r. This representation of a finite-segment p-adic expansion of a p-adic number α is also known as a Hensel Code and is denoted by $H(p, r, \alpha)$. The conditions for the construction of such codes are[46]:

1. For the rational number $\alpha = a/b$ to be represented, the numerator and denominator have the prescribed bounds

$$-[(p^r - 1)/2]^{1/2} \leq a, b \leq [(p^r - 1)/2]^{1/2} \tag{9.69}$$

2. The p-adic expansions are terminated at the right such that r is even.

Thus, any rational number α within the prescribed range (9.69) can be represented by a Hensel code such that

$$H(p, r, \alpha) = \sum_{i=-m}^{r} a_i p^i \tag{9.70}$$

where $a_i \in I_p$.

The four basic arithmetic operations are valid in $\hat{Q}_p$ provided that no overflow occurs, even within the computation. We consider two rational numbers α and β. Then

$$H(p, r, \alpha * \beta) = H(p, r, \alpha) * H(p, r, \beta) \tag{9.71}$$

where $*$ represents an arithmetic operation.

Expression (9.71) is valid, provided the absolute values of the numerator and denominator of α, β, and $\alpha * \beta$ do not exceed $[(p^r - 1)/2]^{1/2}$. The arithmetic operations in $\hat{Q}_p$ are almost identical to the p-ary arithmetic, since it is essentially modulo p^r arithmetic realized as a simple recursion of modulo p operations. Also, p-adic arithmetic operations are error-free. For a detailed presentation of p-adic arithmetic operations and algorithms, the reader is referred elsewhere.[46]

9.4.2. The General p-adic Transform

In this section, a number-theoretic-like transform is defined in the segmented p-adic field $\hat{Q}_p$. The PAT of an N-point sequence $x(n)$, represented by the corresponding Hensel Codes $H(p, r, x(n))$, for a given prime p and even length r, is defined by[18]

$$H(p, r, X(k)) = \sum_{n=0}^{N-1} H(p, r, x(n))[H(p, r, \gamma)]^{nk} \qquad \text{for } 0 \leq k \leq N-1 \tag{9.72}$$

where $X(k)$ is the transformed sequence and $H(p, r, \gamma)$ is the Hensel equivalent of the Nth root of unity in $\hat{Q}_p$. The inverse transform is given by

$$H(p, r, x(j)) = H(p, r, 1/N) \sum_{k=0}^{N-1} H(p, r, X(k))[H(p, r, \gamma)]^{-jk} \qquad \text{for } 0 \leq j \leq N-1 \tag{9.73}$$

Equation (9.72) is substituted into equation (9.73) in order to derive the conditions for orthogonality (or existence of the inverse transform). Hence

$$H(p, r, x(j)) = H(p, r, 1/N) \sum_{k=0}^{N-1} \sum_{n=0}^{N-1} H(p, r, x(n))[H(p, r, \gamma)]^{k(n-j)} \tag{9.74}$$

and then, reordering, we obtain

$$H(p, r, x(j)) = H(p, r, 1/N) \sum_{n=0}^{N-1} H(p, r, x(n)) \sum_{k=0}^{N-1} [H(p, r, \gamma)]^{k(n-j)} \tag{9.75}$$

If

$$H(p, r, s) = H(p, r, 1/N) \sum_{k=0}^{N-1} [H(p, r, \gamma)]^{k(n-j)} \tag{9.76}$$

then we get

$$H(p, r, s) = H(p, r, 1) \qquad \text{if } (n-j) \equiv 0 \pmod{N} \tag{9.77}$$

Similarly

$$[H(p, r, \gamma)]^{N(n-j)} = H(p, r, 1) \qquad \text{in } \hat{Q}_p \tag{9.78}$$

Alternatively, if $(n-j) \not\equiv 0 \pmod{N}$, then

$$[H(p, r, \gamma)]^{(n-j)} \neq 1$$

thus

$$[H(p, r, \gamma)]^{(n-j)} - 1 \neq 0 \tag{9.79}$$

Multiplying $H(p, r, s)$ by expression (9.79) yields

$$\begin{aligned} H(p, r, s)\{[H(p, r, \gamma)]^{(n-j)} - 1\} &= H(p, r, 1/N)\{[H(p, r, \gamma)]^{(n-j)} - 1\} \\ &\quad \times \sum_{k=0}^{N-1} [H(p, r, \gamma)]^{k(n-j)} \\ &= H(p, r, 1/N)\{[H(p, r, \gamma)]^{N(n-j)} - 1\} \\ &= 0 \end{aligned} \tag{9.80}$$

and since $[H(p, r, \gamma)]^{(n-j)} - 1 \neq 0$, then equation (9.80) implies that

$$H(p, r, s) = 0 \tag{9.81}$$

In conclusion

$$H(p, r, s) = \begin{cases} H(p, r, 1) & \text{if } (n-j) \equiv 0 \pmod{N} \\ 0 & \text{if } (n-j) \not\equiv 0 \pmod{N} \end{cases} \tag{9.82}$$

Thus the necessary and sufficient conditions for the PAT to have the DFT structure, or the properties for cyclic convolution, are

1. $H(p, r, \gamma)$ is a root of unity of order N in the field of $\hat{Q}_p$:

$$[H(p, r, \gamma)]^N = H(p, r, 1) \tag{9.83}$$

2. $H(p, r, 1/N)$ should exist, or be representable, in $\hat{Q}_p$;
3. $[H(p, r, \gamma)]^{-1}$ should be representable.

9.4.3. The Existence and Derivation of $H(p, r, \gamma)$ in $\hat{Q}_p$

We have shown in the previous section that one of the conditions for the existence of the PAT is that a root of unity, $H(p, r, \gamma)$, of order N exists in $\hat{Q}_p$. Bachman[47] showed that the equation $x^{p-1} = 0$ has exactly $p-1$ distinct roots in $\hat{Q}_p$. Thus, the maximum transform length which can be used is

$$N_{\max} = p - 1 \tag{9.84}$$

which is determined by the chosen prime, p.

We call the root of unity of order $N_{\max}$ the primitive p-adic root $H(p, r, \gamma_\phi)$; the powers of this primitive root will generate all the roots of unity in the field $\hat{Q}_p$.

The number-theoretic properties are satisfied, i.e., the number of primitive roots is given by $\phi(\phi(p))$; the number of roots of order N is given by $\phi(N)$, and the Nth root γ_N can be obtained from the primitive root γ_ϕ as given by

$$\gamma_N = \gamma_\phi^{(p-1)/N} \tag{9.85}$$

In order to find $H(p, r, \gamma_\phi)$, we have to find the solution for

$$[H(p, r, x)]^{p-1} = H(p, r, 1) \quad \text{in } \hat{Q}_p \tag{9.86}$$

or, in general,

$$[H(p, r, x)]^{N} = H(p, r, 1) \quad \text{in } \hat{Q}_p$$

Let $\gamma = \{a_0 \ a_1 \ \cdots \ a_n \ \cdots\}$ be the p-adic representation of the Nth root of unity in $\hat{Q}_p$. By p-adic multiplication N times, we get

$$\gamma^N = \{a_0^N, Na_0^{N-1}a_1, \ldots\} \equiv \{1, 0, 0, \ldots\} \pmod{p} \tag{9.87}$$

Thus

$$a_0^N \equiv 1 \pmod{p} \tag{9.88}$$

and, by the Fermat–Euler theorem,[25] $a_0^{\phi(p)} \equiv 1 \pmod{p}$, and since $\phi(p) = p - 1$, then

$$a_0^{p-1} \equiv 1 \pmod{p} \tag{9.89}$$

Hence, for $N = p - 1$, there exists a root a_0 of order $p - 1$. Once the congruence (9.89) is solved for a_0, the p-adic primitive root $H(p, r, \gamma)$ can be evaluated.[47]

9.4.4. Mersenne and Fermat p-adic Transforms

The basic arithmetic operations in a segmented p-adic field $\hat{Q}_p$ are performed modulo p, where p is a prime number. However, operations mod p are very costly; thus Mersenne and Fermat primes are usually chosen in order to reduce the complexity of the processor's structure.[27]

9.4.4.1. Mersenne p-adic Transform

We consider the segmented p-adic field $\hat{Q}_p$ with p a Mersenne prime, $p = 2^q - 1$, where q is a prime number. Thus, the PAT in this field has the maximum length

$$\begin{aligned} N_{\max} &= p - 1 \\ &= 2^q - 2 \\ &= 2(2^{q-1} - 1) \end{aligned} \tag{9.90}$$

which is not highly composite. Thus a simple radix-2 FFT algorithm cannot be used. However, more complex and efficient techniques, such as prime factorization or Winograd algorithms, can be used. In these algorithms, the transform length is put into prime factorization form as shown in Table 9.1 for several Mersenne primes. Then, by the CRT, this 1-D PAT is mapped into a multidimensional transform such that it is cyclic, with prime transform length in every dimension.[53-55] This multidimensional PAT is then implemented by the Rader algorithm,[55] which is a conversion of prime length p_i transforms into circular convolutions of $p_i - 1$ points which, in turn, are implemented by FFT algorithms, short convolutions, or polynomial product algorithms.[56]

9.4.4.2. Fermat p-adic Transform

Unlike the Mersenne p-adic transform, the Fermat p-adic transform has a transform length which is highly composite; thus a radix-2 FFT algorithm can be employed. This is due to the fact that, given a Fermat prime $F_t = 2^{2^t} + 1$ for $t = 1, 2, 3, 4$, the maximum transform length $N_{\max}$

Table 9.1. Prime Factorization of Transform Lengths based on Mersenne Primes

q	$2(2^{q-1} - 1)$	$p_1\, p_2 \cdots p_i$
2	2	2
3	6	$2 \cdot 3$
5	30	$2 \cdot 3 \cdot 5$
7	126	$2 \cdot 3^2 \cdot 7$
13	8190	$2 \cdot 3^2 \cdot 5 \cdot 7 \cdot 13$

is given by

$$N_{\max} = F_t - 1$$
$$= 2^{2^t} \qquad (9.91)$$

and consequently the Fermat p-adic transform can be directly implemented by FFT algorithms.

9.4.5. Complex p-adic Transform

We consider a complex p-adic sequence $H(p, r, u_n)$ to be filtered by a complex sequence having N terms $H(p, r, b_n)$, in which $H(p, r, y_m)$ is the output p-adic sequence given by

$$H(p, r, y_m) = \sum_{n=0}^{N-1} H(p, r, b_n)H(p, r, u_{m-n}) \qquad \text{for } 0 \le n, m \le N-1 \qquad (9.92)$$

where

$$H(p, r, u_n) = H(p, r, x_n) + jH(p, r, \hat{x}_n)$$

$$H(p, r, b_n) = H(p, r, h_n) + jH(p, r, \hat{h}_n)$$

$$H(p, r, y_m) = H(p, r, Z_m) + jH(p, r, \hat{Z}_m) \qquad (9.93)$$

and $j = \sqrt{-1}$.

The complex convolution (9.92) can be decomposed into four real convolutions as follows:

$$H(p, r, y_m) = \sum_{n=0}^{N-1} [H(p, r, h_n)H(p, r, x_{m-n}) - H(p, r, \hat{h}_n)H(p, r, \hat{x}_{m-n})]$$
$$+ j\sum_{n=0}^{N-1} [H(p, r, \hat{h}_n)H(p, r, x_{m-n})$$
$$+ H(p, r, h_n)H(p, r, \hat{x}_{m-n})] \qquad (9.94)$$

Thus

$$H(p, r, Z_m) = \sum_{n=0}^{N-1} [H(p, r, h_n)H(p, r, x_{m-n}) - H(p, r, \hat{h}_n)H(p, r, \hat{x}_{m-n})] \qquad (9.95)$$

and

$$H(p, r, \hat{Z}_m) = \sum_{n=0}^{N-1} [H(p, r, \hat{h}_n)H(p, r, x_{m-n}) + H(p, r, h_n)H(p, r, \hat{x}_{m-n})] \qquad (9.96)$$

It is seen that expression (9.94) consists of four real convolutions, which can be implemented separately by a fast Mersenne or Fermat p-adic transform.

9.4.6. Complex Convolution Via a Fermat p-adic Transform

We consider a PAT defined in $\hat{Q}_p$, where p is chosen to be a Fermat prime. With this selection of p, $\mathrm{j} = \sqrt{-1}$ has a special representation in Q_p.[19] In other words, since the equation $x^2 + 1 = 0$ is solvable in Q_p, the square root of -1 can be represented by a p-adic sequence, say $H(p, r, \mathrm{j})$ where p is a Fermat prime, and we can rewrite equation (9.92) in terms of an N-point real p-adic convolution

$$H(p, r, y_m) = \sum_{n=0}^{N-1} [H(p, r, h_n) + H(p, r, \mathrm{j})H(p, r, \hat{h}_n)] \times [H(p, r, x_{m-n}) + H(p, r, \mathrm{j})H(p, r, \hat{x}_{m-n})] \tag{9.97}$$

where

$$H(p, r, y_m) = H(p, r, Z_m) + H(p, r, \mathrm{j})H(p, r, \hat{Z}_m) \tag{9.98}$$

In order to recreate the in-phase and quadratic components $H(p, r, Z_m)$ and $H(p, r, \hat{Z}_m)$ of the output sample given by equation (9.98), we consider the auxiliary convolution

$$H(p, r, v_m) = \sum_{n=0}^{N-1} [H(p, r, h_n) - H(p, r, \mathrm{j})H(p, r, \hat{h}_n)] \times [H(p, r, x_{m-n}) - H(p, r, \mathrm{j})H(p, r, \hat{x}_{m-n})] \tag{9.99}$$

where

$$H(p, r, v_m) = H(p, r, Z_m) - H(p, r, \mathrm{j})H(p, r, \hat{Z}_m) \tag{9.100}$$

On combining equations (9.98) and (9.100), we get

$$H(p, r, Z_m) = H(p, r, 1/2)[H(p, r, y_m) + H(p, r, v_m)] \tag{9.101}$$

and

$$H(p, r, \hat{Z}_m) = H(p, r, 1/2\mathrm{j})[H(p, r, y_m) - H(p, r, v_m)] \tag{9.102}$$

Thus, in the p-adic field $\hat{Q}_p$ where p is a Fermat number, an N-point complex p-adic convolution can be achieved by two N-point real p-adic convolutions (9.99), instead of a conventional approach with four real p-adic convolutions.

9.4.7. Quadratic p-adic Transform in Extension Fields of Q_p with p a Mersenne Prime

The p-adic field Q_p has infinitely many algebraic extension fields, all the fields being generated by roots of algebraic equations $X^n - p = 0$ ($n = 2, 3, 4, \ldots$).[19] For simplicity, only quadratic extension fields of Q_p are considered. Mahler[57] has shown that, for $p \geq 3$, there are exactly three distinct quadratic extensions of Q_p, and these may be represented by $Q_p(\sqrt{N_p})$, $Q_p(\sqrt{p})$, and $Q_p(\sqrt{pN_p})$, where N_p is the smallest positive quadratic nonresidue; otherwise, the equation $X^2 - N_p = 0$ is said to have no solution in Q_p. By Euler's theorem,[28] this is further equivalent to

$$\left[\frac{N_p}{p}\right] = (N_p)^{(N_p-1)/2} = -1 \tag{9.103}$$

where $[\frac{m}{p}]$ is the Legendre symbol defined in equation (9.18).

Any one of the three quadratic extension fields can be denoted by $K_p = Q_p(\sqrt{d})$. Every element Z of K_p can be written in the form

$$Z = x + jy$$

where $x, y \in Q_p$ and $j = \sqrt{d}$.

In the following sections we define the p-adic transforms in the three extension fields of Q_p with p a Mersenne prime.

9.4.7.1. *p-adic Transforms in* $Q_p(\sqrt{N_p})$

We consider the quadratic extension field $K_p = Q_p(\sqrt{N_p})$ with p a Mersenne prime, $p = 2^q - 1$, and q a prime ≥ 2. Since $x^2 \equiv -1 \pmod{p}$ has no solution in Q_p, or by Euler's theorem,

$$\begin{aligned}(-1/p) &= (-1)^{(p-1)/2} \\ &= (-1)^{(2^q-2)/2} \\ &= (-1)^{(2^{q-1}-1)} \\ &= -1\end{aligned} \tag{9.104}$$

Thus $N_p = -1$ is the smallest quadratic nonresidue modulo a Mersenne prime, and the polynomial $P(x) = x^2 + 1$ is said to be irreducible in Q_p. Let us informally represent $\mathrm{i} = \sqrt{-1}$ as a root of the polynomial

$P(x) = x^2 + 1$, satisfying $i^2 = -1$, where i is an element of the extension field $Q_p(\sqrt{-1})$, which is composed of the set

$$K_p = Q_p(\sqrt{-1}) = \{A + iB\} \tag{9.105}$$

where $A, B \in Q_p$. We note that $i \in K_p$ plays here the same role over the field Q_p as $i = \sqrt{-1}$ plays over the field of rationals Q, and the basic arithmetic operations in K_p are similar to complex arithmetic in Q.

A complex PAT in the extension field $Q_p(\sqrt{-1})$ is defined as[19]

$$\langle X_k \rangle = \sum_{n=0}^{d-1} \langle x_n \rangle \langle \gamma \rangle^{nk}, \qquad 0 \le k \le d-1 \tag{9.106}$$

and its inverse by

$$\langle x_n \rangle = \left\langle \frac{1}{d} \right\rangle \sum_{k=0}^{d-1} \langle X_k \rangle \langle \gamma \rangle^{-nk}, \qquad 0 \le n \le d-1 \tag{9.107}$$

where $\langle x_n \rangle$, $\langle X_k \rangle$, and $\langle \gamma \rangle \in Q_p(\sqrt{-1})$ and p is a Mersenne prime.

The conditions for the existence of this transform are similar to those of a complex NTT

1. γ is a root of unity of order d in $Q_p(\sqrt{-1})$ in order to have the cyclic convolution property, thus $\gamma^d = 1$ in $Q_p(\sqrt{-1})$.
2. $1/d$ exists in $Q_p(\sqrt{-1})$.
3. $d \mid (p^2 - 1)$.

The number of elements in $Q_p(\sqrt{-1})$ is p^2, as in a Galois field of p^2 elements, GF(p^2).[33] The maximum order t of the multiplicative group with α as the generator is

$$\begin{aligned} t = p^2 - 1 &= (2^q - 1)^2 - 1 \\ &= 2^{q+1}(2^{q-1} - 1) \end{aligned} \tag{9.108}$$

Thus, $\alpha^{p^2-1} = 1$ in $Q_p(\sqrt{-1})$.

Quantity γ as the generator of order d can be obtained from

$$\gamma = \alpha^{(p^2-1)/d} \tag{9.109}$$

as in the NTT. This yields the third condition given above. Also, for $d = 2^{q+1}$, a radix-2 FFT can be used.

When considering processors of finite word length, and as we previously showed in the segmented p-adic field $\hat{Q}_p$, in order to perform arithmetic in the extension fields, it is necessary to form a finite-segment extension field

$\hat{K}_p = \hat{Q}_p(\sqrt{-1})$ which is a subset of K_p. This is obtained by truncating the p-adic series representation of the elements in K_p to r digits. This produces a finite set $\hat{K}_p$ with elements represented by

$$H(p, r, Z) = H(p, r, A) + jH(p, r, B) \tag{9.110}$$

where $H(p, r, A)$ [or $H(p, r, B)$] has the canonic expansion

$$H(p, r, A) = \sum_{n=-m}^{r} a_n p^n \tag{9.111}$$

and $a_n \in I_p$. Thus

$$H(p, r, Z) = \sum_{n=-m}^{r} Z_n p^n \tag{9.112}$$

where $Z_n = a_n + jb_n$.

The PAT pair defined in $\hat{K}_p$ must satisfy the same conditions as the PAT defined in K_p, with an additional dynamic-range constraint such that if

$$\left(\frac{a}{b} + j\frac{c}{d}\right) \in K_p,$$

then, also,

$$-[(p^r - 1)/2]^{1/2} \le a, b, c, d \le [(p^r - 1)/2]^{1/2} \tag{9.113}$$

Thus, to compute the transform over $\hat{K}_p$ of a d-point sequence of complex numbers, $H(p, r, x_n) \in \hat{K}_p$, we have to find the root of unity γ of order d. The complex convolution is now implemented by two forward complex PATs and one inverse transform.

9.4.7.2. p-adic Transforms in $Q_p(\sqrt{p})$ and $Q_p(\sqrt{pN_p})$

Having defined the p-adic transforms in $Q_p(\sqrt{N_p})$ where $N_p = -1$ and p is a Mersenne prime, we now define the PATs in $Q_p(\sqrt{p})$ and $Q_p(\sqrt{-p})$ for the same conditions. These two extension fields are known as ramified extension fields,[57] because the elements in these fields have the special representation

$$Z = x + \sqrt{p}y = \sum_{n=-m}^{\infty} Z_n \hat{p}^n \tag{9.114}$$

where $Z_n \in I_p$, and $\hat{p} = \sqrt{(p/E)}$ or $p = E\hat{p}^2$ such that

$$\begin{aligned} E &= 1 \qquad \text{if } K_p = Q_p(\sqrt{p}), \qquad p > 3 \\ &= -1 \qquad \text{if } K_p = Q_p(\sqrt{-p}), \qquad p > 3 \end{aligned}$$

Thus, an element in these two fields is represented by a series of power $\hat{p}$ with real coefficients $Z_n \in I_p$, as given by expression (9.114). The transform pair in $Q_p(\sqrt{p})$ is defined as

$$X_k = \sum_{n=0}^{d-1} x_n \beta^{nk}, \qquad 0 \leq k \leq d-1 \tag{9.115}$$

$$x_n = \frac{1}{d} \sum_{k=0}^{d-1} X_k \beta^{-nk}, \qquad 0 \leq n \leq d-1 \tag{9.116}$$

where x_n, X_k, $\beta \in Q_p(\sqrt{p})$ and p is a Mersenne prime. The condition given above for the existence of a transform must also be satisfied in this extension field. Quantity β is a root of unity of order d in $Q_p(\sqrt{p})$ and the arithmetic in $Q_p(\sqrt{p})$ is similar to that in a complex p-adic field.

9.4.8. Quadratic p-adic Transform in Extension Fields of Q_p with p a Fermat Prime

By considering the three distinct extension fields given by

$$K_p = Q_p(\sqrt{N_p}),\ Q_p(\sqrt{p}),\ Q_p(\sqrt{pN_p})$$

where N_p is the smallest integer nonresidue modulo a Fermat prime, it can be proved[14] that a complex PAT, where $N_p = -1$, does not exist with p a Fermat prime. Similarly to the previous analysis, a transform pair can be defined in K_p. The maximum transform length N_{max} is given by $N_{max} = 2^{2^t}$ where $t = 1, 2, 3, 4$. Such a transform is highly composite and hence a radix-2 FFT algorithm can be used for its implementation.

9.4.9. p-adic Transform Defined in a g-adic Ring Over a Direct Sum of Several p-adic Fields

Any rational number α can be represented in a g-adic ring Q_g by an infinite series[52]

$$\alpha = \sum_{n=-m}^{\infty} a_n g^n \tag{9.117}$$

Such an infinite series is convergent with respect to the g-adic norm.

Also, Q_g can be defined as the direct sum of the p-adic fields Q_{p_k} for $k = 1, 2, \ldots, n$. Thus

$$Q_g = Q_{p_1} \oplus Q_{p_2} \oplus \cdots \oplus Q_{p_n} \tag{9.118}$$

Defining a Fourier-like transform over Q_g is analogous to defining it over the direct sum of the fields Q_{p_k} [equation (9.118)]. Thus, any element A of Q_g can be represented by its p-adic components

$$A = \langle A_1, A_2, \ldots, A_n \rangle \tag{9.119}$$

and the basic arithmetic operations can be performed in each p-adic field. For example, if two g-adic numbers are composed of n p-adic components,

$$A = \langle A_{p_1}, A_{p_2}, \ldots, A_{p_n} \rangle \qquad \text{and} \qquad B = \langle B_{p_1}, B_{p_2}, \ldots, B_{p_n} \rangle$$

then

$$AB = \langle A_{p_1} B_{p_1}, A_{p_2} B_{p_2}, \ldots, A_{p_n} B_{p_n} \rangle \tag{9.120}$$

Therefore, the arithmetic in each field can be performed in parallel, provided that we retain the ability to reconstruct the number from its components. Thus, a Fourier-like transform defined over Q_g for a g-adic sequence a_n, $n = 0, 1, \ldots, d-1$ is given by

$$A_k = \sum_{n=0}^{d-1} a_n \gamma^{nk}, \qquad k = 0, 1, \ldots, d-1 \tag{9.121}$$

and

$$a_n = \frac{1}{d} \sum_{k=0}^{d-1} A_k \gamma^{-nk}, \qquad n = 0, 1, \ldots, d-1 \tag{9.122}$$

where a_n, A_k, and $\gamma \in Q_g$, or

$$a_n = [(a_n)_{p_1}, (a_n)_{p_2}, \ldots, (a_n)_{p_i}]$$

and

$$A_k = [(A_k)_{p_1}, (A_k)_{p_2}, \ldots, (A_k)_{p_i}]$$

The condition for orthogonality is that

$$\gamma^d = 1 \qquad \text{in } Q_g \tag{9.123}$$

and since γ is given by its components as

$$\gamma = [\gamma_{p_1}, \gamma_{p_2}, \ldots, \gamma_{p_i}] \tag{9.124}$$

where $\gamma_{p_i} \in Q_{p_i}$, then each component of γ should satisfy

$$\gamma_{p_j} = 1 \quad \text{in } Q_{p_j}, \quad j = 1, 2, \ldots, i \tag{9.125}$$

This shows that the transform over Q_g is the same as a direct sum of transforms over $\{Q_{p_1}, Q_{p_2}, \ldots, Q_{p_n}\}$, in parallel, yielding a larger dynamic range.

9.5. THE USE OF TRANSFORMS BASED ON NUMBER THEORETIC ALGORITHMS

In this chapter, we have discussed orthogonal transforms which have the DFT structure, but which are based on number theoretic approaches. Number theoretic transforms, in particular, are attractive in that they are capable of providing error-free computation; in addition, all computations are carried out in the real domain. The main drawback is that all calculations are performed over a finite field (or a finite ring) of integers, and this imposes a limit on the maximum possible transform length constrained by the word length of the machine. The net effect of these constraints is that 2-D block recursive processing, for instance, becomes computationally very inefficient using NTT. Despite this, it may be of value in that no roundoff error is introduced.

On the other hand, for block dimensions which are an integral power of two, polynomial transforms which have an FFT-type structure may be used. These avoid the need to use trigonometric functions and complex arithmetic. For instance, in the case of image-processing applications the number of multiplications per output pixel required, N_{PT}, when using polynomial transform algorithms to implement the circular convolution, is given in Table 9.2. By comparison with Tables 8.3 and 8.4, we see that this method of implementation is the most efficient in terms of number of multiplications, especially for block sizes greater than about 8×8. Additionally, the roundoff noise is much lower than that of the FFT technique.[58]

Jullien *et al.*[59] have reviewed extensively the use of quadratic residue number systems in the implementation of FIR and IIR filter structures. In particular, practical algorithms defined over finite fields or rings and based on complex number theoretic transforms have been considered, together with their use in the indirect computation of digital convolution and a

Table 9.2. Number of Multiplications Required to Compute the Circular Convolution for Various Block Sizes (K, L) using Polynomial Transform Algorithms

(K, L)	N_{PT}
(2, 2)	27.5
(4, 4)	40.6
(8, 8)	60.8
(16, 16)	74.3
(32, 32)	107.2
(64, 64)	121.7
(128, 128)	141.2

recursive FIR filter structure. Furthermore, the hardware implementation details for complex number theoretic transforms have been discussed by Jullien *et al.*[59] Two schemes were considered in this implementation: a massively parallel and a linear systolic structure.

In the case of p-adic transforms it is seen that, for digital signal processing purposes, they have very attractive features, due to the possibility of undertaking completely error-free computation over a p-adic field. Furthermore, they provide a larger dynamic range and a longer transform length, which makes their use more advantageous than the NTT. Their most attractive feature, however, lies in the ability to map, exactly, the field of real numbers onto the integer-like p-adic field and carry out different operations within this domain prior to the final conversion to the real number equivalent.

p-adic number algorithms are currently being practically investigated as viable tools for implementing circular convolution and as a means of eliminating noise sources in digital filters.

REFERENCES

1. J. H. McClellan, Hardware realization of a Fermat number transform, *IEEE Trans. Acoust., Speech, Signal Process.* **ASSP-24,** 216–225 (1976).
2. A. Baraniecka and G. A. Jullien, Hardware implementation of convolution using number theoretic transforms, *Proc. Int. Conf. Acoust., Speech, Signal Process., Washington, D.C.,* 490–492 (1979).
3. R. C. Agarwal and C. S. Burrus, Number theoretic transforms to implement fast digital convolution, *Proc. IEEE* **63,** 550–560 (1975).
4. I. S. Reed and T. K. Truong, A fast DFT algorithm using complex integer transforms, *Electron. Lett.* **14,** 191–193 (1978).
5. H. J. Nussbaumer, Digital filtering using pseudo Fermat number transform, *IEEE Trans. Acoust., Speech, Signal Process.* **ASSP-25,** 79–83 (1977).

6. H. J. Nussbaumer, Linear filtering technique for computing Mersenne and Fermat number transforms, *IBM J. Res. Dev.* **21,** 334–339 (1977).
7. H. J. Nussbaumer, Relative evaluation of various number theoretic transforms for digital filtering applications, *IEEE Trans. Acoust., Speech, Signal Process.* **ASSP-26,** 88–93 (1978).
8. H. J. Nussbaumer, Complex convolutions via Fermat number transforms, *IBM J. Res. Dev.* **20,** 282–284 (1976).
9. H. J. Nussbaumer, Digital filtering using complex Mersenne transforms, *IBM J. Res. Dev.* **20,** 498–504 (1976).
10. H. J. Nussbaumer, *Fast Fourier Transform and Convolution Algorithms,* Springer-Verlag, Berlin and New York (1981).
11. M. C. Vanwormhoudt, On number theoretic Fourier transforms in residue class rings, *IEEE Trans. Acoust., Speech, Signal Process.* **ASSP-25,** 585–586 (1977).
12. M. C. Vanwormhoudt, Structural properties of complex residue rings applied to number theoretic Fourier transforms, *IEEE Trans. Acoust., Speech, Signal Process.* **ASSP-26,** 99–104 (1978).
13. K. M. Henein, *Number Theoretic Transforms in Digital Signal Processing,* MSc. Thesis, Imperial College of Science and Technology, University of London (1983).
14. N. M. Nasrabadi, *Orthogonal Transforms and their Applications to Image Coding,* PhD. Thesis, Imperial College of Science and Technology, University of London (1984).
15. R. N. Gorgui-Naguib and A. Leboyer, Comment on "Determination of p-adic transform bases and lengths," *Electron. Lett.* **21,** 905–906 (1985).
16. A. Leboyer, *P-adic Numbers—P-adic Transform,* MSc. Communications report, Imperial College of Science and Technology (1985).
17. V. Loahakosol and W. Surakampontorn, P-adic transforms, *Electron. Lett.* **20,** 726–727 (1984).
18. N. M. Nasrabadi and R. A. King, Fast digital convolution using p-adic transforms, *Electron. Lett.* **19,** 266–267 (1983).
19. N. M. Nasrabadi and R. A. King, Complex number theoretic transform in p-adic field, *Proc. IEEE—Int. Conf. Acoust., Speech, Signal Process.* **1984,** pp. 28A.4.1–28A.4.3 (1984).
20. S.-C. Pei and J.-L. Wu, Determination of p-adic transform bases and lengths, *Electron. Lett.* **21,** 431–432 (1985).
21. W. W. Adams and L. J. Goldstein, *Introduction to Number Theory,* Prentice-Hall, Englewood Cliffs, NJ (1976).
22. T. A. Apostol, *Introduction to Analytic Number Theory,* Springer-Verlag, Berlin (1976).
23. Z. I. Borevich and I. R. Shafarevich, *Number Theory,* Academic Press, New York (1973).
24. W. J. Leveque, *Fundamentals of Number Theory,* Addison-Wesley, Reading, Massachusetts (1977).
25. I. Niven and H. S. Zuckerman, *An Introduction to the Theory of Numbers,* 4th edn., John Wiley and Sons, New York (1980).
26. I. N. Herstein, *Topics in Algebra,* 2nd edn., John Wiley and Sons, New York (1975).
27. J. H. McClellan and C. M. Rader, *Number Theory in Digital Signal Processing,* Prentice-Hall Signal Processing Series (1979).
28. G. H. Hardy and E. M. Wright, *An Introduction to the Theory of Numbers,* 5th edn., Oxford University Press (1979).
29. C. M. Rader, Discrete convolutions via Mersenne transforms, *IEEE Trans. Comput.* **C-21,** 1269–1273 (1972).
30. R. C. Agarwal and C. S. Burrus, Fast convolution using Fermat number transforms with applications to digital filtering, *IEEE Trans. Acoust., Speech, Signal Process.* **ASSP-22,** 87–97 (1974).
31. S. Winograd, On computing the discrete Fourier transform, *Math. Comput.* **32,** 175–199 (1978).

32. D. P. Kolba and T. W. Parks, A prime factor FFT algorithm using high speed convolution, *IEEE Trans. Acoust., Speech, Signal Process.* **ASSP-25,** 90–103 (1977).
33. I. S. Reed and T. K. Truong, Convolutions over residue classes of quadratic integers, *IEEE Trans. Inf. Theory* **IT-22,** 468–475 (1976).
34. H. J. Nussbaumer and P. Quandalle, Fast computation of discrete Fourier transforms using polynomial transforms, *IEEE Trans. Acoust., Speech, Signal Process.* **ASSP-27,** 169–181 (1979).
35. H. J. Nussbaumer and P. Quandalle, Computation of convolutions and discrete Fourier transforms by polynomial transform, *IBM J. Res. Dev.* **22,** 134–144 (1978).
36. H. J. Nussbaumer, DFT computation by fast polynomial transform algorithms, *Electron. Lett.* **15,** 701–702 (1979).
37. B. Arambepola and P. J. W. Rayner, Discrete transform over polynomial rings with applications in computing multidimensional convolutions, *IEEE Trans. Acoust., Speech, Signal Process.* **ASSP-28,** 407–414 (1980).
38. H. J. Nussbaumer, Digital filtering using polynomial transforms, *Electron. Lett.* **13,** 386–387 (1977).
39. I. S. Reed, H. M. Shao, and T. K. Truong, Fast polynomial transform and its implementation by computer, *IEE Proc.* **128,** Part E, No. 1, 50–60 (1981).
40. B. Arambepola and P. J. W. Rayner, Efficient transforms for multidimensional convolutions, *Electron. Lett.* **15,** 189–190 (1979).
41. B. Arambepola and P. J. W. Rayner, Multidimensional fast Fourier transform algorithms, *Electron. Lett.* **15,** 382–383 (1979).
42. H. J. Nussbaumer, New polynomial transform algorithms for multidimensional DFT's and Convolutions, *IEEE Trans. Acoust., Speech, Signal Process.* **ASSP-29,** 761–783 (1981).
43. H. J. Nussbaumer, Inverse polynomial transform algorithms for DFT's and convolutions, *Proc. IEEE—Int. Conf. Acoust., Speech, Signal Process.* **1981,** 315–318 (1981).
44. K. Hensel, *Theorie der Algebraischen Zahlen*, Teubner, Leipzig (1908).
45. K. Hensel, *Zahlentheorie*, Goschen, Berlin and Leipzig (1913).
46. R. N. Gorgui-Naguib, *P-adic Number Theory and its Applications in a Cryptographic System*, PhD. Thesis, Imperial College of Science and Technology, University of London (1986).
47. G. Bachman, *Introduction to P-adic Numbers and Valuation Theory*, Academic Press, New York (1964).
48. E. V. Krishnamurthy, T. Mahadeva Rao, and K. Subramanian, Finite-segment p-adic number systems with applications to exact computation, *Proc. Indian Acad. Sci.* **81A,** 58–79 (1975).
49. E. V. Krishnamurthy, T. Mahadeva Rao, and K. Subramanian, *P*-adic arithmetic procedures for exact matrix computations, *Proc. Indian Acad. Sci.* **82A,** 165–175 (1975).
50. E. V. Krishnamurthy, Matrix processors using p-adic arithmetic for exact linear computations, *IEEE Trans. Comput.* **C-26,** 633–639 (1977).
51. R. N. Gorgui-Naguib and R. A. King, Comments on matrix processors using p-adic arithmetic for exact linear computation, *IEEE Trans. Comput.* **C-35,** 928–930 (1986).
52. N. Koblitz, *P-adic Numbers, P-adic Analysis and Zeta-functions*, Graduate Texts in Mathematics, Springer-Verlag, Berlin (1977).
53. I. S. Reed and T. K. Truong, Fast Mersenne-prime transforms for digital filtering, *Proc. IEEE* **125,** 433–440 (1978).
54. K. Y. Liu, I. S. Reed, and T. K. Truong, Fast number theoretic transforms for digital filtering, *Electron. Lett.* **12,** 644–646 (1976).
55. C. M. Rader, Discrete Fourier transforms when the number of data samples is prime, *Proc. IEEE* **56,** 1107–1108 (1968).
56. R. C. Agarwal and J. W. Cooley, New algorithms for digital convolution, *IEEE Trans. Acoust., Speech, Signal Process.* **ASSP-25,** 392–410 (1977).

57. K. Mahler, *P-adic Numbers and their Functions*, Cambridge University Press (1980).
58. H. J. Nussbaumer, Fast polynomial algorithms for digital convolution, *IEEE Trans. Acoust., Speech, Signal Process.* **ASSP-28**, 205–215 (1980).
59. G. A. Jullien, R. Krichman, and W. C. Miller, Complex digital signal processing over finite rings, *IEEE Trans. Circuits Syst.* **CAS-34**, 365–377 (1987).
60. G. A. Jullien, Implementation of multiplication modulo *a prime number*, with applications to number theoretic transforms, *IEEE Trans. Comput.* **C-29**, 899–905 (1980).

10

Image Modeling

In the preceding chapters, the design and analysis of recursive and nonrecursive one-dimensional (1-D) and two-dimensional (2-D) digital filters have been discussed for the case of deterministic input signals. However, in situations such as satellite and radar imaging, knowledge of the original image is not deterministically available. Each pixel is considered as a random variable and the image is thought of as a sample of an ensemble of images. For simplicity in modeling, the image is assumed to have a Gaussian distribution which can be specified uniquely by its first- and second-order moments (mean and covariances). To perform the estimation and filtering process, the image should be represented by an appropriate statistical model.

Two-dimensional finite-order autoregressive (AR) or autoregressive moving average (ARMA) models have found numerous applications in parametric representations of image fields. For example, they can be used in the areas of image restoration,[1-4] image coding,[5] and texture analysis.[6]

The problem of image modeling involves fitting an appropriate AR or ARMA model to image data which would realize the given covariance function or, equivalently, the spectral density function (SDF). The procedure in the 2-D case differs from that in the 1-D case in two major ways. First, in the 1-D case, where time is normally the independent variable, causality is a rigid requirement for physical realizability of the model, while in the 2-D case, where variables are normally spatial, the region of support of the model can be causal, semicausal, or noncausal, leading to initial-value, initial-boundary-value, or boundary-value problems,[1] respectively. Second, in contrast to the 1-D case where finite-order models can realize any process with a given rational SDF of finite order, in the 2-D case finite-order difference equations in general may not be capable of representing any 2-D rational SDF, due to the lack of general 2-D spectral factorization.

The aim of this chapter is to study the problem of image modeling and the related issues on causality, stability, and spectral factorization. In this context, first a review of 1-D causal and noncausal models and their properties is provided. Then, several 2-D AR and ARMA models with different prediction geometries are introduced. Examples are included to illustrate the applications of these models to image processing.

10.1. ONE-DIMENSIONAL AUTOREGRESSIVE, MOVING AVERAGE, AND AUTOREGRESSIVE MOVING AVERAGE MODELS

The parametric approach to spectral estimation involves three steps. In the first step an appropriate parametric time-series model, such as an autoregressive (AR), moving average (MA) or combined autoregressive moving average (ARMA) model, is to be selected to represent the measured data. In step two, an estimate of the parameters of the model should be made. In step three, the estimated parameters should be used to match the given SDF. The choice of one of the three models is dictated by the shape and characteristics of the spectra. However, the principle of parsimony precludes the use of a model having a large number of parameters. The computational burden required to estimate the parameters of an AR model is often significantly less than those of MA and ARMA models. Thus, AR models are preferred choices, even though they may not be the most parsimonious representation. In this section, these three models are introduced for modeling 1-D processes.

10.1.1. Causal Representation

If $\{u_k\}$ is a discrete-time, zero mean stationary Markov process, then a causal representation of the type

$$u_k = -\sum_{i=1}^{M} a_i u_{k-i} + \sum_{i=0}^{N} b_i e_{k-i} \tag{10.1}$$

is called a (one-sided) autoregressive moving average (ARMA) model. The driving sequence $\{e_k\}$ is a white noise process of zero mean and variance σ_e^2, i.e.,

$$E[e_k] = 0$$

$$E[e_k\ e_l] = \sigma_e^2 \delta(k-l) \tag{10.2}$$

where $\delta(\cdot)$ represents the Kronecker delta function and E is the expectation operator. The parameters a_i form the autoregressive portion of the ARMA model, while the parameters b_i form the moving-average portion of the model. The notation ARMA(M, N) is often used to indicate an ARMA model of order M autoregressive parameters and order N moving-average parameters. The system function has the rational form

$$H(z) = \frac{B(z)}{A(z)} \tag{10.3}$$

in which the polynomials are

$$A(z) = 1 + \sum_{i=1}^{M} a_i z^{-i} \tag{10.4a}$$

$$B(z) = 1 + \sum_{i=1}^{N} b_i z^{-i}, \qquad b_0 = 1 \tag{10.4b}$$

$$H(z) = 1 + \sum_{i=1}^{\infty} h_i z^{-i} \tag{10.4c}$$

where $\{h_k\}$ is the impulse response sequence.

Both $A(z)$ and $B(z)$ would be assumed to have all of their zeros within the unit circle in the z plane to ensure that $H(z)$ is a BIBO stable, minimum-phase causal filter. If stationarity is assumed, the SDF of $\{u_k\}$ is given by

$$S_u(z) = \sum_{n=-\infty}^{\infty} r_u(n) z^{-n} \tag{10.5}$$

where $r_u(n)$ is the autocorrelation function defined by

$$r_u(n) = E[u_{i+n}\ u_i] \tag{10.6}$$

The autocorrelation function of $\{u_k\}$ is related to that of $\{e_k\}$ by

$$r_u(n) = r_e(n) * h(-n) * h(n) \tag{10.7}$$

where $*$ stands for the convolution operation, and $r_e(n) = E[e_{i+n}\ e_i]$.

Now taking the z transform of both sides of equation (10.7) yields

$$S_u(z) = S_e(z) H(z) H(z^{-1}) \tag{10.8}$$

or

$$S_u(z) = S_e(z) \frac{B(z)B(z^{-1})}{A(z)A(z^{-1})} \tag{10.9}$$

where $S_e(z)$ for the white noise process $\{e_k\}$ with statistics given in equations (10.2) is

$$S_e(z) = \sigma_e^2 \tag{10.10}$$

If all the autoregressive parameters are zero, except $a_0 = 1$, then

$$u_k = \sum_{i=0}^{N} b_i e_{k-i} \tag{10.11}$$

is strictly a moving-average (MA) process of order N, or simply MA(N). The SDF for the model is

$$S_u(z) = S_e(z)B(z)B(z^{-1}) \tag{10.12}$$

If all the moving-average parameters are zero, except $b_0 = 1$, then

$$u_k = -\sum_{i=1}^{M} a_i u_{k-i} + e_k \tag{10.13}$$

is strictly an autoregressive (AR) process of order M, or simply AR(M). The SDF in this case is

$$S_u(z) = \frac{S_e(z)}{A(z)A(z^{-1})} \tag{10.14}$$

We note that in equations (10.9) and (10.14) the denominator $A(z)A(z^{-1})$ is a polynomial in z of order $2M$. The zeros of this polynomial occur in reciprocal pairs, i.e., if z_k is a root of $A(z)$, then $1/z_k$ will be a root of $A(z^{-1})$. Given the SDF of the data, the aim of the spectral factorization is to decompose the poles and zeros of the SDF into those that are within the unit circle, and those that are outside, leading to minimum-phase and maximum-phase factors, respectively. This guarantees the existence of a stable, minimum-phase causal realization. The general techniques to accomplish this include the Wiener–Doob method and the linear prediction method.[7] The linear prediction method fits successively higher-order AR models to the given autocorrelation function by solving a finite set of Toeplitz equations. Under some mild conditions on the given SDF, the SDFs realized by the models can be shown to converge uniformly to the given SDF.

10.1.1.1. Parameter Estimation for Autoregressive Moving-Average Processes[7]

Let us consider the ARMA model in equation (10.1). To determine the model parameters, multiply both sides by u_{k-n} and then take expectation.

This yields

$$E[u_k \; u_{k-n}] = -\sum_{i=1}^{M} a_i E[u_{k-i} \; u_{k-n}] + \sum_{i=0}^{N} b_i E[e_{k-i} \; u_{k-n}] \qquad (10.15)$$

or

$$r_u(n) = -\sum_{i=1}^{M} a_i r_u(n-i) + \sum_{i=0}^{N} b_i r_{eu}(n-i) \qquad (10.16)$$

where $r_{eu}(n)$ is the cross correlation between e and u, and can be expressed in terms of h_k samples using the convolution sum, i.e.,

$$\begin{aligned} r_{eu}(n) &= E[e_{i+n} \; u_i] \\ &= E\left\{ e_{i+n} \left[\sum_{k=0}^{\infty} h_k \; e_{i-k} \right] \right\} \\ &= \sum_{k=0}^{\infty} h_k r_e(n+k), \qquad h_0 = 1 \end{aligned} \qquad (10.17)$$

Since e_k is assumed to be a white noise process, then

$$r_{eu}(n) = \begin{cases} 0 & n > 0 \\ \sigma_e^2 & n = 0 \\ \sigma_e^2 h_{-n} & n < 0 \end{cases} \qquad (10.18)$$

Thus from equation (10.16) the autocorrelation function $r_u(n)$ becomes

$$r_u(n) = \begin{cases} -\sum_{i=1}^{M} a_i r_u(n-i), & n > N \\ -\sum_{i=1}^{M} a_i r_u(n-i) + \sigma_e^2 \sum_{i=n}^{N} b_i h_{i-n}, & 0 \leq n \leq N \\ r_u(-n), & n < 0 \end{cases} \qquad (10.19)$$

The AR parameters can be obtained separately from the MA parameters by solving a system of linear equations for $N + 1 \leq n \leq N + M$. This gives the following matrix expression, which is known as the ARMA Yule–Walker normal equation[7]:

$$\begin{bmatrix} r_u(N) & r_u(N-1) & \cdots & r_u(N-M+1) \\ r_u(N+1) & r_u(N) & \cdots & r_u(N-M+2) \\ \vdots & & & \vdots \\ r_u(N+M-1) & r_u(N+M-2) & \cdots & r_u(N) \end{bmatrix} \begin{bmatrix} a_1 \\ a_2 \\ \vdots \\ a_M \end{bmatrix} = -\begin{bmatrix} r_u(N+1) \\ r_u(N+2) \\ \vdots \\ r_u(N+M) \end{bmatrix} \tag{10.20}$$

We note that this autocorrelation matrix is Toeplitz; thus a number of efficient algorithms[8] may be used to solve this system of equations without performing matrix inversions. The MA parameters can be estimated using an innovation algorithm, or, more efficiently, by using maximum-likelihood and least-squares estimations.[7,9] Owing to the difficulties in estimating the MA parameters and the fact that ARMA and MA processes can be represented generally by AR(∞) models, we develop most of our results for AR models. For these models, the autocorrelation function (10.19) becomes

$$r_u(n) = \begin{cases} -\sum_{i=1}^{M} a_i r_u(n-i), & n > 0 \\ -\sum_{i=1}^{M} a_i r_u(n-i) + \sigma_e^2, & n = 0 \\ r_u(-n), & n < 0 \end{cases} \tag{10.21}$$

Equation (10.21) can be expressed in matrix form:

$$\begin{bmatrix} r_u(0) & r_u(-1) & \cdots & \cdots & r_u(-M) \\ r_u(1) & r_u(0) & \cdots & \cdots & r_u(-M+1) \\ \vdots & & \ddots & & \vdots \\ \vdots & & & \ddots & \vdots \\ r_u(M) & r_u(M-1) & \cdots & \cdots & r_u(0) \end{bmatrix} \begin{bmatrix} 1 \\ a_1 \\ \vdots \\ \vdots \\ a_M \end{bmatrix} = \begin{bmatrix} \sigma_e^2 \\ 0 \\ 0 \\ \vdots \\ 0 \end{bmatrix} \tag{10.22}$$

This matrix equation is known as the AR Yule–Walker normal equation.[7] The autocorrelation matrix in this case is both Toeplitz and Hermitian.

10.1.1.2. Convergence and Stability Properties

An important consequence of the above result is that any given SDF can be approximated arbitrarily closely by a finite-order AR spectrum. In other words, if the SDF, $S_u(z)$, is positive analytic and if $\{a_i(M), i = 1, \ldots, M\}$, and $\sigma_e^2(M)$ where

$$\sigma_e^2(M) = r_u(0) - \sum_{i=1}^{M} a_i(M) r_u(i) \tag{10.23}$$

denote the solution of equation (10.22), then

$$\lim_{M \to \infty} \left[\frac{\sigma_e^2(M)}{A_M(z) A_M(z^{-1})} \right] = S_u(z) \tag{10.24}$$

where $A_M(z)$ is given by expression (10.4a) with a_i replaced by $a_i(M)$. As a result, the solution of equation (10.22) as $M \to \infty$ provides the spectral factorization of $S_u(z)$. If $S_u(z)$ happens to be a rational all-pole function, then $a_i(M)$ will be zero when $i > M$ for some $M < \infty$. In general, equation (10.22) can be used to find a rational, all-pole spectral approximation of an arbitrary positive SDF.

10.1.1.3. Minimum Variance Prediction[10]

The conditional mean estimate of u_k, conditioned on the past u_i, $\forall i \le k-1$, i.e.,

$$\begin{aligned} \hat{u}_k &\triangleq E[u_k \mid u_i, \forall i \le k-1] \\ &= \sum_{i=1}^{M} a_i u_{k-i} \end{aligned} \tag{10.25}$$

is the best mean-square predictor of u_k based on all of its past and depends only on the past M samples. This corresponds to the minimum variance predictor, which minimizes the mean-square error

$$\varepsilon = E[(u_k - \hat{u}_k)^2] \tag{10.26}$$

This minimization results in the following orthogonal property:

$$\frac{\partial \varepsilon}{\partial a_i} = -2E[(u_k - \hat{u}_k)\; u_{k-i}] = 0, \qquad 1 \le i \le M$$

or

$$E[(u_k - \hat{u}_k)\; u_{k-i}] = E[e_k\; u_{k-i}] = 0 \tag{10.27}$$

i.e., $(u_k - \hat{u}_k)$ or e_k is orthogonal to u_{k-i}. Since $\hat{u}_k$ is expressed linearly in terms of u_{k-i} via relation (10.25), then

$$E[e_k \, \hat{u}_k] = 0 \tag{10.28}$$

This orthogonality condition implies the normal equations

$$r_u(k) - \sum_{i=1}^{M} a_i r_u(k-i) = \sigma_e^2 \delta(k), \qquad 0 \leq k \leq M \tag{10.29}$$

which gives the AR Yule–Walker normal equation (10.22).

EXAMPLE 10.1. We consider a zero-mean stationary process with covariance function

$$r_u(n) = E[u_{i+n} \, u_i] = \sigma_u^2 \rho^{|n|}$$

The SDF is given by

$$\begin{aligned} S_u(z) &= \sum_{n=-\infty}^{\infty} r_u(n) z^{-n} \\ &= \sigma_u^2 \left(\frac{1}{1-\rho z^{-1}} + \frac{1}{1-\rho z} - 1 \right) \\ &= \frac{\sigma_u^2(1-\rho^2)}{(1-\rho z^{-1})(1-\rho z)} \end{aligned}$$

from which we can derive an AR model that represents the data exactly, i.e.,

$$u_k = \rho u_{k-1} + e_k \qquad \text{and} \qquad \sigma_e^2 = E[e_k^2] = \sigma_u^2(1-\rho^2)$$

10.1.2. Noncausal Representation[10,11]

In this section, some results on 1-D noncausal representation and the related issues of stability, correlation match, etc. are presented. The intent it to provide better insight into the problem of noncausality which commonly occurs in the 2-D image models. It must be mentioned that the 1-D noncausal models are important in their own right, as they also have applications in image restoration and data compression.

Let us consider a noncausal minimum variance representation (MVR) for a zero-mean Gaussian random process $\{u_k\}$ given by

$$u_k = \hat{u}_k + e_k \tag{10.30}$$

where

$$\hat{u}_k = \sum_{i \neq 0} a_i u_{k-i} \tag{10.31}$$

is the minimum variance prediction estimate of u_k based on all the past and future, and $\{e_k\}$ is the prediction error process. The associated SDF of $\{u_k\}$ can be written as

$$S_u(z) = \sigma_e^2 / A(z), \qquad \sigma_e^2 = \min E[e_k^2] \tag{10.32}$$

where

$$A(z) \triangleq 1 - \sum_{i \neq 0} a_i z^{-i} \tag{10.33}$$

If $\{u_k\}$ is an Mth-order Markov process, i.e., all the information needed for minimum variance noncausal prediction of $\{u_k\}$ is contained in the immediate M past and future samples of $\{u_k\}$, then

$$E[u_k \mid u_{k-i}, \forall i \neq 0] = E[u_k \mid u_{k-i}, 1 \leq |i| \leq M] \tag{10.34}$$

In this case, the Mth-order noncausal model (MVR)

$$u_k = \sum_{\substack{i=-M \\ i \neq 0}}^{M} a_i(M) u_{k-i} + e_k(M) \tag{10.35}$$

represents an exact realization of the process. This is due to the fact that the SDF generated by this model, i.e.,

$$S_u^{(M)}(z) = \sigma_e^2(M) / A_M(z) \tag{10.36}$$

where

$$A_m(z) \triangleq 1 - \sum_{\substack{i=-M \\ i \neq 0}}^{M} a_i(M) z^{-i} \tag{10.37}$$

matches exactly the SDF of a finite-order process, i.e., $S_u^{(M)}(z) = S_u(z)$. The Mth-order minimum variance predictor

$$\hat{u}_k = \sum_{\substack{i=-M \\ i \neq 0}}^{M} a_i(M) u_{k-i} \tag{10.38}$$

is used to predict u_k with prediction error $e_k(M)$. In practice, however, one needs to approximate an infinite-order process by a finite-order minimum variance prediction model. By choosing $M > M_0$, the difference between the actual covariances and the covariances generated by the model can be made arbitrarily small. The correlation match is exact either at $M = \infty$ or at $M = M_0$ if $\{u_k\}$ is an M_0th-order process.

REMARK 10.1. Realization of noncausal models does not require spectral factorization.

REMARK 10.2. Unlike the causal AR models where $\{e_k\}$ is a white process, the noncausal MVR is driven by colored noise. This can be shown by multiplying both sides of equation (10.30) by $e(k - n)$ and taking expectation, which yields

$$r_e(n) = \sigma_e^2 \delta(n) - \sigma_e^2 \sum_{i \neq 0} a_i \delta(n - i) \tag{10.39}$$

or

$$S_e(z) = \sigma_e^2 A(z) \tag{10.40}$$

This implies that $\{e_k\}$ is a moving-average process.

10.1.2.1. *Convergence and Stability Properties*[11]

The following results hold for a Gaussian random process $\{u_k\}$ with positive analytic SDF:

1. An Mth-order noncausal model is stable if and only if

$$A_M(z) \neq 0, \qquad |z| = 1 \tag{10.41}$$

There exists some $M_0 < \infty$ such that

$$A_M(z) \neq 0, \qquad |z| = 1 \qquad \text{and} \qquad M > M_0 \tag{10.42}$$

We note that, unlike the causal AR model, it is not true in general for the noncausal models that stability of the Mth-order model guarantees stability of the $(M + 1)$th-order model.

2. The sequence

$$A_M(z) = 1 - \sum_{\substack{i=-M \\ i \neq 0}}^{M} a_i(M) z^{-i} \tag{10.43}$$

converges uniformly in some neighborhoods of $|z| = 1$ to an analytic limit:

$$\lim_{M \to \infty} A_M(z) = A(z) \tag{10.44}$$

The sequence of prediction error variances converges to a positive lower limit σ_e^2, where

$$\lim_{M \to \infty} \sigma_e^2(M) = \sigma_e^2 = 1 \Big/ \frac{1}{2\pi} \int_{-\pi}^{\pi} S_u^{-1}(\omega)\, d\omega \tag{10.45}$$

The derivations of the above results are given elsewhere.[11]

10.1.2.2. Parameter Estimation for Noncausal Models[11]

The orthogonality condition $E[\hat{u}_k\, e_k] = 0$ can be used to yield the normal equations

$$r_u(n) = \sum_{\substack{i=-M \\ i \neq 0}}^{M} a_i(M) r_u(n-i) + \sigma_e^2(M)\delta(n), \qquad |n| \leq M \tag{10.46}$$

where

$$\sigma_e^2(M) = \min E[e_k^2(M)]$$

The above equation can be written in the matrix form

$$\begin{bmatrix} r_u(0) & r_u(-1) & \cdots & r_u(-2M) \\ r_u(1) & r_u(0) & \cdots & r_u(-2M+1) \\ \vdots & & & \vdots \\ r_u(2M) & r_u(2M-1) & \cdots & r_u(0) \end{bmatrix} \begin{bmatrix} -a_{-M}(M) \\ -a_{-M+1}(M) \\ \vdots \\ -a_{-1}(M) \\ 1 \\ -a_1(M) \\ \vdots \\ -a_M(M) \end{bmatrix} = \begin{bmatrix} 0 \\ 0 \\ \vdots \\ 0 \\ \sigma_e^2(M) \\ 0 \\ \vdots \\ 0 \end{bmatrix} \tag{10.47}$$

or in the compact notation

$$R_M \alpha_M = \sigma_e^2(M) 1_M \tag{10.48}$$

where

$$1_M = [0 \ \cdots \ 0 \ 1 \ 0 \ \cdots \ 0]^t$$

[we note that 1 appears in the $(M+1)$th position]

$$\alpha_M = [-a_{-M}(M) \ \cdots \ -a_{-1}(M) \ 1 \ -a_1(M) \ \cdots \ -a_M(M)]^t$$

and R_M is the autocorrelation matrix. If R_M is positive definite [when $S_u(\omega) > 0$], a unique solution exists. Here

$$a_i(M) = a_{-i}(M), \qquad i = 1, 2, \ldots, M \tag{10.49}$$

This is so, since R_M^{-1} is persymmetric (symmetric about the cross diagonal) and α_M is proportional to the middle column of R_M^{-1}.

EXAMPLE 10.2. We consider the covariance function given in Example 10.1. Let us rewrite the SDF in the form

$$\begin{aligned} S_u(z) &= \frac{\sigma_u^2(1-\rho^2)}{1+\rho^2-\rho(z+z^{-1})} \\ &= \frac{\beta^2}{1-\alpha(z+z^{-1})} \end{aligned}$$

where $\alpha \triangleq \rho/(1+\rho^2)$ and $\beta^2 \triangleq \sigma_u^2(1-\rho^2)/(1+\rho^2)$.

Denoting $A(z) = 1 - \alpha(z + z^{-1})$ and multiplying the denominator and numerator by $A(z^{-1})$ gives

$$S_u(z) = \frac{\beta^2 A(z^{-1})}{A(z)A(z^{-1})}$$

On comparing this with the general expression

$$S_u(z) = \frac{S_e(z)}{A(z)A(z^{-1})}$$

we obtain the following noncausal realization:

$$u_k = \alpha(u_{k+1} + u_{k-1}) + e_k$$

It can be seen that $\{e_k\}$ is an MA process with

$$S_e(z) = \beta^2 A(z^{-1}) = \beta^2 A(z)$$

or

$$r_e(n) = \beta^2 \delta(n) - \beta^2 \alpha[\delta(n-1) + \delta(n+1)]$$

10.2. TWO-DIMENSIONAL LINEAR PREDICTION MODELS[10-12]

As pointed out earlier, one major difference between 1-D models and 2-D models is that, for an image, causality is not inherent in the data. Moreover, it may not be possible to achieve a finite-order causal realization for a given rational SDF owing to the lack of general factorization for 2-D polynomials. In general, the prediction (support) region of 2-D models can take three types of geometry, leading to causal, semicausal, and noncausal models. These representations are the discrete versions of the classical 2-D linear-system categories that are: initial value (or hyperbolic), initial-boundary value (or parabolic), and boundary value (or elliptic), represented by partial differential equations.[1]

Most of the previous work on image modeling and the related issues has been confined to causal models. In recent years, semicausal and noncausal models have found numerous applications in a wide variety of image processing problems. Semicausal models are used in image data compression, image filtering, and spectral estimation. Noncausal models have been found useful in developing block processing algorithms with applications in image data compression using block coders. In this section, these models and their properties will be studied. We begin our discussion with the familiar 2-D causal models.

10.2.1. Causal Representation[10,11]

We consider an image $\{u_{i,j}\}$ represented by a zero-mean, 2-D stationary Gaussian process with autocorrelation function $r_u(k, l)$, i.e.,

$$r_u(k, l) \triangleq E[u_{i+k,j+l} \; u_{i,j}] \tag{10.50}$$

Let us assume that the image field is modeled by an MVR with prediction region W_k,

$$u_{i,j} = \hat{u}_{i,j} + e_{i,j} \tag{10.51}$$

where $\hat{u}_{i,j}$ is the linear prediction estimate of $u_{i,j}$ given by

$$\hat{u}_{i,j} = \sum_{m,n \in W_k}\sum a_{m,n} u_{i-m,j-n}, \qquad k = 1, 2, 3 \tag{10.52}$$

The quantities $a_{m,n}$ are called reflection coefficients. The geometry of W_k determines the type of the estimator, namely causal, semicausal, and noncausal estimators. By considering a scanning scheme which scans the image line by line, top to bottom, and left to right on each line, the three prediction regions associated with these models are defined as follows:

Causal prediction

$$W_1 = \{(m, n): \{n \geq 1, \forall m\} \cup \{n = 0, m \geq 1\}\} \tag{10.53a}$$

Semicausal prediction

$$W_2 = \{(m, n): \{n \geq 1, \forall m\} \cup \{n = 0, \forall m \neq 0\}\} \tag{10.53b}$$

Noncausal prediction

$$W_3 = \{(m, n): \{\forall (m, n) \neq (0, 0)\}\} \tag{10.53c}$$

The above prediction regions are shown respectively in Figure 10.1a, c, and d. In practice, since only finite-order realizations are of interest, a finite number of nearest-neighbor pixels from the prediction windows $\hat{W}_i \subset W_i$ should be used in the prediction. These prediction windows have rectangular support and are respectively given by

$$\hat{W}_1 = \{(m, n): 1 < n \leq N, -M \leq m \leq M \text{ and } n = 0, 1 \leq m \leq M\} \tag{10.54a}$$

$$\hat{W}_2 = \{(m, n): 0 \leq n \leq N, -M \leq m \leq M, (m, n) \neq (0, 0)\} \tag{10.54b}$$

$$\hat{W}_3 = \{(m, n): -N \leq n \leq N, -M \leq m \leq M, (m, n) \neq (0, 0)\} \tag{10.54c}$$

We note that the prediction region W_1 represents a general causal region, which is also called the nonsymmetric half-plane (NSHP).[2] A special case of the causal prediction region is the familiar quarter-plane region, which is depicted in Figure 10.1b. In this section, we restrict ourselves to causal models.

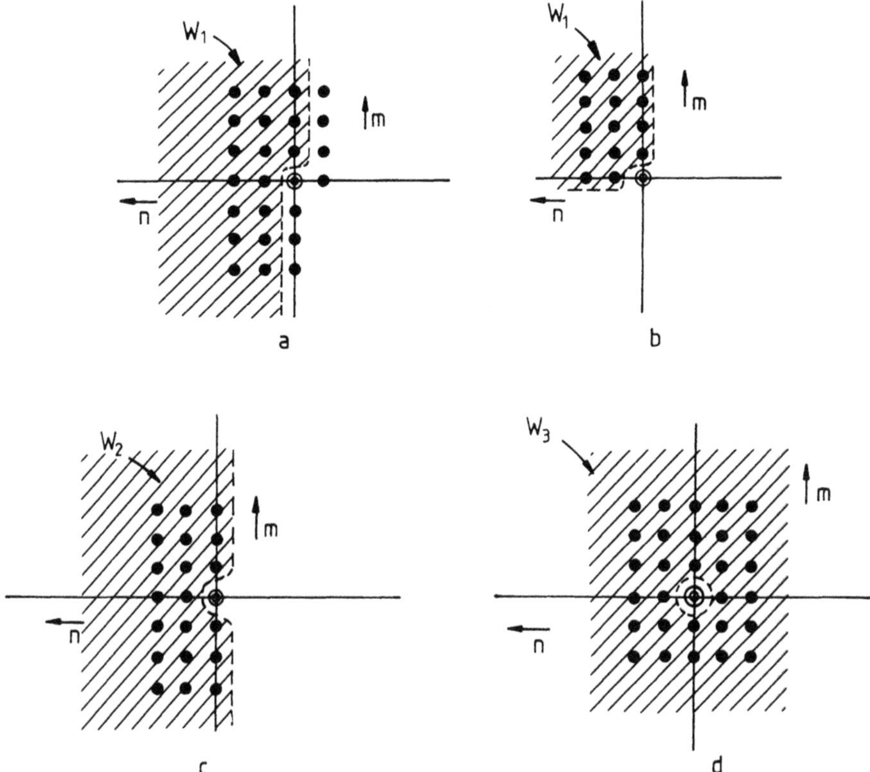

Figure 10.1. Prediction regions and windows for (a) causal, (b) causal quarter-plane, (c) semicausal, and (d) noncausal geometries.

The 2-D transfer function has a rational form

$$H(z_1, z_2) = \frac{1}{A(z_1, z_2)} \tag{10.55a}$$

where

$$A(z_1, z_2) \triangleq 1 - \sum_{m,n \in W_1}\sum a_{m,n} z_1^{-m} z_2^{-n} \tag{10.55b}$$

It is noteworthy that, in contrast to the 1-D case where the poles and zeros of the system function occur at isolated points in the z plane, in the 2-D case these singularities form a continuous surface or manifold in a four-dimensional space. As a result, it is not generally possible to decompose a 2-D polynomial into isolated poles and zeros (i.e., first-order polynomials). The stability of system (10.55) can be tested using the techniques in Chapter 3.

The SDF of the image $\{u_{i,j}\}$ is defined by

$$S_u(z_1, z_2) = \sum_{k=-\infty}^{\infty} \sum_{l=-\infty}^{\infty} r_u(k, l) z_1^{-k} z_2^{-l} \tag{10.56}$$

It is assumed that $S_u(z_1, z_2)$ is positive analytic,[11] i.e.,

(a) $S_u(\omega_1, \omega_2) > 0$, $z_1 = e^{j\omega_1}$, $z_2 = e^{j\omega_2}$;
(b) $S_u(z_1, z_2)$ is analytic in a neighborhood of $|z_1| = |z_2| = 1$.

The consequences of the above assumption are:

1. $S_u(z_1, z_2)$ and $S_u^{-1}(z_1, z_2)$ will have Laurent expansions in some neighborhoods.
2. $\{r_u(m, n)\}$ is exponentially bounded.
3. $\{r_u(m, n)\}$ is positive definite.

The autocorrelation function $r_u(m, n)$ is related to that of the prediction error $r_e(m, n)$ by

$$r_u(m, n) = r_e(m, n) * [h(m, n) * h(-m, -n)] \tag{10.57}$$

where $\{h(m, n)\}$ represents the impulse response sequence of the system. Taking the z transform of both sides of equation (10.57) yields

$$S_u(z_1, z_2) = \frac{S_e(z_1, z_2)}{A(z_1, z_2)A(z_1^{-1}, z_2^{-1})} \tag{10.58}$$

where $S_e(z_1, z_2)$ is the SDF of the prediction error $\{e_{i,j}\}$. In general, $\{e_{i,j}\}$ would be modeled as an MA process so that expression (10.58) is a rational function expressed as a ratio of 2-D polynomials in z_1 and z_2. However, it can be easily shown that the causal MVRs are white-noise-driven representations.[10] Since causal models are recursive in both dimensions, realization of these models from the SDF $S_u(z_1, z_2)$ requires factorization in both z_1 and z_2 variables. In general, for a given positive analytic SDF there exists a unique causal realization of infinite order. However, it is possible to find finite-order causal MVRs which would come arbitrarily close to the infinite-order causal MVRs of a given SDF. The corresponding SDF for this finite-order realization will then have the form

$$S_u^{(M,N)}(z_1, z_2) = \frac{\sigma_e^2(M, N)}{A_{M,N}(z_1, z_2)A_{M,N}(z_1^{-1}, z_2^{-1})} \tag{10.59a}$$

where

$$A_{M,N}(z_1, z_2) = \sum_{m,n \in \hat{W}_1 \cup (0,0)}\sum a_{m,n}(M, N) z_1^{-m} z_2^{-n} \tag{10.59b}$$

A method for finding the reflection coefficients $a_{m,n}(M, N)$ associated with a finite-order polynomial $A_{M,N}(z_1, z_2)$ and also the variance $\sigma_e^2(M, N)$ of the white noise sequence $\{e_{i,j}(M, N)\}$ is given in the following section.

10.2.1.1. Parameter Estimation of Causal MVRs[11]

In this section, we examine the problem of finite-order causal models, given the covariances on a finite window. The normal equation for an (M, N)-order causal model is

$$\begin{aligned} r_u(k, l) = &\sum_{m=1}^{M} a_{m,0}(M, N) r_u(k-m, l) \\ &+ \sum_{m=-M}^{M} \sum_{n=1}^{N} a_{m,n}(M, N) r_u(k-m, l-n) \\ &+ \sigma_e^2(M, N)\delta(k, l) \end{aligned} \tag{10.60}$$

for $(k, l) \in \hat{W}_1 \cup (0, 0)$.

In matrix notation, the above equation can be written as

$$R_{M,N}\alpha_{M,N} = \sigma_e^2(M, N) 1_{0,0} \tag{10.61}$$

where

$$R_{M,N} \triangleq \begin{bmatrix} \rho'_0 & \rho'_{-1} & \rho'_{-2} & \cdots & \cdots & \rho'_{-N} \\ \rho'_1 & \rho_0 & \rho_{-1} & & & \rho_{-N+1} \\ \vdots & & & & & \vdots \\ \vdots & & & & & \rho_{-1} \\ \rho'_N & \rho_{N-1} & & \cdots & \cdots & \rho_0 \end{bmatrix} \tag{10.62a}$$

Quantities ρ_k and ρ'_k are defined by

$$\rho_k \triangleq \begin{bmatrix} r_u(0, k) & r_u(-1, k) & \cdots & r_u(-2M, k) \\ r_u(1, k) & & & \vdots \\ \vdots & \ddots & & \\ \vdots & & \ddots & \vdots \\ r_u(2M, k) & \cdots & r_u(1, k) & r_u(0, k) \end{bmatrix} = \rho_{-k}^t \tag{10.62b}$$

and

$$\rho'_0 = \begin{bmatrix} r_u(0,0) & r_u(-1,0) & \cdots & r_u(-M,0) \\ r_u(1,0) & r_u(0,0) & \cdots & r_u(-M+1,0) \\ \vdots & & & \vdots \\ r_u(M,0) & r_u(M-1,0) & \cdots & r_u(0,0) \end{bmatrix}$$

$$\rho'_{-k} = \begin{bmatrix} r_u(M,-k) & r_u(M-1,-k) & \cdots & r_u(-M,-k) \\ r_u(M+1,-k) & r_u(M,-k) & \cdots & r_u(-M+1,-k) \\ \vdots & & & \vdots \\ r_u(2M,-k) & r_u(2M-1,-k) & \cdots & r_u(0,-k) \end{bmatrix}$$

$$\rho'_k = \rho''_{-k} \tag{10.62c}$$

$$\alpha^t_{M,N} \triangleq [\alpha^t_0 \ \alpha^t_1 \ \cdots \ \alpha^t_N]$$

$$\alpha^t_0 \triangleq [1 \ -a_{10}(M,N) \ \cdots \ -a_{M0}(M,N)], \qquad a_{00}(M,N) \triangleq 1$$

$$\alpha^t_i \triangleq [-a_{-M,i}(M,N) \ \cdots \ -a_{0i}(M,N) \ \cdots \ -a_{M,i}(M,N)] \tag{10.62d}$$

$$1^t_{0,0} \triangleq [1^t_0 \ 0^t \ \cdots \ 0^t], \qquad 1^t_0 \triangleq [1 \ 0 \ \cdots \ 0] \tag{10.62e}$$

Given that $S_u(\omega_1, \omega_2) > 0$, a unique solution to the system of equations (10.62) exists. For 2-D causal models with quarter-plane region of support, the autocorrelation matrix $R_{M,N}$ becomes doubly Toeplitz, i.e., a block Toeplitz matrix with Toeplitz blocks. Thus, efficient algorithms can be utilized for the solution of the corresponding 2-D Yule–Walker equation.[7,8]

10.2.1.2. Convergence and Stability Properties[11]

Let $\{u_{i,j}\}$ be a Gaussian random process with positive analytic SDF; the following results then hold:

1. It can be shown[11] that

$$A_{M,N}(z_1, z_2) = 1 - \sum_{m=1}^{M} a_{m,0}(M,N) z_1^{-m} - \sum_{m=-M}^{M} \sum_{n=1}^{N} a_{m,n}(M,N) z_1^{-m} z_2^{-n} \tag{10.63}$$

   converges uniformly to the minimum phase limit $A(z_1, z_2)$ in the region

$$\Gamma_1 = \{|z_1| = 1, |z_2| \geq 1\} \cup \{|z_1| \geq 1, z_2 = \infty\} \tag{10.64}$$

2. The variance of the prediction error converges to a positive lower limit, i.e.,

$$\lim_{M,N\to\infty} \sigma_e^2(M, N) = \sigma_e^2$$

$$= \exp\left[\frac{1}{4\pi^2}\int_{-\pi}^{\pi}\int_{-\pi}^{\pi} \log S_u(\omega_1, \omega_2)\, d\omega_1\, d\omega_2\right] \tag{10.65}$$

3. For $(M, N) > (M_0, N_0)$, the correlation match can be made arbitrarily small. Moreover, $A_{M,N}(z_1, z_2) \neq 0$ in Γ_1, i.e., the causal model of order $(M, N) > (M_0, N_0)$ is stable.

EXAMPLE 10.3. We consider an image modeled by a 2-D stationary separable Markov process with autocorrelation function

$$r_u(k, l) = E[u_{i+k,j+l}\; u_{i,j}] = \sigma_u^2 \rho^{|k|+|l|}, \qquad 0 < \rho < 1$$

and zero mean

$$\mu_u = E[u_{i,j}] = 0$$

σ_u^2 is the variance of $u_{i,j}$. We are interested in identifying a causal realization for the above autocorrelation function. Since the image is assumed to be stationary, the SDF can be obtained as

$$S_u(z_1, z_2) = \sum_{m=-\infty}^{\infty} \sum_{n=-\infty}^{\infty} r_u(m, n) z_1^{-m} z_2^{-n}$$

$$= \sigma_u^2 \left(\sum_{m=-\infty}^{\infty} \rho^{|m|} z_1^{-m}\right)\left(\sum_{n=-\infty}^{\infty} \rho^{|n|} z_2^{-n}\right)$$

On using the result in Example 10.1, we obtain

$$S_u(z_1, z_2) = \frac{\sigma_u^2(1-\rho^2)^2}{(1-\rho z_1^{-1})(1-\rho z_2^{-1})(1-\rho z_1)(1-\rho z_2)}$$

It is known that the causal models require spectral factorization in both directions. Therefore, we can write

$$A(z_1, z_2) = (1-\rho z_1^{-1})(1-\rho z_2^{-1}) \qquad \text{and} \qquad S_e(z_1 z_2) = \sigma_u^2(1-\rho^2)^2$$

which would yield the following 2-D causal model:

$$u_{i,j} = \rho u_{i-1,j} + \rho u_{i,j-1} - \rho^2 u_{i-1,j-1} + e_{i,j}$$

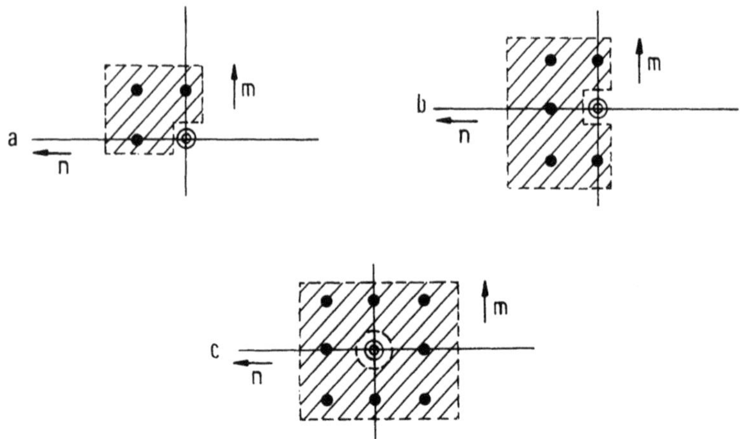

Figure 10.2. Regions of the causal, semicausal, and noncausal model used in Examples (a) 10.3, (b) 10.4, and (c) 10.5.

This model, which has realized the above SDF function exactly, has a quarter-plane region of support as shown in Figure 10.2a. The statistics of the white-noise-driving sequence $\{e_{i,j}\}$ are given by

$$\mu_e = 0 \qquad \text{and} \qquad r_e(k, l) = \beta_1^2 \delta(k, l)$$

where $\beta_1^2 \triangleq \sigma_u^2(1 - \rho^2)^2$.

REMARK 10.3. (a) Causal models require factorization in both z_1 and z_2 variables.

(b) Causal MVRs are white noise driven.

(c) Causal models are recursive in both directions.

(d) The causal MVRs are necessarily quarter-plane for the separable autocorrelation functions, although the converse is not true.

10.2.2. Semicausal Representation[10,11]

Semicausal models are recursive in one of the dimensions and nonrecursive in the other. Unlike causal MVRs, which require spectral factorization in both variables, semicausal MVRs require factorization of $S_u(z_1, z_2)$ only in one of the variables which corresponds to the causal direction. In addition, semicausal models are not white-noise-driven models. As in the causal case, it is possible to find finite-order semicausal MVRs which would realize a given positive analytic SDF arbitrarily closely.

Let us consider a finite-order semicausal MVR with a prediction window $\hat{W}_2$ defined by expression (10.54b). Then we can write

$$u_{i,j} = \hat{u}_{i,j} + e_{i,j}(M, N)$$

$$\hat{u}_{i,j} = \sum\sum_{m,n \in \hat{W}_2} a_{m,n}(M, N) u_{i-m,j-n}$$

and

$$\sigma_e^2(M, N) = \min E[e_{i,j}^2(M, N)] \tag{10.66}$$

The SDF of this semicausal MVR is given by[10]

$$S_u^{(M,N)}(z_1, z_2) = \frac{S_e^{(M,N)}(z_1, z_2)}{A_{M,N}(z_1, z_2) A_{M,N}(z_1^{-1}, z_2^{-1})} \tag{10.67a}$$

where

$$A_{M,N}(z_1, z_2) = \sum\sum_{m,n \in \hat{W}_2 \cup (0,0)} a_{m,n}(M, N) z_1^{-m} z_2^{-n} \tag{10.67b}$$

and

$$\begin{aligned} S_e^{(M,N)}(z_1, z_2) &= \sigma_e^2(M, N) \left[1 - \sum_{m=-M}^{M} a_{m,0}(M, N) z_1^{-m} \right] \\ &= \sigma_e^2(M, N) A_0(z_1) \end{aligned} \tag{10.67c}$$

i.e., $\{e_{i,j}(M, N)\}$ is an MA process in the "i" variable and is white in the "j" variable. Thus, semicausal models are driven by colored noise. It can be seen from system (10.67) that the finite-order semicausal MVRs realize SDFs which contain both numerator and denominator polynomials. However, the numerator polynomial is one-dimensional, and that is in the noncausal direction.

10.2.2.1. Parameter Estimation of Semicausal MVRs[11]

The normal equations for model (10.66) can be written as

$$\begin{aligned} r_u(k, l) = &\sum_{\substack{m=-M \\ m \neq 0}}^{M} a_{m,0}(M, N) r_u(k - m, l) \\ &+ \sum_{m=-M}^{M} \sum_{n=1}^{N} a_{m,n}(M, N) r_u(k - m, l - n) \\ &+ \sigma_e^2(M, N) \delta(k, l), \qquad k, l \in \hat{W}_2 \cup (0, 0) \end{aligned} \tag{10.68}$$

By defining $a_{0,0}(M, N) = 1$, we may write the above equation in the matrix form

$$R_{M,N} \alpha_{M,N} = \sigma_e^2(M, N)\, 1_{M,0} \tag{10.69a}$$

where

$$R_{M,N} \triangleq \begin{bmatrix} \rho_0 & \rho_{-1} & \rho_{-2} & \cdots & \rho_{-N} \\ \rho_1 & \rho_0 & & & \rho_{-N+1} \\ \vdots & & \ddots & & \vdots \\ \vdots & & & \ddots & \vdots \\ \rho_N & \cdots & & \cdots & \rho_0 \end{bmatrix} \tag{10.69b}$$

ρ_k was defined earlier in equation (10.62b), and

$$\alpha^t_{M,N} = [\alpha^t_0 \ \alpha^t_1 \ \cdots \ \alpha^t_N]$$

$$\alpha_k = [-a_{-M,k}(M, N) \ -a_{-M+1,k}(M, N) \ \cdots$$
$$-a_{0,k}(M, N) \ \cdots \ -a_{M,k}(M, N)]^t$$

$$1^t_{M,0} = [1^t_M \ 0^t \ \cdots \ 0^t]$$

$$1^t_M = [0 \ 0 \ \cdots \ 0 \ 1 \ 0 \ \cdots \ 0]^t \tag{10.69c}$$

(We note that 1 appears in the $(M+1)$th position.)

Thus $\sigma^2_e(M, N)$ and $a_{m,n}(M, N)$ can be found from

$$\sigma^2_e(M, N) = 1/[R^{-1}_{M,N}]_{M+1,M+1} \tag{10.70a}$$

$$a_{m,n}(M, N) = \sigma^2_e(M, N)[R^{-1}_{M,N}]_{k,M+1}, \qquad k = n(2M+1) + m + M + 1 \tag{10.70b}$$

where $[R^{-1}_{M,N}]_{i,j}$ denotes the (i, j)th element of the $R^{-1}_{M,N}$ matrix. It is seen that, since the covariances in system (10.69) from a window twice the size of $\hat{W}_2$ are needed to calculate $\{a_{m,n}(M, N)\}$, there is no unique correspondence between the reflection coefficients and the covariances realized by the model.

10.2.2.2. *Convergence and Stability Properties*[11]

The following results hold for a Gaussian random field $\{u_{i,j}\}$ with positive analytic SDF:

1. $$A_{M,N}(z_1, z_2) = 1 - \sum\sum_{(m,n)\in \hat{W}_2} a_{m,n}(M, N) z_1^{-m} z_2^{-n} \tag{10.71}$$

   converges uniformly to the analytic semiminimum phase[11] $A(z_1, z_2)$ in the region

   $$\Gamma_2 = \{|z_1| = 1, |z_2| \geq 1\} \tag{10.72}$$

2. The variance of the prediction error converges to a positive lower limit σ_e^2,

$$\lim_{M,N\to\infty} \sigma_e^2(M, N) = \sigma_e^2 = 1 \Big/ \left[\frac{1}{2\pi}\int_{-\pi}^{\pi} D^{-1}(\omega_1)\, d\omega_1\right]$$

$$D(\omega_1) = \sigma_e^2 A_0^{-1}(\omega_1) \tag{10.73}$$

3. An (M, N)-order semicausal model is stable if and only if

$$A_{M,N}(z_1, z_2) \neq 0, \qquad (z_1, z_2) \in \Gamma_2$$

There exists some $(M_0, N_0) < \infty$ such that for $(M, N) > (M_0, N_0)$

$$A_{M,N}(z_1, z_2) \neq 0, \qquad (z_1, z_2) \in \Gamma_2 \tag{10.74}$$

4. By choosing $(M, N) > (M_0, N_0)$, the difference between the actual correlations and the correlations generated by the model can be made arbitrarily small. At $(M, N) = \infty$, the correlation match is exact.

EXAMPLE 10.4. Let us consider the problem of finding a semicausal realization for the separable autocorrelation function in Example 10.3.

If the SDF is rewritten in the form

$$\begin{aligned} S_u(z_1, z_2) &= \frac{\sigma_u^2(1-\rho^2)^2}{[1+\rho^2-\rho z_1^{-1}-\rho z_1](1-\rho z_2)(1-\rho z_2^{-1})} \\ &= \frac{\sigma_u^2(1-\rho^2)^2/(1+\rho^2)}{[1-\alpha_1(z_1^{-1}+z_1)](1-\rho z_2)(1-\rho z_2^{-1})} \end{aligned}$$

with $\alpha_1 \triangleq \rho/(1+\rho^2)$, and both numerator and denominator are multiplied by $[1-\alpha_1(z_1+z_1^{-1})]$, then we obtain

$$S_u(z_1, z_2) = \frac{\sigma_u^2(1-\rho^2)^2[1-\alpha_1(z_1+z_1^{-1})]/(1+\rho^2)}{[1-\alpha_1(z_1^{-1}+z_1)][1-\alpha_1(z_1+z_1^{-1})](1-\rho z_2)(1-\rho z_2^{-1})}$$

From this SDF we can deduce

$$A(z_1, z_2) = [1-\alpha_1(z_1^{-1}+z_1)](1-\rho z_2^{-1})$$

and

$$S_e(z_1, z_2) = \sigma_u^2(1-\rho^2)^2[1-\alpha_1(z_1+z_1^{-1})]/(1+\rho^2)$$

Thus the semicausal model which realizes the given SDF exactly becomes

$$u_{i,j} = \alpha_1(u_{i-1,j} + u_{i+1,j}) - \rho\alpha_1(u_{i+1,j-1} + u_{i-1,j-1}) + \rho u_{i,j-1} + e_{i,j}$$

The region of support of the model is shown in Figure 10.2b. The statistics of the prediction error $e_{i,j}$ are given by

$$\mu_e = 0$$

$$r_e(k, l) = \beta_2^2[\delta(k, l) - \alpha_1\delta(k-1, l) - \alpha_1\delta(k+1, l)]$$

$$\beta_2^2 \triangleq \sigma_u^2(1-\rho^2)^2/(1+\rho^2)$$

which shows an MA process along the "i" variable and a white process along the "j" variable.

The transfer function of the model is

$$H(z_1, z_2) = \frac{1}{A(z_1, z_2)} = \frac{1}{(1-\rho z_2^{-1})(1-\alpha_1 z_1^{-1} - \alpha_1 z_1)}$$

The poles of this transfer function are

$$z_2 = \rho \qquad \text{(stable causal part along "}j\text{" direction)}$$

$$z_1 = \frac{1}{2\alpha_1}(1 \pm \sqrt{1-4\alpha_1^2})$$

One pole is greater than unity. This instability occurs along the noncausal "i" direction. The above transfer function can be decomposed and realized by two separable difference equations:

$$u_{i,j} = \rho u_{i,j-1} + v_{i,j} \qquad \text{(causal part)}$$

$$v_{i,j} = \alpha_1(v_{i+1,j} + v_{i-1,j}) + e_{i,j} \qquad \text{(noncausal part)}$$

These equations form an initial boundary value problem, with the initial value problem in the "j" variable and the boundary value problem in the "i" variable. Thus, the recursive solution of the second equation along the forward direction (i.e., starting from $i = 0$) is unstable. On the other hand, if the boundary conditions v_{i0} and $v_{i,P+1}$ (the image is assumed to be of size $P \times P$) are known, the solution for each j is obtained by solving this equation as a boundary value problem by applying a Riccati transformation technique.[13]

REMARK 10.4. (a) The numerator of the SDF of a semicausal MVR is a 1-D polynomial in z_1, i.e., the noncausal direction.

(b) Semicausal MVRs are driven by colored noise. In fact $\{e_{i,j}\}$ is uncorrelated in the "j" variable and correlated and MA along the "i" variable.

(c) Semicausal models are recursive (or causal) only in one variable, "j." A reindexing in the "i" variable to attempt to represent it as a 2-D causal model would yield an unstable model.

(d) With the properties of $\{e_{i,j}\}$ as above, the autocorrelation of $u_{i,j}$ obtained directly from the semicausal model (in this example) will be the same as that of the causal model.

(e) For a given semicausal MVR, a causal MVR can always be found under some mild restrictions.[11] However, the causal realization may have a higher (even infinite) order.

10.2.3. Noncausal Representation[10-12]

Noncausal models are nonrecursive in both directions. They do not require spectral factorization in any variable and are driven by colored noise. An arbitrary positive analytic SDF can be realized arbitrarily closely by a stationary finite-order noncausal MVR.

We consider a finite-order noncausal MVR with a prediction window $\hat{W}_3$ given by expression (10.54c). Then we can write

$$u_{i,j} = \hat{u}_{i,j} + e_{i,j}(M, N)$$

$$\hat{u}_{i,j} = \sum_{m,n \in \hat{W}_3}\sum a_{m,n}(M, N)u_{i-m,j-n}$$

and

$$\sigma_e^2(M, N) = \min E[e_{i,j}^2(M, N)] \tag{10.75}$$

The SDF for this representation is given by

$$S_u^{(M,N)}(z_1, z_2) = \frac{S_e^{(M,N)}(z_1, z_2)}{A_{M,N}(z_1, z_2)A_{M,N}(z_1^{-1}, z_2^{-1})} \tag{10.76a}$$

where

$$A_{M,N}(z_1, z_2) = \sum_{m,n \in \hat{W}_3 \cup (0,0)}\sum a_{m,n}(M, N)z_1^{-m}z_2^{-n} \tag{10.76b}$$

and

$$S_e^{(M,N)}(z_1, z_2) = \sigma_e^2(M, N)A_{M,N}(z_1, z_2) \tag{10.76c}$$

Thus the autocorrelation of $\{e_{i,j}\}$ is given by

$$r_e(k, l) = \sigma_e^2(M, N)\left[\delta(k, l) - \sum_{m,n \in \hat{W}_3}\sum a_{m,n}(M, N)\delta(k-m, l-n)\right] \quad (10.77)$$

i.e., noncausal models are driven by colored noise. It can be seen that the realized SDF contains a 2-D polynomial of order (M, N) in the denominator and a *zero*th-order polynomial in the numerator. Thus, if the given SDF is a rational function with *zero*th-order numerator polynomial, i.e.,

$$S_u^{(M,N)}(z_1, z_2) = \frac{\sigma_e^2(M, N)}{A_{M,N}(z_1, z_2)} \quad (10.78)$$

then there exists a unique finite-order noncausal MVR realization of this SDF. We note that, similar to the 1-D noncausal models,

$$a_{-m,-n} = a_{m,n} \quad (10.79)$$

10.2.3.1. Parameter Estimation of Noncausal MVRs[11]

The associated normal equation for model (10.75) can be written as

$$r_u(k, l) = \sum_{m,n \in \hat{W}_3}\sum a_{m,n}(M, N) r_u(k-m, l-n) + \sigma_e^2(M, N)\delta(k, l), \qquad k, l \in \hat{W}_3 \cup (0, 0) \quad (10.80)$$

In matrix notation, equation (10.80) may be expressed as

$$R_{M,N}\alpha_{M,N} = \sigma_e^2(M, N)1_{M,N} \quad (10.81a)$$

where in this case $R_{M,N}$ is a doubly Toeplitz symmetric matrix given by

$$R_{M,N} \triangleq \begin{bmatrix} \rho_0 & \rho_{-1} & \cdots & \rho_{-2N} \\ \rho_1 & \rho_0 & & \rho_{-2N+1} \\ \vdots & & \ddots & \vdots \\ \rho_{2N} & \rho_{2N-1} & \cdots & \rho_0 \end{bmatrix} \quad (10.81b)$$

with ρ_k defined in equation (10.62b), and

$$\alpha^{t}_{M,N} \triangleq [\alpha^{t}_{-N} \ \cdots \ \alpha^{t}_{0} \ \cdots \ \alpha^{t}_{N}]$$

$$\alpha_k \triangleq [-a_{-M,k}(M,N) \ -a_{-M+1,k}(M,N) \ \cdots \ -a_{0,k}(M,N) \ \cdots \ -a_{M,k}(M,N)]^{t}$$

$$a_{00}(M,N) \triangleq 1 \tag{10.81c}$$

$$1^{t}_{M,N} \triangleq [0^{t} \ 0^{t} \ \cdots \ 1^{t}_{M} \ 0^{t} \ \cdots \ 0^{t}]$$

[we note that 1^{t}_{M} appears in the $(N+1)$th block position]

$$1^{t}_{M} \triangleq [0 \ 0 \ \cdots \ 0 \ 1 \ 0 \ \cdots \ 0]$$

[we note that 1 appears in the $(M+1)$th position] (10.81d)

If $S_u(\omega_1, \omega_2) > 0$, then $R_{M,N}$ is positive definite and a unique solution exists for the reflection coefficients. This is given by

$$\sigma^2_e(M,N) = 1/[R^{-1}_{M,N}]_{k,k}, \qquad k = N(2M+1) + M + 1 \tag{10.82a}$$

$$a_{m,n}(M,N) = -\sigma^2_e(M,N)[R^{-1}_{M,N}]_{l,k},$$

$$l = (n+N)(2M+1) + m + M + 1 \tag{10.82b}$$

Similar to the 1-D noncausal models, if $\{u_{i,j}\}$ is a Markov field, then a finite-order noncausal model exists which realizes the given SDF exactly. Generally speaking, a noncausal MVR can realize a given SDF only approximately. However, by increasing the order of the model, the correlation match improves asymptotically.

10.2.3.2. Convergence and Stability Properties[11]

The following results hold for a Gaussian random field $\{u_{i,j}\}$ with positive analytic SDF:

1.

$$A_{M,N}(z_1, z_2) = 1 - \sum_{m,n \in \hat{W}_3}\sum a_{m,n}(M,N) z_1^{-m} z_2^{-n} \tag{10.83}$$

converges uniformly on the unit bidisc $|z_1| = 1, |z_2| = 1$ to the analytic limit $A(z_1, z_2)$.

2. The variance of the prediction error converges to a positive lower limit σ_e^2, i.e.,

$$\lim_{M,N\to\infty} \sigma_e^2(M, N) = \sigma_e^2 = \left[\frac{1}{4\pi^2}\int_{-\pi}^{\pi}\int_{-\pi}^{\pi} S_u^{-1}(\omega_1, \omega_2)\, d\omega_1\, d\omega_2\right]^{-1} \tag{10.84}$$

3. An (M, N)-order noncausal model is stable if, and only if, for

$$A_{M,N}(z_1, z_2) \neq 0, \qquad |z_1| = |z_2| = 1$$

there exists some $(M_0, N_0) < \infty$ such that

$$A_{M,N}(z_1, z_2) \neq 0, \qquad |z_1| = |z_2| = 1, \qquad (M, N) > (M_0, N_0) \tag{10.85}$$

4. By choosing $(M, N) > (M_0, N_0)$, the difference between the actual correlations and the correlations realized by the model can be made arbitrarily small. At $(M, N) = \infty$, the correlation match is exact.

EXAMPLE 10.5. In this example, we require a noncausal model for the realization of the SDF given in Example 10.3.

Let us rewrite the relevant SDF in the form

$$S_u(z_1, z_2) = \frac{\sigma_u^2(1-\rho^2)^2[1-\alpha_1(z_1^{-1}+z_1)][1-\alpha_1(z_2^{-1}+z_2)]/(1+\rho^2)^2}{[1-\alpha_1(z_1^{-1}+z_1)]^2[1-\alpha_1(z_2^{-1}+z_2)]^2}$$

By comparing this with system (10.76), we can deduce that

$$A(z_1, z_2) = [1-\alpha_1(z_1^{-1}+z_1)][1-\alpha_1(z_2^{-1}+z_2)]$$

and

$$S_e(z_1, z_2) = \sigma_u^2(1-\rho^2)^2[1-\alpha_1(z_1^{-1}+z_1)][1-\alpha_1(z_2^{-1}+z_2)]/(1+\rho^2)^2$$

The noncausal model which realizes the SDF exactly is

$$\begin{aligned} u_{i,j} &= \alpha_1(u_{i-1,j} + u_{i+1,j} + u_{i,j-1} + u_{i,j+1}) \\ &\quad - \alpha_1^2(u_{i-1,j-1} + u_{i+1,j-1} + u_{i-1,j+1} + u_{i+1,j+1}) + e_{i,j} \end{aligned}$$

The region of support for this model is depicted in Figure 10.2c. The colored noise sequence $\{e_{i,j}\}$ has the following statistics:

$$\mu_e = 0$$

$$\begin{aligned} r_e(k, l) = \beta_3^2[\delta(k, l) &- \alpha_1\delta(k-1, l) - \alpha_1\delta(k+1, l) - \alpha_1\delta(k, l-1) \\ &- \alpha_1(k, l+1) + \alpha_1^2\delta(k-1, l-1) + \alpha_1^2\delta(k+1, l-1) \\ &+ \alpha_1^2\delta(k-1, l+1) + \alpha_1^2\delta(k+1, l+1)] \end{aligned}$$

$$\beta_3^2 \triangleq \sigma_u^2(1-\rho^2)^2/(1+\rho^2)^2$$

REMARK 10.5. (a) Noncausal models do not require spectral factorization for realization in any direction.

(b) Noncausal representations are driven by colored noise.

(c) With the properties of $\{e_{i,j}\}$ as above, the autocorrelation functions of $\{u_{i,j}\}$ generated from this noncausal model are the same separable autocorrelation functions generated by the causal or semicausal models.

10.3. VECTOR MODELS

In the preceding sections, only scalar linear prediction models have been considered. In these models, a given pixel is related to the nearest-neighbor pixels using a causal, semicausal, or noncausal MVR. As a result, at every stage of the algorithm only one pixel estimate is produced. With the growing trends in VLSI technologies, and in particular recent improvements in systolic array processors, parallel and pipeline implementations of image processing algorithms have become increasingly popular. This naturally calls for vector image models which inherently benefit from this parallelism. In these models, the scalar difference equation describing the image generation process is arranged in vector or state-space form, in which the state consists of the whole or part of the row or column of the image. The corresponding models are referred to as "full" or "reduced" order models, respectively. The former type of model[1] is not of practical interest, owing to its enormous computational requirements. The latter models lead to estimators that are only suboptimal. Over the past few years, several new reduced-order vector image models have been introduced[2-4] for 2-D Kalman filtering of images degraded by additive white noise and blur.

The aim of this section is to provide the reader with some exposure to the vector models and their relations with the scalar models in the previous sections. In this regard, only the general vector modeling schemes are discussed. For more recent work, the reader is referred elsewhere.[2-4]

10.3.1. Vector Causal Models

Let us consider an image of size $P \times P$ modeled by 2-D stationary separable Markov statistics with the separable autocorrelation function given in Example 10.3. The causal representation which would generate this autocorrelation exactly is found to be

$$u_{i,j} = \rho u_{i-1,j} + \rho u_{i,j-1} - \rho^2 u_{i-1,j-1} + e_{i,j} \qquad \forall i, j \in [1, P] \tag{10.86a}$$

where $\{e_{i,j}\}$ is a white noise sequence with variance

$$\beta_1^2 = \sigma_u^2(1 - \rho^2)^2 \tag{10.86b}$$

and σ_u^2 is the variance of $\{u_{i,j}\}$. If U_j and E_j of size $P \times 1$ represent the "j" column of the image and the error, respectively, then writing equation (10.86a) for each pixel of the image in this column yields

$$A_0 U_j = A_1 U_{j-1} + E_j + F_j \tag{10.87a}$$

where the Toeplitz matrices A_0 and A_1 are defined by their (i, j)th elements $A_{0\,(i,j)}$ and $A_{1\,(i,j)}$ as

$$A_{0\,(i,j)} = \begin{cases} 1, & i = j \\ -\rho, & i - j = 1 \\ 0, & \text{otherwise} \end{cases} \qquad \text{and} \qquad A_{1\,(i,j)} = \begin{cases} \rho, & i = j \\ -\rho^2, & i - j = 1 \\ 0, & \text{otherwise} \end{cases} \tag{10.87b}$$

Thus $A_1 = \rho A_0$.

The initial condition vector F_j is defined by

$$F_j \triangleq [\rho u_{0j} - \rho^2 u_{0\,j-1},\ 0 \ \cdots \ 0]^t$$

$$U_j = [u_{1j}\ u_{2j}\ \cdots\ u_{Pj}]^t \tag{10.87c}$$

and similarly for E_j. By defining

$$Q_j \triangleq E_j + F_j \tag{10.88}$$

system (10.87) becomes

$$\begin{aligned} U_j &= A_0^{-1} A_1 U_{j-1} + A_0^{-1} Q_j \\ &= \rho U_{j-1} + A_0^{-1} Q_j, \qquad j \in [1, P] \end{aligned} \tag{10.89}$$

This model represents a Markov vector process, provided that $\{Q_j\}$ is a white noise process. Since $E[F_j U_{j-1}^t] = 0$ and, furthermore, $\{E_j\}$ is a white noise process, this condition is met. The correlation matrix of Q_j is

$$R_Q = E[Q_j Q_j^t] = \beta_1^2 I + \rho^2(1-\rho^2)I_1 \tag{10.90a}$$

where

$$I_{k(i,j)} = \begin{cases} 1, & i=j=1,\dots,k \\ 0, & \text{otherwise} \end{cases} \tag{10.90b}$$

The Kalman filtering equations for this model are derived elsewhere.[1]

10.3.2. Vector Semicausal Models

Let us consider the semicausal representation obtained in Example 10.4, which also realizes the relevant separable SDF exactly:

$$u_{i,j} = \alpha_1(u_{i-1,j} + u_{i+1,j}) - \rho\alpha_1(u_{i+1,j-1} + u_{i-1,j-1}) + \rho u_{i,j-1} + e_{i,j} \tag{10.91a}$$

where $\{e_{i,j}\}$ is a white process along the causal direction "j" and is a correlated and MA process along the noncausal direction "i." The statistics for this sequence are given by

$$r_e(k,l) = \beta_2^2[\delta(k,l) - \alpha_1\delta(k-1,l) - \alpha_1\delta(k+1,l)] \tag{10.91b}$$

where

$$\beta_2^2 \triangleq \sigma_u^2(1-\rho^2)^2/(1+\rho^2)$$

If equation (10.91a) is arranged in vector form, similar to the previous case, such that the propagation of the state vector is along the causal direction, we get

$$B_0 U_j = B_1 U_{j-1} + E_j + G_j, \qquad j \in [1, P] \tag{10.92a}$$

where the Toeplitz matrices B_0 and B_1 are defined by their i, jth elements $B_{0\,(i,j)}$, $B_{1\,(i,j)}$ in the form

$$B_{0\,(i,j)} = \begin{cases} 1, & i=j \\ -\alpha_1, & |i-j|=1 \\ 0, & \text{otherwise} \end{cases} \quad \text{and} \quad B_{1\,(i,j)} = \begin{cases} \rho, & i=j \\ -\alpha_1\rho, & |i-j|=1 \\ 0, & \text{otherwise} \end{cases} \tag{10.92b}$$

Thus $B_1 = \rho B_0$. The vectors U_j and E_j are defined in a similar manner to equations (10.87c). The vector G_j consists of initial-boundary values and is defined by

$$G_j = \alpha_1[(u_{0j} - \rho u_{0,j-1}) \ 0 \ 0 \ \cdots \ 0 \ (u_{P+1,j} - u_{P+1,j-1})]^t \tag{10.92c}$$

If system of equations (10.92) is regarded as a state-space equation with state vector U_j, the state matrix in this model becomes

$$A = B_0^{-1} B_1 = \rho I \tag{10.93}$$

Since $0 < \rho < 1$, all the eigenvalues of this matrix are located inside the unit circle in the z plane. We note that these eigenvalues correspond to the stable part of the scalar model in Example 10.4, which is along the causal direction. Thus, the vector AR model (10.92) which propagates along the causal direction is always stable.

The statistics of E_j are defined by

$$E[E_j] = 0 \qquad \text{and} \qquad E[E_j E_k^t] = \beta_2^2 B_0 \delta(j, k) \tag{10.94}$$

The vector Markovian property of the model can similarly be shown by checking

$$E[G_j U_{j-1}^t] = 0 \qquad \text{and} \qquad E[G_j G_k^t] = 0, \qquad j \neq k$$

Therefore, by defining

$$P_j \triangleq E_j + G_j \tag{10.95}$$

system (10.92) becomes a vector Markov process. The correlation matrix of P_j becomes

$$R_P = \beta_2^2 B_0 + \alpha_1(1 - \rho^2)(I_1 + \tilde{I}_P) \tag{10.96}$$

where the (i, j)th element $\tilde{I}_{k(i,j)}$ of $\tilde{I}_k$ is given by

$$\tilde{I}_{k(i,j)} = \begin{cases} 1, & i = j = k, k+1, \ldots, P \\ 0, & \text{otherwise} \end{cases}$$

Now, if we arrange the same scalar equation (10.91) so that the state propagation occurs along the noncausal direction "i," we get

$$B_0' U_i = B_1' U_{i-1} + E_i + G_i' \tag{10.97a}$$

where

$$B'_0 = \begin{bmatrix} I & 0 \\ 0 & C_1 \end{bmatrix}_{2P\times 2P} \quad \text{and} \quad B'_1 = \begin{bmatrix} 0 & I \\ -C_1 & C_0 \end{bmatrix}_{2P\times 2P} \tag{10.97b}$$

while matrices C_0 and C_1 are defined by their i, jth elements,

$$C_{0\,(i,j)} = \begin{cases} -1, & i = j \\ \rho, & i - j = 1 \\ 0, & \text{otherwise} \end{cases} \quad \text{and} \quad C_{1\,(i,j)} = \begin{cases} -\alpha_1, & i = j \\ \alpha_1\rho, & i - j = 1 \\ 0, & \text{otherwise} \end{cases} \tag{10.97c}$$

Thus $C_1 = \alpha_1 C_0$. In this case U_i and E_i consist of two rows of the image pixels and the error, respectively, i.e.,

$$U_i = [u_{i1} \quad \cdots \quad u_{iP} u_{i+1,1} \quad \cdots \quad u_{i+1,P}]^t \tag{10.97d}$$

The boundary vector G'_i is given by

$$G'_i = [-\rho\alpha_1(u_{i+1,0} + u_{i-1,0}) + \rho u_{i,0} \;\; 0 \;\; \cdots \;\; 0]^t \tag{10.97e}$$

The state matrix in this case is

$$A' = B_0'^{-1} B'_1 = \begin{bmatrix} 0 & I \\ I & \alpha_1^{-1} I \end{bmatrix} \tag{10.98}$$

The eigenvalues of this matrix are the roots of

$$\lambda^2 - \alpha_1^{-1}\lambda + 1 = 0$$

namely, $\lambda_{1,2} = (1 \pm \sqrt{1 - 4\alpha_1^2})/2\alpha_1$.

These roots correspond to the unstable part of the scalar model in Example 10.4. As a result, arranging the scalar model in a vector model in which the state propagates along the noncausal direction leads to an unstable model in the forward direction. It should be noted that the noncausal representation in Example 10.5 cannot be implemented in vector recursive form along any direction.

The state-space representation and implementation of 1-D noncausal models has been the topic of a multitude of papers over the past few years. These systems are known as "singular" or "descriptor" systems.[14] The extension of these methods to 2-D noncausal models opens plenty of room for future research.

REFERENCES

1. A. K. Jain and J. R. Jain, Partial difference equations and finite difference methods in image processing—Part 2: image restoration, *IEEE Trans. Autom. Control* **AC-23,** 817-833 (1978).
2. J. W. Woods and C. H. Radewan, Kalman filtering in two dimensions, *IEEE Trans. Inf. Theory* **IT-23,** 473-482 (1977).
3. B. R. Suresh and B. A. Shenoi, New results in two-dimensional Kalman filtering with applications to image restoration, *IEEE Trans. Circuits Syst.* **CAS-28,** 307-319 (1981).
4. M. R. Azimi-Sadjadi and P. W. Wong, Two-dimensional block Kalman filtering for image restoration, *IEEE Trans. Acoust., Speech, Signal Process* **ASSP-35,** 1736-1749 (1987).
5. E. J. Delp, R. L. Kashyap, and O. R. Mitchell, Image data compression using autoregressive time series models, *Pattern Recognition* **11,** 313-323 (1979).
6. R. Chellappa and R. L. Kashyap, Texture synthesis using 2-D noncausal autoregressive models, *IEEE Trans. Acoust., Speech, Signal Process* **ASSP-33,** 194-202 (1985).
7. S. Lawrence Marple, Jr., *Digital Spectral Analysis with Applications,* Prentice-Hall, Englewood Cliffs, NJ (1987).
8. R. Kumar, A fast algorithm for solving a Toeplitz system of equations, *IEEE Trans. Acoust., Speech, Signal Process* **ASSP-33,** 254-267 (1985).
9. P. J. Brockwell and R. A. Davis, *Time Series Theory and Methods,* Springer-Verlag, Berlin (1987).
10. A. K. Jain, Advances in mathematical models for image processing, *Proc. IEEE* **69,** 502-528 (1981).
11. S. Ranganath and A. K. Jain, Two-dimensional linear prediction models—Part 1: Spectral factorization and realization, *IEEE Trans. Acoust., Speech, Signal Process* **ASSP-33,** 280-299 (1985).
12. R. L. Kashyap, Characterization and estimation of two-dimensional ARMA models, *IEEE Trans. Inf. Theory* **IT-30,** 736-745 (1984).
13. A. K. Jain, A semicausal model for recursive filtering of two-dimensional images, *IEEE Trans. Comput.* **C-26,** 343-350 (1977).
14. G. C. Verghese, B. C. Levy, and T. Kailath, A generalized state-space for singular systems, *IEEE Trans. Autom. Control* **AC-26,** 811-831 (1981).

11

Applications

11.1. INTRODUCTION

In the previous chapters, we have considered the major theoretical issues involved in digital filter design, analysis, and implementation. In some cases, examples have been provided to illustrate specific points.

These two final chapters concentrate rather more on specific signal processing problems and examine how digital filters are applied in the solution of these problems.

The range of areas in which filters are applied is extremely diverse. It includes areas such as time series analysis, where the signal is processed in a manner which hides the fact that this is in fact a digital signal filtering process, although it does not involve reference to the fundamental theory of linear digital filters.

We are concerned here with those areas where the theory is well used, such as one-dimensional (1-D) and multidimensional signal processing with application to speech, image, seismic, and biomedical signals. Excellent surveys and reviews, such as Bose,[1] Mersereau and Dudgeon,[2] and Hunt,[3] give some indication of these application areas.

The approach adopted here is to select areas of interest and discuss them in some detail, examining how the specific requirements of the application may be met by utilizing the theory discussed previously. This is followed by a consideration of some case studies in the next chapter.

Digital filters are widely applied to 1-D and multidimensional signals. Applications of 1-D digital filters are well covered in texts such as those by Hamming,[4] Huang,[5] Oppenheim and Shafer,[6] and Rabiner and Gold.[7] Techniques used for two-dimensional (2-D) systems can generally be extended to multidimensional systems, and this chapter concentrates on 2-D applications, specifically image processing. However, the techniques are equally applicable to other 2-D and multidimensional systems.

Examples of 2-D systems include image processing, seismic signal processing,[8] meteorology, gravity field mapping, etc. For image processing both independent variables are spatial, while for seismic signal processing one of the variables is time and the other spatial.

The most common applications of digital filters in image processing are in the areas of digital image enhancement and restoration.[9,10] To enhance images digitally, filters can be designed that reduce noise by smoothing, emphasize some region of the spectrum, sharpen edges, and implement other functions that produce improved images. Such improvement may be intended to provide a subjectively better image for humans to look at, or it may be intended as a preprocessor for higher-level tasks, such as segmentation and pattern recognition, or for machine vision.

Digital restoration filters may be used to invert a degradation process. For example, the effect of known camera defocusing, or some other limitation, can be reduced or removed by developing a restoration modeled on it.[11]

Digital filters may also be applied in postprocessing after image coding, and in preprocessing before image segmentation.

In other applications, filters may be used to extract features such as texture, spectral content, etc., which could be used to describe and recognize scenes. In common with speech signals, image patterns may be detected using matched filtering.

For the purposes of this book, we shall restrict ourselves to linear systems. Linear filtering techniques are mathematically more tractable, but have a fundamental disadvantage when used in image-processing applications. This is because the noise process is frequently nonlinear. Furthermore, edges are very important in images, and consequently filtering techniques may need to perform differently in edge regions; hence shift variance may be required.

Although the analysis of nonlinear systems is extremely complex, it should be noted that many such systems have been developed to produce very good results, in many cases much better than those produced by linear systems. For example, maximum entropy spectral estimation methods have proved extremely good. In addition there are pseudolinear systems, such as Oppenheim's work[12] on homomorphic filtering, that are worth considering.

As discussed in earlier chapters, it is noted that some constraints need to be imposed on digital filters that are used to process images. Of course, the nature of these requirements can differ widely, depending on the restrictions imposed by the intended application. Some requirements, however, are common, being properties of the nature of images and their perception by the human visual system. One such is that of linear phase; another may be the need for circular symmetry of the impulse response.

11.2. DIGITAL IMAGE PROCESSING—MODELS

The subject of digital image processing is concerned with the capture, modification, manipulation, and display of digital data in two or more dimensions. An ordinary photograph is one example of an analogue image that may be converted to digital form by scanning to produce a 2-D digital image with both dimensions being spatial.

Digital television images represent three-dimensional image functions, with the third dimension being the temporal dimension representing sucessive frames. In other situations, such as computer-aided tomography or solid modeling applications, the third dimension may also be spatial.

A general model often used for the image formation process considers the image as resulting from a process that transforms the radiant energy $u(m, n)$ of the original object into an ideal observed image $g(m, n)$. Hence, in the discrete formulation for a linear system,

$$g(m, n) = \sum_{k=1}^{N_1} \sum_{l=1}^{N_2} h(m, n, k, l)u(k, l) \tag{11.1}$$

where $g(m, n)$ is the 2-D image observed, $u(k, l)$ is the object image, and $h(m, n, k, l)$ is the discrete point spread function (PSF) or impulse response of the system. The images are of finite extent and the range of k and l is $1, \ldots, N_1$ and $1, \ldots, N_2$, respectively. The PSF models the blur and smoothing degradation that takes place in the image-formation process. Sources of blur include, among others, defocused systems, motion, the effects of atmospheric turbulence, and sampling with extended pulses.

Equation (11.1) represents a spatially varying linear system. If shift invariance is assumed, this will be modified to the convolution relationship.

$$g(m, n) = \sum_{k=1}^{N_1} \sum_{l=1}^{N_2} h(m - k, n - l)u(k, l) \tag{11.2}$$

Most digital image-processing schemes involve analysis of the above equations. We shall assume throughout that the images are of square finite extent, of dimensions $N \times N$; there is no loss of generality in making this assumption, as rectangular images can always be suitably padded to make them square. Hence, the range of m, n, k, l in equation (11.2) is $1, \ldots, N$.

Apart from the occurrence of blur in the image-formation process, there are other sources of degradation that must be taken into account. For example, the process of recording and storing the image may introduce sensor noise and film (or other media) grain noise. Detailed analyses of these noises and the development of models, and assumptions regarding the models, are not particularly pertinent to this book; these topics have been well treated elsewhere.[13]

For most sensors used in imaging, the associated noise can be modeled as a Gaussian or Poisson distributed random process. The dominant cause is thermal noise resulting from random electron fluctuations. In general, most practical detector circuits can be modeled as being dominated by additive Gaussian noise. Poisson noise occurs when the detector has a large internal electron amplification and the image light level is very low.

Film grain noise refers to the inherent randomness in film grain formation owing to the nondeterministic nature of the process of converting silver halide grains to silver by exposure to a quantity of light. It is assumed that film grain noise is multiplicative in the intensity domain and additive in the density domain.[13]

In addition, there are other forms of noise whose actual nature may vary. An accepted model treats all the various noises as generating an essentially additive component. We therefore have a relationship between the object $u(m, n)$, the noise-free image $g(m, n)$, and the observed image $y(m, n)$,

$$y(m, n) = g(m, n) + v(m, n) \tag{11.3}$$

where $v(m, n)$ is an additive noise process representing the various noise sources. This leads to the representation of images obtained through linear convolutional blurring systems with additive noise by combining equations (11.1) and (11.3) to yield

$$y(m, n) = \sum_k \sum_l h(m, n, k, l)u(k, l) + v(m, n) \tag{11.4}$$

and for time-invariant systems

$$y(m, n) = \sum_k \sum_l h(m - k, n - l)u(k, l) + v(m, n) \tag{11.5}$$

This model is illustrated in the block diagram of Figure 11.1.

A vector representation, as discussed in Section 1.6.3.1, may be obtained by performing a lexicographic scan ordering to give

$$y = Hu + v \tag{11.6}$$

where y, u, and v are $N^2 \times 1$ vectors and H is an $N^2 \times N^2$ matrix representing the PSF.

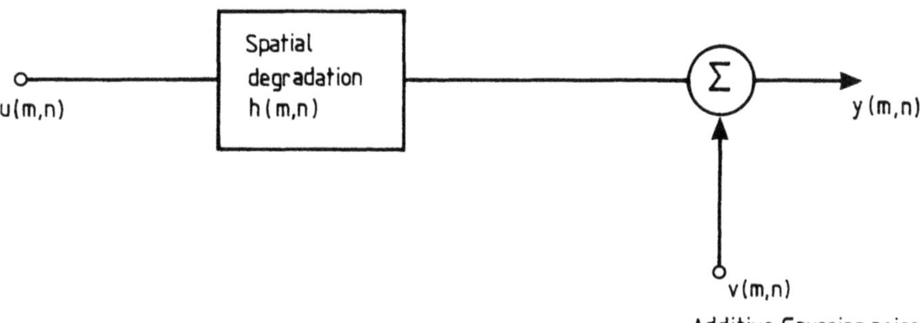

Figure 11.1. Image-formation model.

11.3. IMAGE ENHANCEMENT

Digital filters are applied in image enhancement to try and improve the subjective appearance of images without attempting directly to invert the effect of any degradation. They may also be employed to enhance certain properties of the images and to suppress others.

It is difficult to define what constitutes an improved image, as this is a matter of subjective judgment; as a result, it is not possible to specify enhancement filters in any scientific manner. Most of the enhancement filters used are based on the knowledge that a specific action leads to subjectively improved images. For example, it is generally accepted that improving contrast is desirable, that pictures look better with enhanced edges, etc., and so filters can be designed to perform these functions. The enhancement methods are *ad hoc*, and most of them involve nonlinear filters, hence the application of linear filters in image enhancement is limited.

The approach to enhancement seems to be to find a property or feature of the image that is related to its subjective quality, and to specify a filter to enhance that feature.

Taking the Fourier transform of images provides a spectral decomposition that corresponds to certain components of the image. For example, the lower frequencies tend to be dominated by the average intensities in the image, while the high frequencies correspond to the rapidly changing parts of the image. Consequently, it is possible to enhance certain features of the image by operating on the corresponding parts of the spectrum. This property is utilized in some of the enhancement schemes described below. These schemes consider the application of digital filters in image enhancement for three kinds of applications: as a preprocessor to provide an improved image, as a noise-reduction process in its own right, and as a postprocessor to reduce the effects of noise introduced by some other form of processing.

The most commonly used methods of image enhancement cannot really be viewed as filtering techniques. Examples of these methods include contrast enhancement through histogram equalization, and similar methods based on gray-scale modification.

11.3.1. Enhancement to Reduce the Effect of Noise

Some forms of noise can be reduced by enhancement methods, because they happen to coincide with components whose contribution to the image is reduced compared to other components during enhancement. Two kinds of noise are considered here. The first is impulse noise, and the second is the so-called "blocking effect."

11.3.1.1. Removal of Impulse Noise

Impulse noise appears in images as discrete isolated pixel variations that are not spatially correlated. It may be due to noisy sensors, or be caused by channel transmission errors. In general, the noisy pixels are quite different from their neighbors. This marked variation between image pixels, and the lack of spatial correlation, form the basis of many noise-cleaning algorithms. Some of these are considered below. This is regarded as enhancement, rather than restoration, because there is no attempt to estimate the original image; rather, an effort is made to mitigate the effects of this type of noise.

Impulse noise is essentially a high spatial frequency phenomenon in comparison to the rest of the image, due to its spatial decorrelation. Consequently, low-pass filtering will reduce the effect of noise, and thus lead to a smoother image. It will obviously have the effect of smoothing any other high-frequency phenomena in the image, thus adversely affecting edges and rapidly varying parts of the image.

The spatial filtering operation may be performed by convolving the noisy image with an appropriate low-pass mask, such as the 3×3 mask

$$h = \frac{1}{10}\begin{bmatrix} 1 & 1 & 1 \\ 1 & 2 & 1 \\ 1 & 1 & 1 \end{bmatrix} \tag{11.7}$$

Other forms of low-pass mask may be used of the same or higher orders. The selection of weighting factors and mask size will affect the degree of smoothing obtained.

Alternatively, the filtering can be performed indirectly in the Fourier, or other orthogonal transform, domain which may have computational advantages over the direct convolution for large mask sizes. By this approach, the transform of the image is obtained and multiplied by a filter function, and the filtered image is obtained by taking the inverse transform

$$y(m, n) = T^{-1}U(u, v)H(u, v) \tag{11.8}$$

where $y(m, n)$ is the filtered image; T is the transform; $U(u, v)$ is the transform of $u(m, n)$, the unfiltered image; and $H(u, v)$ is the filter function in the transform domain. The filter function $H(u, v)$ is specified to give the required level of smoothing.

The cutoff frequency of the filter is an important design parameter. If it is set too low, too much smoothing will occur and information in the picture will be lost; at the same time, if it is set too high, there will be insufficient noise cleaning. In practice, it is impossible to set a perfect cutoff, as there is substantial overlap in the frequency domain between the noise spectrum and the image spectrum. Figure 11.2 illustrates impulse noise reduction by linear filtering.

Another linear technique that can be used to reduce impulse noise is linear arithmetic-mean filtering, which tends to reduce impulse noise and to distribute it to its surrounding pixels.

While these approaches give useful results, it is noted that the more successful methods of combating impulse noise have used nonlinear filtering techniques, such as median filtering[14] and generalized mean filtering.[15] These are described briefly below, since they are closely related to linear techniques.

Figure 11.2a. Impulse noise degradation.

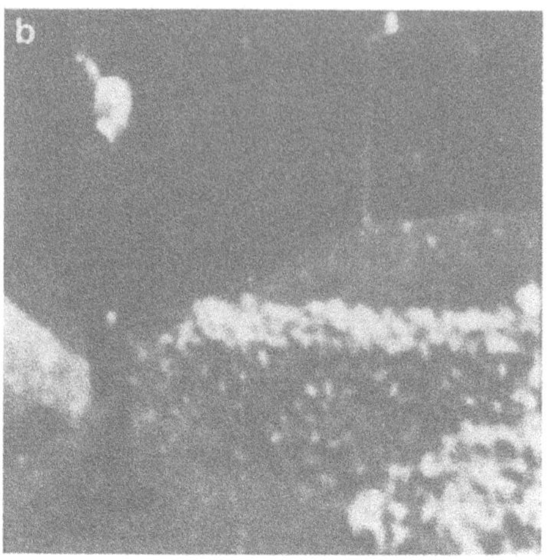

Figure 11.2b. Impulse noise reduction by low-pass filtering.

In median filtering, a sliding window is moved over the image, and the central pixel in the window is replaced by the median of all pixels inside the window.

Thus, for a 3×3 window, a new image is generated by

$$\begin{aligned} y(m, n) = \text{median}\{&u(m-1, n-1), u(m-1, n), u(m-1, n+1), \\ &u(m, n-1), u(m, n), u(m, n+1), u(m+1, n), \\ &u(m+1, n-1), u(m+1, n+1)\} \end{aligned} \tag{11.9}$$

Analysis of this filter[14] indicates that it is a useful tool for impulse-noise suppression.

Another nonlinear filtering scheme has been proposed,[15] where the new image is generated by a generalized mean filter given by the $(2L+1) \times (2L+1)$ sliding window over the $N \times N$ image with

$$y(m, n) = \left\{ \frac{\sum_{i=-L}^{L} \sum_{m=-L}^{L} w(i, j)[u(m+i, n+j)]^p}{\sum_{i=-L}^{L} \sum_{j=-L}^{L} w(i, j)} \right\}^{1/p}, \qquad m, n = 1, \ldots, N \tag{11.10}$$

where $w(i, j)$ are weights and p is a parameter chosen to get the best image, i.e., by trial and error.

This and other filters have been incorporated in an even wider class of nonlinear mean filters by Pitas and Venetsanopoulos.[16] In one dimension

$$y = g^{-1} \frac{\sum_{i=1}^{N} w_i g(u_i)}{\sum w_i} \tag{11.11}$$

where $g(u)$ is a single-valued analytic nonlinear function and w_i are weights.

11.3.1.2. Reduction of the Blocking Effect

This is an application of digital filters to the postprocessing of images which have suffered as a result of processing in block sections; this occurs in applications such as transform coding, block truncation coding, vector quantization, and other methods.

The effect is characterized by noticeable and frequently objectionable degradation near block boundaries, especially for low bit rates.

Usually, this may be considered as quantization noise and is mostly, though not entirely, outside the bandwidth of the image. The noise components that contribute to the blocking effect include:

1. Abrupt changes of intensity between blocks which are themselves smooth within an area of the image that is also smooth.
2. Lack of continuity of edges across blocks. Edges may be "staggered" as each block is coded independently.

This effect appears because the blocks are treated independently, and it could possibly be reduced if global block information was available. However, this would involve an overhead in requiring that such information be sent along with the coded coefficients, thus affecting the bit rate. A number of schemes have been presented for combating this sort of noise[17,18] without requiring more information.

Linear filtering can be used to remove out-of-band noise. It is observed that, within each region, the blocking effect is clearly out of band. For example, in the image of Figure 11.3a, the effect appears as a grid against the smooth background behind the MAN. Furthermore, the effect is less objectionable in some portions of the image than in others, owing to the spatial masking effect of the human visual system. The object of applying linear filtering is to reduce the blocking effect by low-pass filtering. The selection of the cutoff frequency is difficult to optimize, because a global cutoff is required while the noise in each region is out of band, having different cutoffs. Figure 11.3b shows the effect of linear filtering. Better results may be obtained using shift variant filters whose characteristics would be different in the different regions.

Figure 11.3a. Blocking effect degradation.

Figure 11.3b. Noise reduction after linear filtering.

11.3.2. Edge Enhancement

Edges convey the most information about an image, and hence enhancing them tends to lead to a subjective improvement of the image.

There is a variety of methods by which this may be achieved. As in the previous case, linear filtering can be applied (but nonlinear methods are usually better).

Edge enhancement is frequently used in preprocessing, before segmentation or for subjective improvement of images to be presented to a user. A related function, edge detection, enhances the edges without consideration as to what happens to the other image components. Frequently, the goal is to produce an edge-only image; this cannot really be considered under enhancement, and fits better in feature extraction or image analysis.

11.3.2.1. High-Pass Filtering

Enhancement of the edges in a picture can be obtained by high-pass filtering. Again, the rationale for this is that edges dominate the higher spatial frequencies, and a filter function which attenuates the lower frequencies will enhance the effect of the higher frequencies, and hence the edges.

This may be achieved by direct convolution with a high-pass mask, such as

$$h = \begin{bmatrix} -1 & -1 & -1 \\ -1 & 9 & -1 \\ -1 & -1 & -1 \end{bmatrix} \tag{11.12}$$

Other high-pass masks may be used as previously.

Alternatively, the operation may be performed indirectly in the frequency domain, as described below.

In this case, $H(u, v)$ in equation (11.8) represents a high-pass filter function. Similar design considerations apply as for the low-pass filter described previously; the main goal is to obtain good edge enhancement without sacrificing the nonedge information.

11.4. IMAGE RESTORATION

Some systems may be modeled by a linear shift-invariant impulse response with additive noise; images which are degraded by such systems can be restored by linear filtering.

In the digital filtering context outlined in Chapter 1, this involves the design of a filter whose transfer function is the effective inverse of the degradation process, and its application to the observed image $f(m, n)$ to give an estimate $\bar{u}(m, n)$ of the object image. This is illustrated in Figure 11.4. As discussed previously, indirect computational techniques based on transforms may be utilized.[19]

Obtaining a model of the PSF can be a problem in itself. This could be obtained analytically from the physical nature of the problem if there

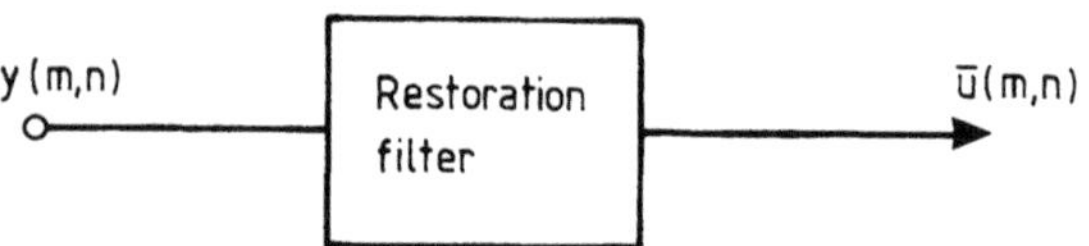

Figure 11.4. Image restoration model.

were sufficient *a priori* information. This is not possible in most cases, and the information has to be determined from the observed image.

A common approach is to assume a functional form for the blur—for example, that the PSF is smooth or uniform. The problem then becomes the determination of the parameters of this function from the Fourier spectrum of the blurred image. This technique has its limitations, as only certain classes of the PSF can be so identified. Detailed discussion of this problem lies outside the scope of this book.

Here we shall begin by considering essentially nonrecursive restoration filters. This is followed by a discussion of recursive filters and of some nonlinear restoration filters.

11.4.1. Linear Nonrecursive Filtering

The earliest image-restoration techniques used nonrecursive filters, some of the most common being discussed below.

11.4.1.1. Inverse Filtering

The basic inverse filter is aimed purely at the removal of blur, and does not take into account the presence of additive noise. It is based on the concept of inverting the degradation transfer function to yield a restored image. The ideal restoration filter will act on the degraded output to yield the original image in the form

$$u = H_R y \tag{11.13}$$

and, using equation (11.6), this leads to

$$u = H_R[Hu + v] \tag{11.14}$$

where H_R is the impulse response of the linear shift-invariant restoration filter, and H is the impulse response of the degrading system.

In the Z domain

$$U(z_1^{-1}, z_2^{-1}) = H_R(z_1^{-1}, z_2^{-1}) Y(z_1^{-1}, z_2^{-1}) \tag{11.15}$$

or, taking the DFT,

$$U(\omega_1, \omega_2) = H_R(\omega_1, \omega_2) Y(\omega_1, \omega_2) \tag{11.16}$$

For the inverse filtering operation, the restoration filter is chosen such that its transfer function corresponds to the inverse of that of the degradation process:

$$H_R(\omega_1, \omega_2) = \frac{1}{H(\omega_1, \omega_2)} \tag{11.17}$$

and, taking the DFT of equation (11.8), the transform of the restored image $\hat{U}(\omega_1, \omega_2)$ now becomes

$$\hat{U}(\omega_1, \omega_2) = U(\omega_1, \omega_2) + \frac{V(\omega_1, \omega_2)}{H(\omega_1, \omega_2)} \tag{11.18}$$

Performing the inverse DFT gives the restored image

$$\hat{u}(m, n) = u(m, n) + \sum_{\omega_1=0}^{N} \sum_{\omega_2=0}^{N} \frac{V(\omega_1, \omega_2)}{H(\omega_1, \omega_2)} W_N^{-\omega_1 m - \omega_2 n} \tag{11.19}$$

where $W_N = \exp(j2\pi/N)$ or, in vector notation,

$$\hat{u} = u + H^{-1} v \tag{11.20}$$

The above solution suggested by the inverse filter leads to certain comments. In the end, one is attempting to invert integral equations which, in the discrete domain, imply that the matrix H will be ill-conditioned. Even if the inverse H^{-1} defined by equation (11.20) exists, the image estimate may be dominated by noise. This noise is an additive reconstruction error, which may become large for spatial frequencies at which H^{-1} is low. In the absence of noise, reconstruction would be perfect and the effects of blur essentially removed. This, however, is rarely the case.

Thus, inverse filtering performs poorly in the presence of noise and, in addition, has several other related problems. For example, the image blur and formation process may not be invertible, and consequently H_R may not exist; this problem may be avoided to some extent by using a pseudoinverse in equation (11.17). Another problem is that the frequency response H usually falls off at high frequencies and, if high-frequency noise is present, this may lead to severe noise amplification, especially in high-detail regions of the image.

Finally, the inverse transfer function becomes infinite at the zeros of H. Consequently, if the transfer function contains zeros in its pass band, the inverse filter is not physically realizable, and this causes severe difficulties, as the filter response must be approximated by a large value at these frequencies.

To overcome the difficulties caused by ignoring the noise process, one must explicitly take the noise into account; this leads to the discrete Wiener formulation considered in Section 11.4.1.2.

11.4.1.2. Wiener Filtering

Wiener filtering techniques incorporate *a priori* statistical knowledge of the noise field.

To develop the discrete Wiener formulation, we consider equation (11.6) in which the noise is independent of the image and such that

$$E(uu^{t}) = P, \qquad E(vv^{t}) = R, \qquad \text{and} \qquad E(uv) = 0$$

where $E(\ )$ is the expectation operator and ^{t} denotes transpose.

The impulse response of the filter is chosen to minimize the mean-square error between the restoration estimate $\hat{u}$ and the original image u, namely,

$$\min_{u} E[(u - \hat{u})^{t}(u - \hat{u})] \tag{11.21}$$

The optimal estimate for linear transformations subject to Gaussian statistics is given by[20]

$$\hat{u} = PH^{t}(HPH^{t} + R)^{-1}y \tag{11.22}$$

On introducing the useful assumption of space invariance, zero mean, and stationarity, the transfer function $H_{R}(\omega_1, \omega_2)$ of the restoration filter may be obtained from

$$U(\omega_1, \omega_2) = H_{R}(\omega_1, \omega_2)\, Y(\omega_1, \omega_2)$$

where

$$\hat{H}_{R}(\omega_1, \omega_2) = \frac{H^{*}(\omega_1, \omega_2)}{|H(\omega_1, \omega_2)|^2 + W_v(\omega_1, \omega_2)/\, W_u(\omega_1, \omega_2)} \tag{11.23}$$

where $*$ denotes complex conjugate; W_v is the power spectral density of the noise, namely, the 2-D DFT of its covariance; and W_u is the power spectral density of the image, namely, the 2-D DFT of its covariance. Since the effect of noise has been explicitly included, equation (11.23) will not become infinite at any frequency in the pass band.

The main limitation of the Wiener filter is that it is not particularly well suited to the way in which the human visual system works, largely because of its reliance on the criterion of minimizing the mean-square error. It is overly concerned with noise suppression, and the stationarity assumptions that must be introduced to make the filter computationally feasible tender it insensitive to abrupt changes, so that it tends to smooth edges and reduce contrast. Thus, the filter leads to noise suppression at the expense of resolution. However, a variety of other filters have been proposed which alleviate some of the shortcomings of the Wiener filter.[21]

The *a priori* information required for the Wiener filter is fairly substantial. In addition to the PSF, which is also required by the inverse filter, we need the second-order statistics of the original image and the noise. For certain types of blur, it may be reasonable to assume that the PSF is known, even though in many cases the entire PSF, or several of its parameters, may not be known and may have to be estimated.

11.4.1.3. Parametric Estimation Filters

As seen in the previous section, the inverse filter is characterized by a potentially high resolution, though poor noise performance, while the Wiener filter has lower resolution with good noise performance. There is therefore scope for devising restoration schemes that attempt to trade off the advantages of each of the above filters, and several schemes have been reported and investigated by research workers.

Stockham *et al.*[22] consider a filter, known as the image power spectrum filter, that is the geometric mean of the inverse and Wiener filters. This relates the restored image with the input by

$$\hat{U}(\omega_1, \omega_2) = [H(\omega_1, \omega_2)]^{-s}\left[\frac{H^*(\omega_1, \omega_2)}{|H(\omega_1, \omega_2)|^2 + W_v(\omega_1, \omega_2)/W_u(\omega_1, \omega_2)}\right]^{1-s} Y(\omega_1, \omega_2) \tag{11.24}$$

where $0 \leq s \leq 1$ is a design parameter.

If $s = \frac{1}{2}$ and $H = H^*$, we have

$$\hat{U}(\omega_1, \omega_2) = H_R(\omega_1, \omega_2) Y(\omega_1, \omega_2) \tag{11.25a}$$

where

$$|H_R(\omega_1, \omega_2)| = \left[\frac{1}{|H(\omega_1, \omega_2)|^2 + W_v(\omega_1, \omega_2)/W_u(\omega_1, \omega_2)}\right]^{1/2} \tag{11.25b}$$

The power spectrum of the reconstructed image $W_{\hat{u}}(\omega_1, \omega_2)$ is thus related to the power spectrum of the observed image $W_y(\omega_1, \omega_2)$ by

$$W_{\hat{u}}(\omega_1, \omega_2) = |H_R(\omega_1, \omega_2)|^2 W_y(\omega_1, \omega_2) \tag{11.26}$$

and $W_y(\omega_1, \omega_2)$ is in turn related to the power spectrum of the ideal object image $W_u(\omega_1, \omega_2)$ by

$$W_y(\omega_1, \omega_2) = |H(\omega_1, \omega_2)|^2 W_u(\omega_1, \omega_2) + W_v(\omega_1, \omega_2) \tag{11.27}$$

where $W_v(\omega_1, \omega_2)$ is the power spectrum of the noise.

Thus it is seen from equations (11.25b), (11.26), and (11.27) that the power spectrum of the reconstructed image is identical to that of the ideal object image. By contrast, it may be shown that, using the Wiener filter, the power spectra only become identical for noise-free observations. The equivalence of the power spectra does not necessarily lead to better images, and none of the transfer functions considered above has shown the phase effect of the restoring filters. This is an important consideration, given the importance of phase in image processing and viewing.[23,24] This point is considered further in Case Study 3 in Chapter 12.

Another filter intermediate between the Wiener and inverse filters is the constrained least-squares filter. Here, $\hat{u}$ is chosen to minimize

$$J(\hat{u}) = \hat{u}C^T C\hat{u} \tag{11.28}$$

subject to the constraint

$$(H\hat{u} - y)^T(H\hat{u} - y) = e \tag{11.29}$$

The matrix C gives some flexibility in selecting the properties of the filter.

In the original 1-D formulation, Phillips[25] asserted that the vector C must be chosen so as to compute the second difference of a vector multiplied by it. In the 2-D work[13,14] this second-difference property is retained or approximated. Two ways of constructing such a matrix are given. The first method creates a Laplacian-like matrix of appropriate size. The second method constructs a radially symmetric representation of the second-difference operator; it starts from a 1-D second-difference sequence, transforms it to the frequency domain, and builds up a 2-D matrix.

Quantity C is thus a finite-difference matrix, which can minimize some measure of the rate of fluctuation in the estimate.

The solution to equations (11.28) and (11.29) is

$$\hat{u} = (H^T H + \gamma C^T C)^{-1} H^T y \tag{11.30}$$

where γ is a Lagrange multiplier found by iteration to satisfy the constraint (11.29), and e is a parameter which enables control of the level of noise suppression.

On taking transforms for the shift-invariant case, we obtain the transfer function

$$\hat{H}_{R}(\omega_1, \omega_2) = \frac{H^*(\omega_1, \omega_2)}{|H(\omega_1, \omega_2)|^2 + \gamma|C(\omega_1, \omega_2)|^2} \tag{11.31}$$

This filter has eliminated the necessity of knowing the second-order statistics of u and v and, moreover, it provides some control over the filter action through modification of the parameters γ, e, and C.

As an example we consider the case $C = I$; comparison of equations (11.31) and (11.23) shows that this results in a pseudoinverse filter. Again, if $|C(\omega_1, \omega_2)|^2$ is set equal to the ratio of the image to the noise power spectra, we have the parametric Wiener filter.

The techniques considered so far have used explicit models of the degradation, and have in some cases utilized specific information about the original image and the noise process. For example, some filtering schemes assume knowledge of the first and second moments of the image.

A way of improving restoration performance is to apply constraints such as positivity, upper bound, specification of individual pixel values, etc. Constrained restoration provides a framework for imposing some forms of *a priori* knowledge, which can often lead to a reduction in the ill-conditioning of the problem. The selection and application of constraints, the computational techniques, and their effects are all areas of current research.

A number of methods have been proposed for constrained restoration in which the constraints are incorporated into the restoration process, rather than on the final restored image. For example, using a least-squares measure with positivity and upper-bound constraints, the restoration problem reduces to a quadratic programming problem. Furthermore, iterative methods have been proposed where the constraints are applied to intermediate estimates. Huang's projection method of constrained restoration[26] uses numerical techniques to solve the set of equations representing the degradation model, applying constraints in each iteration.

11.4.2. Linear Recursive Restoration Filters

Early techniques in image restoration, such as those described in the previous section, concentrated on frequency domain implementations of nonrecursive algorithms. As discussed above, they assume a stationary image model and the technique cannot perform satisfactorily on real-world images that are nonstationary or on images that are degraded by space-invariant blur.

Recent research has concentrated on 2-D recursive filtering algorithms extended from the 1-D Kalman filtering algorithm. The main advantage of recursive filtering algorithms is that they require less computation time and can be easily adapted to handle nonstationary images and those degraded by space-variant blur.

Early approaches to the problem of 2-D recursive filtering of images used lexicographic scan ordering to convert the 2-D image into a 1-D signal, which could then be filtered using recursive Kalman filtering[27] (see Section 10.3). No 2-D model for the image was used, other than the usual covariance description. These filters are considered briefly in Section 11.4.2.1.

The next approach was to utilize 2-D recursive models of the images and to realize 2-D optimal Kalman filters.[28,29] Various models were used, including separable autocorrelation functions. Owing to the large storage and computation requirements of optimal Kalman filtering, suboptimal schemes have been developed.[29]

Since the Kalman filter is based on the criterion of minimizing the mean-square error, it will sacrifice resolution in favor of noise suppression, and relies heavily on *a priori* knowledge of the image covariance. If the image model is erroneous, the approach may not give accurate results.

Recursive filtering has not been limited to causal image models. A semicausal model based on Woods's[30] 2-D Markov fields has been used by Jain and Angel[31] to model images in recursive filtering. For further discussion on these models we refer the reader to Chapter 10.

11.4.2.1. Recursive Scan Techniques

These involve the 1-D processing of the scan-ordered image. We consider the original image $u(m, n)$, which for simplicity is assumed to be zero mean with stationary covariance $r(k, l)$, where

$$r(k, l) = E[u(m, n) \ u(m + k, n + l)] \tag{11.32}$$

E being the expectation operator.

The observed image $y(m, n)$ is corrupted with additive noise, which we assume to be zero mean and white Gaussian. Initially, assuming that there is no blur,

$$y(m, n) = u(m, n) + v(m, n) \tag{11.33}$$

The noise covariance is

$$r_n(k, l) = E[v(m, n) \ v(m + k, n + l)] \tag{11.34}$$

After lexicographic scan ordering, and using vector notation, the above equations become

$$y(k) = u(k) + v(k) \tag{11.35}$$

$$r_u(j) = E[u(k)\ u(k+j)] \tag{11.36}$$

$$r_n(j) = E[v(k)\ v(k+j)] \tag{11.37}$$

Given the above model, Kalman filtering may be used to suppress the noise.

Developing a state-space model for u requires stochastic realization, which should be updated at every stage when $r_u(j)$ is not stationary. Various approximations have been proposed. Franks[32] proposes a stationary covariance

$$r_u(k) = \frac{1}{N} \sum_{m=1}^{N} u(m, m+k) \tag{11.38}$$

When the covariance has been defined, a realization procedure can be used to find the state-space model for u and apply direct Kalman filtering.

The stationary approximation is clearly inappropriate for the line-scan-ordered function and various methods have been suggested to reduce its effect. Vector scanning[33,34] simultaneously scans a number of lines. These schemes lead to a better approximation of the covariance.

One popular method[32,33] considers the image autocorrelation function to be separable, of the exponential form

$$r_u(m, n) = \rho_1^{|m|} \rho_2^{|n|} \tag{11.39}$$

The autocorrelation function of $u(k)$ can then be determined[33] and Kalman filtering performed. The resulting recursive estimate of the ith sample of $u(k)$ is then given by

$$\hat{u}(i) = a(i-1)\hat{u}(i) + b(i-1)y(i-1) \tag{11.40}$$

where a and b are weighting terms dependent on the correlation function of $u(k)$.[35]

11.4.2.2. Two-Dimensional Optimal and Suboptimal Kalman Filtering

The models used in the scan-ordered techniques discussed in the preceding section are basically one-dimensional. The use of 2-D recursive models has been examined by a number of authors and still remains the subject of research.[36] We present a few of these methods here, and discuss them from the standpoint of estimating from a noisy-image observation. As before, we use a model that initially assumes no blur [see equation (11.33)]. Of particular interest in these methods is the size of the resulting Kalman filter, and the question of whether the methods can be extended to estimate in the presence of a nontrivial PSF.

One of the earliest methods[28] considered images described by the exponential autocorrelation function (11.39). This function can be modeled by the 2-D autoregressive process (see Example 10.3)

$$u(m+1, n+1) = \rho_2 u(m+1, n) + \rho_1 u(m, n+1) - \rho_1\rho_2 u(m, n) + [(1-\rho_1^2)(1-\rho_2^2)]^{1/2} w(m, n) \tag{11.41}$$

where $w(m, n)$ is a white, zero-mean process with the same variance as elements of the image, while ρ_1 and ρ_2 are horizontal and vertical adjacent image-point correlations.

If the observed image $y(m, n)$ is modeled as in equation (11.33),

$$y(m, n) = u(m, n) + v(m, n)$$

where $u(m, n)$ is the ideal image and $v(m, n)$ is additive noise.

The 2-D recursive estimate $\hat{u}(m+1, n+1)$ found by Habibi[28] is a one-step predictor along the northeast direction and gives

$$\hat{u}(m+1, n+1) = \rho_2 \hat{u}(m+1, n) + \rho_1 \hat{u}(m, n+1) - \rho_1\rho_2 \hat{u}(m, n) + K(m, n)[y(m, n) - \hat{u}(m, n)] \tag{11.42}$$

Quantity $K(m, n)$ is the Kalman gain function, which is dependent on factors such as model matrices, covariance matrices of the model noise, and additional noise, found[28] to be a complicated function of the covariance of the estimation error. This estimator has a simple appealing structure, but is not optimal because it is not based on the global state.

Optimal 2-D Kalman filtering of images has been studied by several workers. Woods and Radewan[29] assume a nonsymmetrical half-plane (NSHP) model given below, and show that a rather large optimal Kalman filter can be written in the form

$$u(m, n) = \sum_k \sum_l b(k, l) u(m-k, n-l) + \sum_l b(0, l) u(m, n-l) + w(m, n)$$

A more efficient estimator is obtained by processing one line of data at a time. This method can be modified to perform optimal Kalman filtering in the presence of blur but, as before, it results in a giant Kalman filter.

Other optimal line-by-line Kalman filtering schemes have been considered by various workers, such as Attasi,[37] but all seem to develop large Kalman filters.

The large amounts of storage required by 2-D optimal Kalman filtering algorithms have prompted work on suboptimal estimators, which require less computation. One suboptimal method[29] involves processing the picture in strips, or sections, so as to reduce the dimension of the global state. Edge effects caused by incorrect boundary conditions between strips are minimized by overlapping the strips.

Another suboptimal method is the reduced-update Kalman filter, which uses a finite-order NSHP model.[26,29] Here, only those estimates of the state near the point being processed are updated, rather than all the components of the state. The proposed block Kalman filter[38] uses a vector scanning scheme in two adjacent strips, providing smoother estimates in blocks and eliminating edge effects.

This is an area of active research, and suboptimal methods other than these two have also been proposed.

Restoration quality in linear filters is necessarily imperfect, in the sense that local features such as edges may be obscured or smoothed out, or that noise levels are too objectionable in nonedge regions. High noise suppression image restoration is in practice difficult to achieve using linear filtering techniques, which are subject to the well-known resolution versus noise dilemma.

Furthermore, the human visual system (HVS) has a spatial masking effect which enables the eye to tolerate relatively high noise levels in edge regions, but not in flat or smoothly varying regions of an image. The criteria of minimizing the mean-square error, implicit in these filtering techniques, are rather nonsubjective and may not necessarily correspond to the HVS. Consequently, many of the more successful restoration techniques utilize nonlinear processes, which are more adaptive and responsive to the HVS criteria of image intelligibility and are able to utilize constraints and available *a priori* information.

A description of some current research work, and a brief mention of this nonlinear work, are provided.

11.4.3. Nonlinear and Space-Variant Restoration Filtering

The image model used in previous sections has assumed that the image is a wide-sense stationary field. This is not strictly correct and, furthermore, there are many nonlinearities in image sensing that lead to nonlinear blur. Consequently, the basic assumption in using linear processing, i.e., that the degradation is linear, is not realistic.

It is therefore of interest to consider nonlinear restoration filters. The main problem with nonlinear schemes is the difficulty in implementation and the excessive computational requirements. One method which can be implemented is homomorphic filtering.

11.4.3.1. Homomorphic Filtering

Although this book is concerned mostly with the applications of linear systems, there are applications in nonlinear systems where linear filtering can be used.

In common with linear filtering, homomorphic filtering is based on the validity of a superposition rule relating input and output functions. Essentially it consists of a linear filter preceded and succeeded by characteristic functions

$$H = D_y^{-1} L D_x \tag{11.43}$$

where D_y and D_x are called characteristic systems, while L is a standard linear system. The first characteristic system operates on the signal, enabling a linear filter to be used. This is followed by a complementary system to yield the desired output.

This is a useful technique for image restoration when an observed image is subject to multiplicative degradation. An example could be a nonuniform field of illumination or multiplicative noise,

$$y(m, n) = u(m, n)E(m, n) \tag{11.44}$$

The homomorphic restoration is illustrated in Figure 11.5 and involves first taking the logarithm of equation (11.44) to give

$$\log[y(m, n)] = \log[u(m, n)] + \log[E(m, n)] \tag{11.45}$$

Conventional linear filtering can then be used to estimate $\log[\hat{u}(m, n)]$ and exponentiation provides the required image. Researchers are pursuing the design, analysis, and use of more complex nonlinear filters than the one discussed above.

The use of linear space-variant filters is also an open research problem. These techniques are more realistic and will be of more research interest as the computational problems are reduced by better hardware.

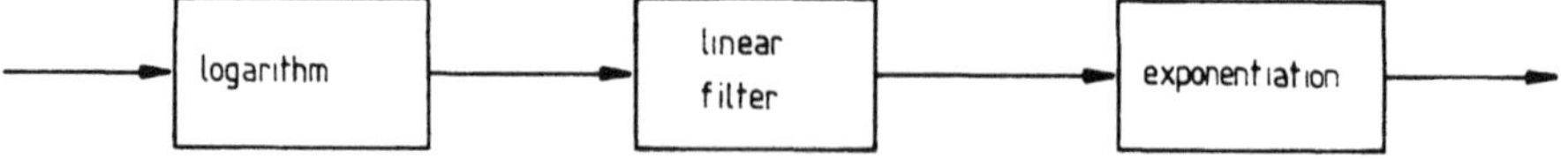

Figure 11.5. Homomorphic filter.

11.5. IMAGE RECONSTRUCTION

Reconstruction refers to that class of inverse problems where the object image function is reconstructed from information or measurements in an observation space.

In the general reconstruction problem, the nature of the observation space can vary. For example, in tomography it consists of object projections, and in sampling it consists of image samples on a grid. Consequently, we either have observations of the image or of some function related to the desired image field.

Restoration could be viewed as a special form of reconstruction where the observation is related to the image by a degradation function, which could consist of blur and noise.

Reconstruction is usually assumed to be from full or complete information when the available observations are sufficient for reconstruction of the image. The major problem of reconstruction from incomplete information occurs when there are fewer observations than are normally required to reconstruct the image, or the observations are incomplete.

11.5.1. Reconstruction from Complete Information

Applications where the image can be considered to be reconstructed from complete information include image sampling and reconstruction, and tomographic image reconstruction.

A common model of the tomographic problem involves the reconstruction from projection measurements of the linear attenuation coefficient integrated along the path of a collimated X-ray beam. The basic reconstruction techniques are based on Radon's rigorous solution to the problem. Reconstruction of the image from the projections is also known as the "inverse scattering problem." This involves reconstructing the image function on a uniform grid. The reconstruction is possible when the distribution of the projections is known at finite and suitably sampled points in the projection domain. A common reconstruction method is the so-called filtered back projection technique.

11.5.2. Reconstruction from Incomplete Information

This problem is found in quite a number of applications of image processing, and has been the subject of a great deal of research effort by many workers on various subproblems.

One common application is image reconstruction from incomplete projection data. Here, parts, or the whole, of some projections may be missing, due to physical constraints on the system or the object being imaged. In the limited-angle problem, projections are limited in terms of angular number or angular range.

Another application is the reconstruction of images when only partial information about their Fourier spectrum is available. This may be due to physical constraints, such as when only the Fourier phase, or only the magnitude, can be measured. A lot of research effort has recently been directed at reconstruction from slightly artificial problems, such as reconstruction given one bit of phase information; this leads to a better understanding of this problem area.

Other significant applications include band-limited spectrum extrapolation, synthetic aperture radar imaging, and many others.

Some of these application areas are considered in more detail in Case Study 4 in the next chapter. In the remainder of this section we briefly review some of the techniques available for reconstruction from incomplete information.

Research workers in this area have approached the problem from two points of view. The first has been the study of uniqueness conditions under which the original image should be recoverable from the given partial information. The second has been the development and study of algorithms for reconstruction from incomplete information.

The techniques fall into two broad categories—iterative and non-iterative.

For many incomplete information problems, explicit inversion techniques are not available or are difficult to use. Consequently, iterative reconstruction algorithms are widely used. Many of the techniques described in the literature to solve incomplete information problems are variations of the 2-D Gerchberg–Papoulis iterative reconstruction algorithm.[39,40] The use of iterative algorithms also introduces the need to study and prove convergence of the algorithm.

Papoulis considered the band-limited spectrum extrapolation problem for continuous signals. He wished to extrapolate a finite segment of data, given that it is band-limited. The algorithm proposed for solving the problem iterated between band-limiting the estimated signal, and then replacing the known segment by its correct value. Convergence was proved by exploiting the properties of prolate spheroid wave functions.

Gerchberg studied the same problem with the frequency and time domains reversed. He estimated the high frequencies of a finite-length signal from the given low frequencies, using a similar iterative algorithm.

These approaches determined the structure of many of the algorithms that have been applied to similar problems of reconstruction from incomplete information. They alternate between forcing constraints in one domain, such as time, and then in another domain, such as frequency.

The presentation and proof of the convergence of these and similar algorithms have been discussed in a unified framework by a number of authors, for example, Tom *et al.*[41]

REFERENCES

1. N. K. Bose (ed.), Special issue on multidimensional systems, *Proc. IEEE* **65,** 819–922 (1977).
2. R. M. Mersereau and D. E. Dudgeon, Two dimensional digital filtering, *Proc. IEEE* **63,** 610–623 (1975).
3. B. R. Hunt, Digital image processing, *Proc. IEEE* **63,** 693–708 (1975).
4. R. W. Hamming, *Digital Filters,* Prentice-Hall, Englewood Cliffs, NJ (1983).
5. T. S. Huang (ed.), *Picture Processing and Digital Filtering,* Springer-Verlag, Berlin (1979).
6. A. V. Oppenheim and R. W. Shafer, *Digital Signal Processing,* Prentice-Hall, Englewood Cliffs, NJ (1975).
7. L. R. Rabiner and B. Gold, *Theory and Application of Digital Signal Processing,* Prentice Hall, Englewood Cliffs, NJ (1975).
8. J. W. Tukey, *Exploratory Data Analysis,* Addison-Wesley, Reading, Massachusetts (1971).
9. A. V. Oppenheim, A. S. Willsky, with I. T. Young, *Signals and Systems,* Prentice-Hall, Englewood Cliffs, NJ (1983).
10. A. S. Kwabwe, *Image Reconstruction from Incomplete Information,* Ph.D. Thesis, Imperial College (1984).
11. J. S. Lim, Image restoration by short space spectral subtraction, *IEEE Trans. Acoust., Speech, Signal Process.* **ASSP-28,** 191–197 (1980).
12. A. V. Oppenheim, R. W. Schafer, and T. G. Stockham, Nonlinear filtering of multiplied and convolved signals, *Proc. IEEE* **56,** 1264–1291 (1968).
13. H. C. Andrews and B. R. Hunt, *Digital Image Restoration,* Prentice-Hall, Englewood Cliffs, NJ (1977).
14. W. K. Pratt, *Digital Image Processing,* John Wiley and Sons, New York (1978).
15. A. Kundu, S. K. Mitra, and P. P. Vaidyanathan, Application of two-dimensional generalized mean filtering for removal of impulse noises from images, *IEEE Trans. Acoust., Speech, Signal Process.* **ASSP-32,** 600–609 (1984).
16. I. Pitas and A. N. Venetsanopoulos, Nonlinear mean filters in image processing, *IEEE Trans. Acoust., Speech, Signal Process.* **ASSP-34,** 573–584 (1986).
17. B. Ramamurthi and A. Gersho, Nonlinear space-variant postprocessing of block coded images, *IEEE Trans. Acoust., Speech, Signal Process.* **ASSP-34,** 1258–1268 (1986).
18. D. C. Munson and J. S. Lim, Reduction of the Blocking Effect in Images, IEEE International Conference on Acoustics, Speech and Signal Processing, San Diego (1984).
19. C. W. Helstrom, Image restoration by the method of least squares, *J. Opt. Soc. Am.* **57,** 297–303 (1967).
20. G. M. Robbins and T. S. Huang, Inverse filtering for linear shift-variant imaging systems, *Proc. IEEE* **60,** 862–872 (1972).
21. W. K. Pratt, Generalized Wiener filtering computation techniques, *IEEE Trans. Comput.* **C-21,** 636–641 (1972).
22. T. G. Stockham Jr., T. M. Cannon, and R. B. Ingebretsen, Blind deconvolution through digital signal processing, *Proc. IEEE* **63,** 678–692 (1975).
23. A. V. Oppenheim and J. S. Lim, The importance of phase in signals, *Proc. IEEE* **69,** 529–541 (1981).
24. T. S. Huang, J. W. Burnett, and A. G. Deczky, The importance of phase in image processing filters, *IEEE Trans. Acoust., Speech, Signal Process.* **ASSP-23,** 529–542 (1975).

25. D. L. Phillips, A technique for numerical solution of certain integral equations of the first kind, *J. Assoc. Comput. Mach.* **9,** 97–101 (1962).
26. M. S. Murphy and L. M. Silverman, Image model representation and line-by-line recursive restoration, *IEEE Trans. Autom. Control* **AC-23,** 809–816 (1978).
27. S. R. Powell and L. M. Silverman, Modeling of two-dimensional covariance functions with application to image restoration, *IEEE Trans. Autom. Control* **AC-19,** 8–13 (1974).
28. A. Habibi, Two-dimensional Bayesian estimate of images, *Proc. IEEE* **60,** 878–883 (1972).
29. J. W. Woods and C. H. Radewan, Kalman filtering in two dimensions, *IEEE Trans. Inf. Theory* **IT-23,** 473–482 (1977).
30. J. W. Woods, Two-dimensional discrete Markovian fields, *IEEE Trans. Inf. Theory* **IT-18,** 232–240 (1972).
31. A. K. Jain and E. Angel, Image restoration, modelling, and reduction of dimensionality, *IEEE Trans. Comput.* **C-23,** 470–476 (1974).
32. L. E. Franks, A model for the random video process, *Bell Syst. Tech. J.* **45,** 609–630 (1966).
33. N. E. Nahi and C. A. Franco, Recursive image enhancement—vector processing, *IEEE Trans. Commun.* **21,** 305–311 (1973).
34. B. R. Suresh and B. A. Shenoi, New results in two-dimensional Kalman filtering with applications to image restoration, *IEEE Trans. Circuits Syst.* **CAS-28,** 307–319 (1981).
35. N. E. Nahi, Role of recursive estimation in statistical image enhancement, *Proc. IEEE* **60,** 872–877 (1972).
36. A. M. Tekalp, H. Kaufman, and J. W. Woods, Identification of image and blur parameters for the restoration of noncausal blurs, *IEEE Trans. Acoust., Speech, Signal Process.* **ASSP-34,** 963–972 (1986).
37. S. Attasi, Modelling and recursive estimation for double indexed sequences, in *System Identification: Advances and Case Studies* (R. K. Mehra and D. G. Lainiotis, eds.), Academic Press, New York (1977).
38. M. R. Azimi-Sadjadi and P. W. Wong, Two-dimensional block Kalman filtering for image restoration, *IEEE Trans. Acoust., Speech, Signal Process.* **ASSP-35,** 1736–1749 (1987).
39. R. W. Gerchberg, Super-resolution through error energy reduction, *Opt. Acta* **21,** 709–729 (1974).
40. A. Papoulis, A new algorithm in spectral analysis and band-limited extrapolation, *IEEE Trans. Circuits Syst.* **CAS-22,** 735–742 (1975).
41. V. T. Tom, T. F. Quatieri, M. H. Hayes, and J. H. McClellan, Convergence of iterative non-expansive signal reconstruction algorithms, *IEEE Trans. Acoust., Speech, Signal Process.* **ASSP-29,** 1052–1058 (1981).

12

Case Studies

12.1. INTRODUCTION

This chapter describes some case studies that demonstrate the techniques discussed in Chapter 11. The scope, however, is restricted to applications of linear digital filters only. In any case study, digital filtering forms only a part, nevertheless an important part, of the processes which must be applied to the image in order to achieve the desired objective; as such it is described within the context of the application.

Four case studies illustrating these applications are considered. These are chosen from a few of the major application areas for digital filters in image processing. The first examines a medical application, while the second is concerned with an industrial problem. Case studies 3 and 4 are of a more general nature, considering image restoration and reconstruction problems which form the backbone of many applications in image processing, communications, and astronomy. The intention is not only to show digital filters at work, but also to get a feel for the overall context in which they may be applied, some of the constraints under which they operate, and how these translate into physical requirements.

12.2. CASE STUDY 1: X-RAY IMAGE ANALYSIS

Pictures such as radiographs, angiographs, tomographs, etc., are widely used in medicine for diagnosis, treatment monitoring, and research. Consequently, it is not surprising that image processing and digital filters find considerable application in this field.

Image processing has two related functions here. First, it may be used to enhance and improve images which have been formed by X radiography or other means. Such pictures can be quite poor for a variety of reasons, and it is necessary to process and improve them. For example, there may be degradation blur and noise, occurring during the image formation process or as a result of uneven illumination. The nature of the X-radiographic process itself may lead to distortion. In another variation, image processing may actually be required to produce an image that physicians then study.

Perhaps the most publicized application of image processing in medicine is the formation of X-ray images via computerized tomography (CT). The basic radiographic problem involves reconstruction from projection measurements of the linear attenuation coefficient integrated along the path of a collimated X-ray beam. Digital filters are widely applied during the reconstruction process that produces the image.

Alternatively, one may wish to reduce the radiation dosage to which a patient is exposed. This may lead to underexposed or low-contrast images, which may need to be enhanced.

The second function of image processing is as a preprocessor for machine diagnosis, recognition, or measurement. In such an application, the output of the image processor is aimed at a subsequent set of algorithms, rather than for subjective human viewing.

This case study takes an example of the various stages of image processing required in such a system, and outlines the use of digital filters in effecting the required processes.

12.2.1. Background

Human growth can be linked to the development of certain bones in the hand which go through invariant event sequences covering the entire developmental span. It is thus possible to determine maturity, measure tempo of growth, and predict adult height, by examining these bones according to some established rules. Ordinarily, such determination is carried out by human experts studying the images, but this may be time-consuming and imprecise, as well as nonquantitative. Since such a study could form the basis on which hormones and other treatment may be administered and monitored, there is a fair amount of interest in obtaining an objective and quantitative rating method.

The features used in analyzing the images can be extracted and compared using image-processing techniques, and if all knowledge of the rating methods can be encoded into a suitable expert system, then these can form the basis of a machine classification. Alternatively, image processing can be used to enhance those features that enable a human expert to make a more effective comparison between the current X ray and a "standard" X ray.

Two distinct problems arise in such an application and are concerned not only with image processing itself, but also with analysis, measurement, and classification of images. The first, which we shall dwell on in some detail in this study, is the problem of operating on the X-ray images to produce some description that can be used by a system that possesses expert knowledge of interpreting such X-ray images. The image-processing problems that have to be solved may include the following, depending upon how the problem is approached:

- Compensation for uneven illumination and poor contrast of the original X-ray images.
- Region enhancement and noise smoothing.
- Consistent segmentation of the image.
- Bone contour extraction and contour following.
- Extraction of features to describe the image properties (such as shape).

The second problem is creating such a system. This involves the actual choice of the method of representing the knowledge and of applying it. This is an interesting problem, which mostly falls outside the scope of this work. The subject of knowledge representation is itself an active research area, with workers looking at the most appropriate methods for each application. Knowledge-based signal processing provides a structure for the incorporation of application-independent knowledge into the solution. It has yet to be effectively applied to digital filters.

The methodology used here is essentially a two-stage process. The first stage broadly involves using image-processing knowledge, and knowledge about the nature and content of X-ray images of the hand, to produce a segmented image with only the bones outlined on the background. These bones are then identified and labeled.

The second stage involves taking the images of some of these bones, chosen according to the method of rating in use, and processing them to generate suitable descriptions of certain preselected and defined features. These features are then used, together with expert knowledge of how they relate to development, in order to assess each of the bones. The overall block diagram in Figure 12.1 shows the processes involved.

It is in the first stage that digital filters find most utilization.

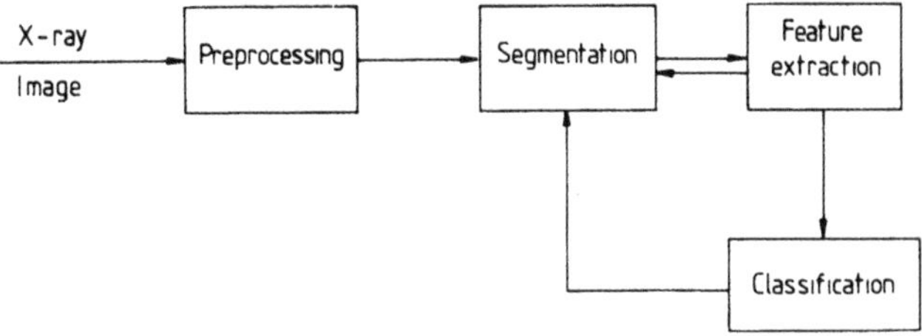

Figure 12.1. X-ray image analysis.

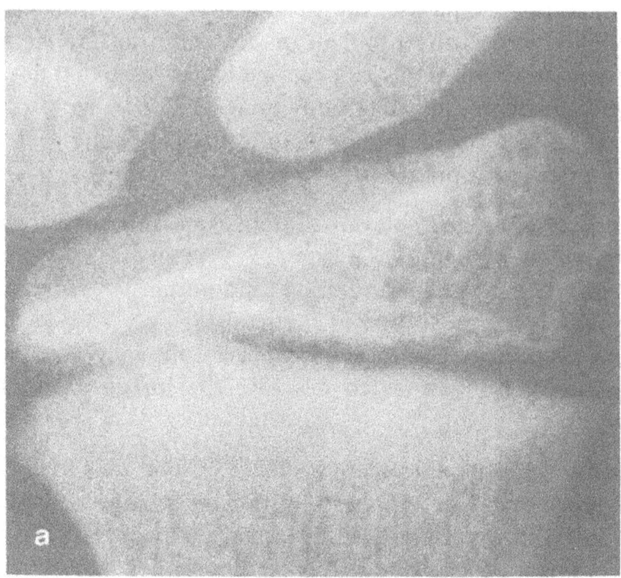

Figure 12.2a. Original X-ray image.

12.2.2. X-ray Image Filtering Examples

Various operations are performed on images before they are segmented. This is partly because it is necessary to standardize the images so as to enable higher-level algorithms to operate on reasonably standard data. The preprocessing is intended to compensate for known degradations and to minimize interpicture variation.

The methods available to achieve this include various forms of gray scale modification, such as histogram equalization, as well as various forms of filtering.

The original X-ray image is shown in Figure 12.2a. This has very poor contrast and edge definition. The edge-enhanced image shown in Figure 12.2b is obtained by high-pass filtering. This forms the input to the segmentation algorithm and leads to much better segmentation than can be obtained without the filtering.

Another use of digital filters is in the process of smoothing. This may be applied after use of a nonlinear operator, such as segmentation or edge detection.

Some edge-detection operators (such as the Sobel or Laplacian) enhance edges, but may also lead to noise enhancement. Typically, it is found that the combination of histogram equalization and edge detection leads to well-defined edges, which are important for the application in hand.

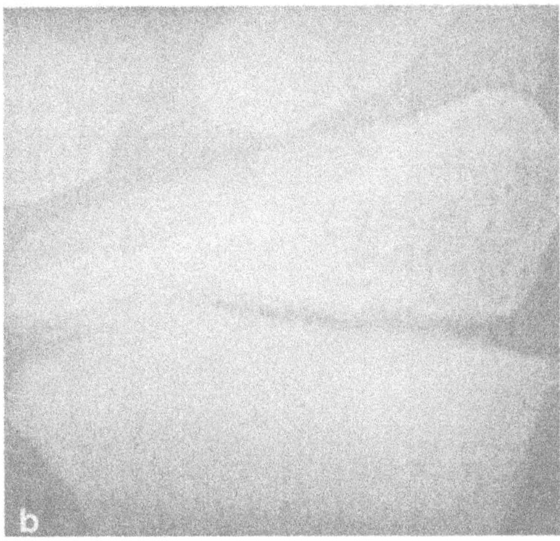

Figure 12.2b. Enhanced image obtained by high-pass filtering of Figure 12.2a.

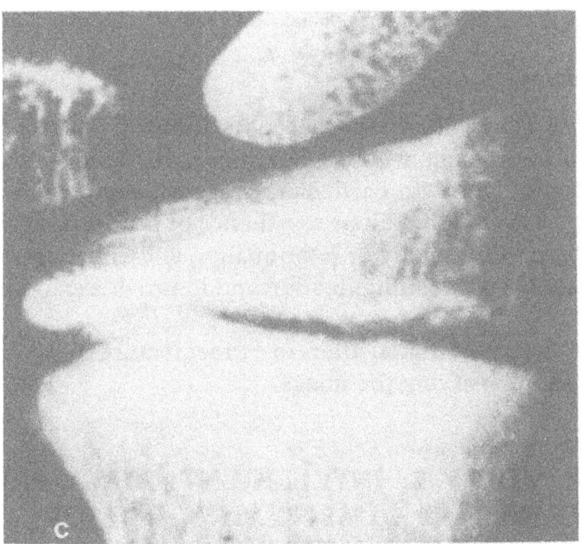

Figure 12.2c. Noisy image obtained by compensation and enhancement of Figure 12.2a.

As the example in Figure 12.2c shows, there is a drawback in that isolated noise is enhanced. This may be reduced by filtering, using a low-pass filter, and the result is illustrated in Figure 12.2d.

Filters may also be used to extract information about specific portions of the spectrum. These are measures which can be used to give some indication of the texture of the scene.

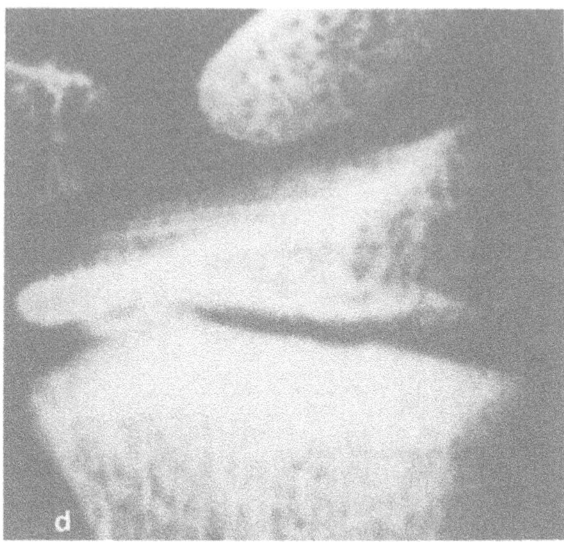

Figure 12.2d. Smoothed image obtained by low-pass filtering of Figure 12.2c.

12.2.3. Summary

This case study illustrates the application of image processing to a specific medical problem, whose solution involves, among other techniques, the use of digital filters. In particular, two areas are identified:

First, the use of digital filters to enhance or restore X-ray images. These images are by nature usually of poor quality, and enhancement is essential to compensate for uneven illumination and to provide smoothing and noise reduction.

Second, the use of digital filters to extract features from an image which may be used in classifying the image.

12.3. CASE STUDY 2: INTELLIGENT IMAGE-PROCESSING SYSTEMS FOR QUALITY CONTROL OF MANUFACTURED PARTS

This case study considers the application of two-dimensional (2-D) digital filters to problems of automated inspection for quality control of manufactured parts. These applications include image deblurring, edge enhancement for contour detection problems, and feature extraction.

The filters considered in this study are linear-phase recursive filters designed from frequency-domain specifications. The linear-phase constraint enables computationally efficient algorithms to be employed.

12.3.1. Background

In industrial quality control, the basic inspection task involves examining manufactured parts and identifying any that may have defects. When digital image processing is applied to this task, an image of the part under inspection is formed and features extracted from it are compared in some way with expected features. When the variation between the expected and determined features is significant, the part is considered to be defective and is rejected.

Figure 12.3 shows an overall block diagram of the type of inspection and quality-control system considered in this case study. The problems involved in implementing such a system include image quality that is marred by nonuniform reflectance from metallic surfaces, and this problem can be solved using linear filtering. In addition, some of the feature extraction and matching techniques can be implemented using linear filtering techniques. A major consideration is the ability to perform the various computations and reach a decision at real-time speeds.

Automated inspection is the largest manufacturing application of image processing and machine vision in use at the present time. Applications range from quality control, robot control and vision, counting, assembly line operations, etc.

In this case study, we consider three specific inspection problems involved in the quality control of the manufacture of rather simple components. These are: car piston heads, differential gear wheels, and strain

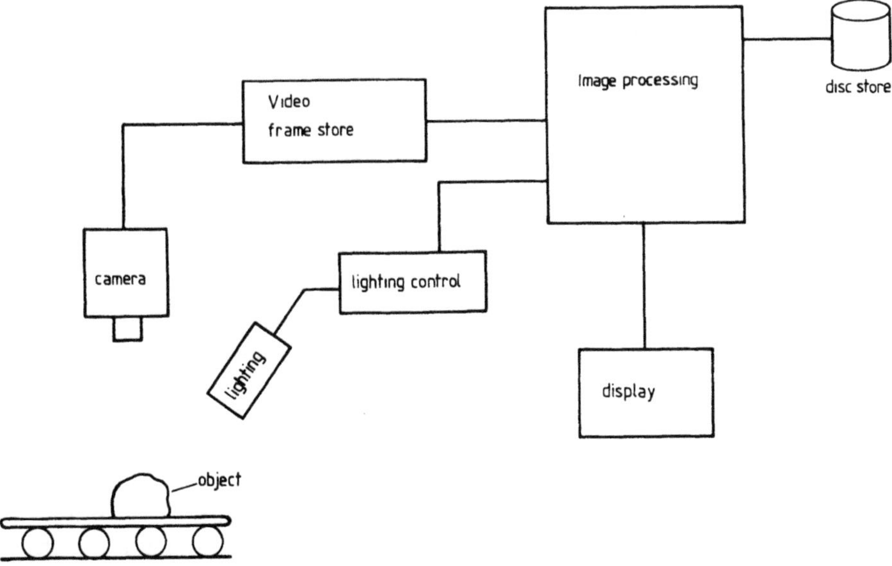

Figure 12.3. Inspection and quality-control system.

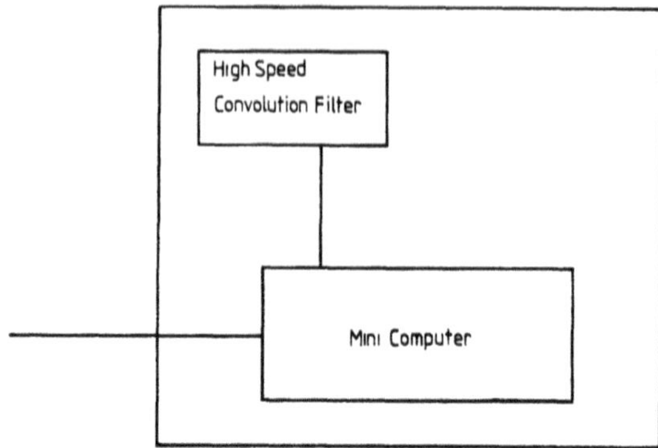

Figure 12.4. Image processor components.

gauges for integrated chip manufacture. We consider a generalized quality-control system that was developed to handle just such problems. It has the overall structure outlined in Figure 12.3, and the image-processing function is shown in more detail in Figure 12.4. It consists of the main computational element (an SEL 32/27) and a high-speed convolution filter developed to perform the filtering and template matching required in processing the images.

The other components of the system shown in Figure 12.3 include a camera and illumination control, an image display, disk memory, image memory, EPROM programmer, and system console. While this system utilizes a video camera for image capture, many industrial systems use other imaging modalities, such as CCD camera, laser illumination, polarized light, etc. The two components of the image-processing function illustrate the use of a general-purpose computer to oversee the quality control, with specialized hardware to perform the time-consuming computation.

The SEL 32/27 minicomputer is based on a 32-bit processor and has 512 K of MOS memory, 32 M hard-disk drives, and 1.2 M floppy-disk drives. It runs a multitasking executive (MPX-32); image-processing algorithms are developed using FORTRAN 77 and 8085 assembler. The SEL supports a high-speed data interface and other general-purpose communications interfaces. The EPROM programmer is used to program the EPROMs required for implementation of the residue arithmetic operations within the convolution filter. It is interfaced to the image-processing system via an RS-232 port.

12.3.2. Filter

The second computational element is the high-speed convolutional filter, which is used to preprocess the image of the manufactured part. It is also used for template matching during the flaw-detection procedures.

The design of the convolution filter is based on number-theoretic hardware techniques and is implemented via a 2-D radix-2 Number Theoretic Transform (NTT) computational element. The entire filter is constructed with EPROMs, adders, registers, two 128 × 128 word memory buffers, and a 128 × 128 word coefficient memory.

The filtering operation is performed by taking the NTT of the image, multiplying this by the NTT of the filter impulse response, and then taking the inverse NTT of the product. The filter hardware is organized to compute the circular convolution of a 128 × 128 image with a 128 × 128 spatial filter kernel. The linear convolution of different sizes of the images and the filter kernels is implemented via software, using the overlap-add technique of block convolutions (see Section 8.1.2).

The filter is interfaced to the SEL via the high-speed data interface and can process an image of size 128 × 128 bytes in 83.5 ms. Therefore, if the overhead of the interface handler and the software for the overlap-add technique is included, a time of 0.8 s is required to filter a 256 × 256 image by a 17 × 17 spatial filter kernel.

Filters are required for preprocessing for the following reasons. The complexity of the flaw-detection algorithm depends on the quality of the image of the part under inspection. For metal parts, uneven surface reflection results in poor quality images; this effect can be reduced by filtering. Furthermore, the flaw-detection algorithms can be greatly simplified if edges of features within the image are well defined. Another justification for the use of filters is that one of the most frequent operations performed during feature extraction is template matching, which can be implemented using filters.

The specification stipulates that a time slot of 4 s is available for determination of flaws on a given part. Given the implementation speed of a convolution algorithm implemented in software, it is clear that, to achieve inspection in the available time slot, a special hardware convolution filter must be used.

Furthermore, implementation of the convolution filter via the indirect transform method results in a lower-cost high-speed system than by using other hardware configurations.

12.3.2.1. Filter Specification

The factors to be considered in determining a kernel size include the results of flaw-detection algorithms using various kernels, the size of the templates that may be required for feature extraction, the size of the spatial filter kernels that are required to remove the effects of nonuniform illumination and edge enhancement, and the size of the images of the parts under consideration.

Given these considerations, a kernel size of 17 × 17 was determined as the requirement. By considering the time required for the implementation of different flaw-detection algorithms, it was concluded that the system needed to filter a 256 × 256 8-bit image is less than one second.

Based on the simulation of NTT filters and the dynamic range required for implementing the different flaw-detection algorithms, it was concluded that the convolution filter should have a dynamic range of at least 17 bits.

The above constraints determined the design and development of the convolution filter. Operation of the inspection system is illustrated by the results obtained from the three examples considered.

12.3.3. Car Piston Head Casting

Figure 12.5a shows a car piston head. The purpose of inspection is to check the quality of the casting of the surfaces of the piston head. Casting defects may lead to pores, which appear as dark spots on the face of the

Figure 12.5a. Original car piston head image.

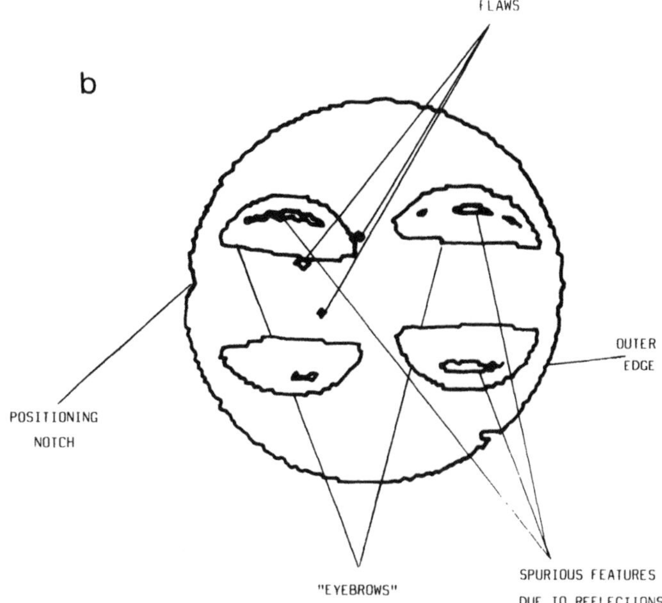

Figure 12.5b. Eyebrows and defects on car piston head.

piston head. The image-processing problem is to detect the occurrence of these dark spots and isolate them from the image. It is complicated by the fact that some parts of the piston also appear dark, even though they have been cast properly. These are the so-called "eyebrows" which are shown in Figure 12.5b.

Consequently, an important part of the inspection problem is to separate the defects from the "eyebrows." This is achieved by template matching using fast correlation schemes to produce a defects-only image. A measure of these defects may then be used to determine whether the car piston head should be accepted or rejected.

12.3.4. Differential Gear Wheels

Another application is in the selection or rejection of differential gear wheels. Figure 12.6a shows a differential gear. The dark spots on the gear teeth indicate the areas of contact when the gear is run with another wheel. The area of contact is obtained by coating the gear with a white lubricant and running it against another gear. With correct usage, the dark spots should lie in the center of the teeth; this therefore gives the criterion by which gear wheels should be selected for use, or rejected.

Thresholding the image produces an image with these dark spots isolated. Figure 12.6b shows the filtered and thresholded image. The centroid

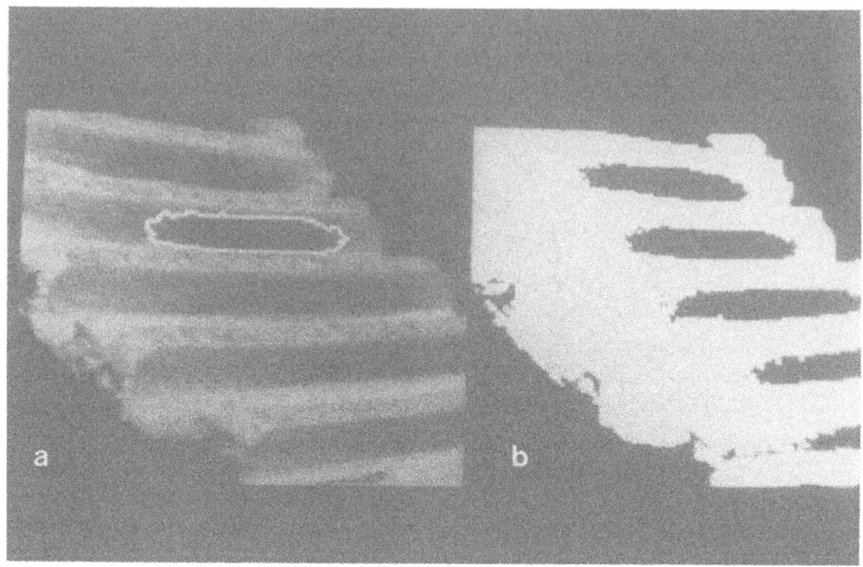

Figure 12.6. (a) Original differential gear image. (b) Filtered and thresholded image of differential gear.

of the perimeter of the contact area is computed after following the contour of these spots. Using the corner of the tooth as a reference point enables determination of whether or not the centroid is located at the center of the tooth and, consequently, of whether the gear should be selected or not.

12.3.5. Strain Gauges

Figure 12.7a shows the image of a strain gauge. The inspection problem is to determine whether there are any broken or missing connections in the integrated chip. This may be achieved by processing the image to produce only the border outlines, as shown in Figure 12.7b.

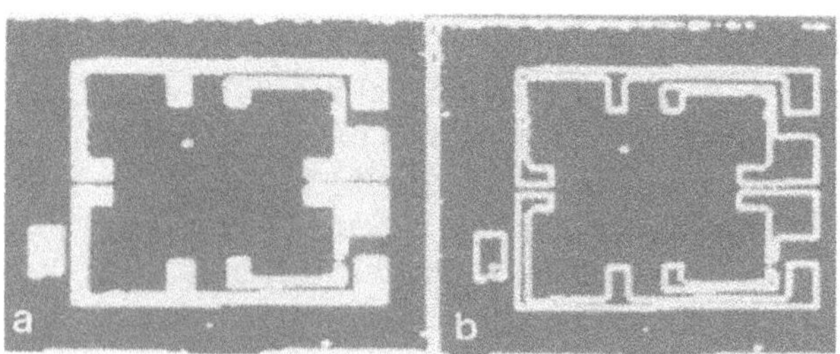

Figure 12.7. (a) Original strain gauge image. (b) Outline of strain gauge.

This processing is performed by filtering, enhancement, and thresholding, as described previously. A boundary-following algorithm is then applied to extract the contours of the strain gauge. The final stage in inspection is the processing to detect broken or missing segments; this is achieved through template matching.

12.3.6. Summary

This case study has illustrated the application of digital filters in quality-control inspection problems. There is a need for filtering, in order to reduce the effects of uneven-surface reflection and to improve the performance of template matching.

Owing to the requirement that the inspection be performed in real time, it is necessary to implement the filter using hardware convolution. The factors influencing the kernel size and the required word length are considered.

12.4. CASE STUDY 3: RESTORATION OF LINEARLY DEGRADED IMAGES WITH ADDITIVE NOISE

12.4.1. Background

Image restoration is applied to obtain the best estimate of an image which has been degraded in some way. Applications of image restoration are to be found in virtually all areas where images are used or formed, because degradation occurs in image acquisition, transmission, recording, handling, etc. Examples include: images blurred due to motion of the camera platform or the object; nonuniform illumination; variations in the transmission medium (e.g., atmospheric turbulence, data transmission noise, channel noise); nonlinearities in sensing devices, etc.

In some of these areas there is adequate information to model the degradation, while in other cases little or no information is available. Most restoration schemes, including those examined in this work, assume a linear degradation model, as discussed in Chapter 11. More realistic models are complex and computationally expensive to implement. The main problem in trying to obtain a "best" estimate is that there can be no authoritative definition of "best." This means that it is difficult to establish a meaningful criterion by which the level of restoration is judged. Examples of criteria used include minimum mean-square error, maximum likelihood, maximum *a posteriori* probability, and maximum entropy. Using these, the best estimate is chosen to satisfy the selected criterion.

In general, the success of an image-restoration technique depends on three factors: the level of degradation, the restoration criterion and degradation model assumed, and the amount of *a priori* information used.

First, the higher the degradation, the more unlikely it is that any method will succeed. Clearly, better restoration is obtained if the degradation model assumed is close to the actual degradation. In this sense, the filters discussed in Chapter 11 differ in performance, with the more realistic ones (such as the Wiener filter) giving better overall performance than, say, the inverse filter. Another difference between these filters is in the amount of knowledge that they assume about the original image and the distortion. For example, the Wiener filter requires knowledge of the covariance of the original image and the noise, while the constrained least-squares method assumes a smooth original signal and only requires the noise variance.

Second, better restoration may also be obtained if the restoration criterion is close to some measure that humans perceive to be important in image quality.

The third factor follows to some extent from the above. Better restorations result from restoration methods that incorporate more *a priori* information. The price to be paid for improved results is in computation time. The best estimate should agree with all *a priori* information, but sometimes the information required simply is not available. Furthermore, even if a lot of *a priori* information is available, the restoration methods may not have an adequate structure by which to impose this knowledge.

The most popular method adopted by research workers for improving *a priori* knowledge in the restoration process has been the use of iterative techniques. These are fairly flexible and, as described in Chapter 11, enable a fair amount of *a priori* knowledge to be incorporated into the restoration process. The main limitation of these iterative techniques is that not all *a priori* knowledge can be modeled and represented this way. Specifically, the *a priori* information is modeled as fixed points of an operator; however, the operator may not be simple and it may be difficult to investigate properties (such as nonexpansivity) in order to determine a feasible solution.

A lot of research effort in this field is being spent at the present time on developing better-constrained restoration techniques.[1] Alternative methods have included projection onto closed convex sets,[2,3] and fuzzy sets,[4] which allow more forms of *a priori* information to be included. At the same time, the field of knowledge-based signal processing is developing and should be well applied in restoration.

12.4.2. Restoration Example

This case study is concerned with the application of digital filters to the problem of restoring images that have been subjected to blur caused

Figure 12.8. Original image.

by linear convolutional systems with additive noise. An example of such an application is sensor noise and quantization noise degradation in low bit-rate data-transmission systems.

The case study here describes a restoration scheme that illustrates some of the points discussed above with regard to restoration. The original image shown in Figure 12.8 can be treated as a single frame of a sequence of images that are subjected to linear blur and transmitted along a noisy channel that degrades the image further with additive noise.

One approach frequently utilized for such restoration problems is Wiener filtering. Another similar approach is spectral subtraction filtering. As discussed previously, these schemes are aimed at estimating the power spectrum; they are based on minimum mean-square error criteria and often sacrifice resolution in favor of noise suppression. It is therefore of interest to improve the performance of such systems by estimating the phase as well as the magnitude. The use of *a priori* information, as well as other information abstracted from the noisy image, is illustrated. The case study considers a hybrid restoration scheme, in which the first stage involves estimation of the power spectrum. The second stage takes the output of the first stage as input for the phase-estimation procedure. This procedure is illustrated in Figure 12.9. It shows the use of a filtering technique based on minimum mean-square error criteria to produce noise reduction. This is followed by a constrained iteration scheme to estimate the full spectrum, where the

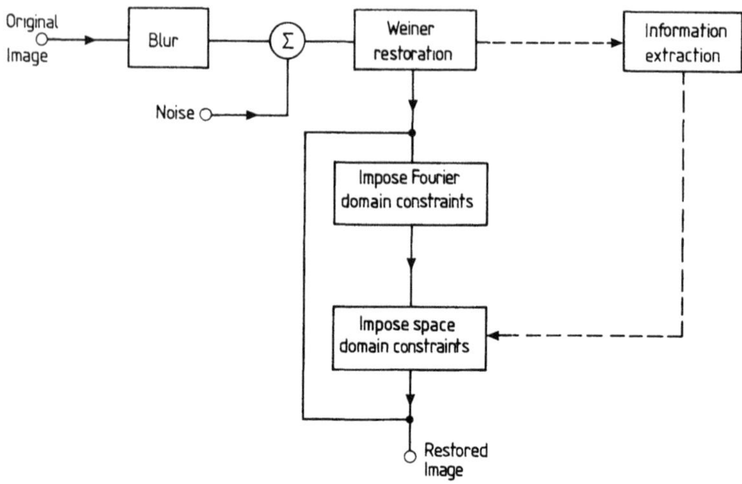

Figure 12.9. Hybrid restoration scheme.

a priori information is utilized in the iteration. The constraints selected are related to the subjective perception of image quality.

The estimation of a phase function from a given magnitude function is discussed in more detail elsewhere.[5] The basic facts to note are that an iterative scheme based on the Gerchberg–Papoulis algorithm[6] improves the phase estimate at each iteration by applying constraints alternately in the space and frequency domains. Given adequate constraints and a good first phase estimate, this method will improve the phase.

Both blind deconvolution and image restoration may be improved by using reconstruction techniques to estimate the phase, in addition to the magnitude estimation that is conventionally performed. In blind deconvolution, the signal has been blurred by a blurring function about which complete knowledge is unavailable.

In some special cases, the distorting signal is known to have a phase function that is approximately zero, and consequently the phase of the degraded image is very similar to that of the original. In these cases, the problem becomes one of reconstruction from phase only, for example, wavelet reconstruction in seismic signal processing.

The problem considered in this case study can be approached in a similar way. The filters conventionally used do not restore the phase. At the same time, it can be assumed that the phase degradation is not too serious. The method presented utilizes a Wiener filter to provide an estimate of the Fourier magnitude. Given such an estimate, an iterative restoration scheme is used to estimate the phase. This may fail to converge if a good first estimate of the phase function is not available.

One way of specifying further constraints is suggested by the interrelationship between edge structure and the phase of an image. If information pertaining to the spatial edge activity of the image is known, then successive spatial domain estimates can be constrained to conform in edge structure to *a priori* known edge information, and so indirectly constrain the phase function. Such information can be obtained by an edge region detection on the output of the Wiener restoration filter. Information on the rough location of edges is often available from the restored image.

The phase-estimation algorithm is based on edge-preserving space-domain constraints.

Figure 12.10a shows the "MAN" image degraded by added noise, the image having a signal-to-noise ratio of 10 dB. The linear blur has a Gaussian point spread function. The image is 128×128 pixels and the impulse response of the blur is 9×9 square.

This image was Wiener filtered to give the restored image of Figure 12.10b. We note that, although noise levels in the picture are well suppressed, this is at the expense of somewhat poor resolution and edge quality.

The final restored image is obtained by the constrained iteration scheme described previously. This image is shown in Figure 12.10c. The "blocking effect" present in the restored image arises from the fact that the reconstruction algorithm is implemented by initially partitioning the image into 32×32 blocks to minimize memory requirements. Despite this effect, close examination of Figure 12.10c reveals additional detailed edge structure absent in the Wiener-filtered image of Figure 12.10b.

Figure 12.10a. Degraded image of Figure 12.8 with a 10 dB signal-to-noise ratio.

Figure 12.10b. Wiener restoration of Figure 12.10a.

Figure 12.10c. Restored image using 30 iterations of the hybrid scheme.

12.4.3. Summary

This study illustrates the use of a nonrecursive digital filter for image restoration. The importance of models of the degradation criteria used and of the *a priori* information utilized is noted.

The use of constrained iteration to implement a restoration filter that enables constraints to be included in the restoration process is observed.

Image restoration research is utilizing such methods where *a priori* information may be applied. The scheme described here implements a method of applying information which is directly relevant to image quality in the restoration criteria and constraints. The limitations of iterative methods lie in their inability to include all forms of information, in convergence problems, and in the fact that sometimes a fixed point of the iteration may not exist. Other schemes for imposing knowledge and which are being studied include projection on convex sets, fuzzy sets, and knowledge-based signal processing.

12.5. CASE STUDY 4: IMAGE RECONSTRUCTION FROM INCOMPLETE INFORMATION

12.5.1. Background

The problem of image reconstruction from incomplete information is an important one, and occurs in a number of applications of digital image processing.

In principle, any digital image can be viewed as quantized samples of an analogue image function. For our purposes, an image produced by sampling above the Nyquist rate and quantizing to a useful number of levels (say 256) can be considered as a full or complete information image.

There are some problem areas in which it is required to reconstruct an image from much less information than this, and it is such applications which form the subject of this case study.

The importance of this subject may be illustrated by the breadth of applications in which it occurs. These fall into three general categories.

The first category covers those applications in which only partial information is available and there is no other way of forming an image. This is the largest category. An example is tomographic reconstruction from limited projections where fewer projections than usual are available for reconstruction. In other applications, only some of the information required can be obtained, due to time, equipment, or other limitations. Examples include cases in which only part of the Fourier spectrum can be measured, such as the phase or magnitude only, high-resolution synthetic aperture radar, and many others.[7]

The second category covers cases in which full information is present, but is noisy or erroneous. An example is the case treated in the previous case study, where reconstruction is based on that part of the data known to be correct.

The third category consists of cases in which full information is present, but some is selectively discarded for purposes of data compression. An example is the zonal sampling methods used in transform coding. The received image from such a system could be treated as a limited data image.

The examples considered in this case study fall into the first category. Let us consider the image formation model

$$y(m, n) = \mathrm{D}u(m, n) \tag{12.1}$$

where $y(m, n)$ is the observed image, $u(m, n)$ is the input image, and D is an operator that acts on $u(m, n)$ to produce the observed output $y(m, n)$.

This model is utilized to describe three general classes of problems that occur in digital systems where digital filters are applied.[8] These classes are system identification, system realization, and inverse problems.

The system identification problem occurs when both the input and output are known and the operator D is to be determined.

System realization is the more usual problem in digital filter design. Here, the input and the operator are known, while the output is to be determined.

In inverse problems, the output is known and it is required to determine the input. Image restoration, reconstruction, and construction fall into this classification.

In restoration, D is known or assumed, and an attempt is made to invert the effects of D and to reach some estimate of $u(m, n)$ which is close according to a selected criterion.

In reconstruction, D is recognized as an operator that removes some components of $u(m, n)$ in some domain, be it spatial, frequency, or some other. Generally, information may be missing in the time domain or in a transform domain, and the reconstruction process may use either or both of the domains.

In the transform domain, we have several applications where it is necessary to reconstruct an image from incomplete information. In the Fourier phase problem, only the Fourier transform magnitude is available for measurement and the phase must be reconstructed. In image transform coding, only a few samples of the transform coefficients are transmitted. These are chosen so that a reconstruction algorithm at the receiving end can give reasonable reconstruction of the image.

Another application in the transform domain is image reconstruction from projections—especially when imaging fast-moving organs such as the beating heart. In this case, we are interested both in reducing the total number of projections needed for reconstruction of static scenes, and in producing images of reasonable quality from the few projections that can be taken within the very short "static" viewing period of a fast-moving object.

Figure 12.11. Image construction model.

In the time domain, we may have insufficient information due to decimation. Another application may be sub-Nyquist sampling. Images represented by their contours or polygonal approximations are another example of partial information.

An attempt is made to describe, to the maximum extent possible, the full information image $u(m, n)$ from which the partial information image $y(m, n)$ was obtained, subject to (1) acceptance of all available data, (2) ensuring that all estimated data are consistent with *a priori* knowledge, and (3) being neutral about data that are not measured and cannot be estimated.

Image construction is similar to reconstruction, but the emphasis is on changing the input to the system rather than the output. It occurs in environments in which the input can be controlled to some extent, but not D (see Figure 12.11) or the output; for example, when sending images through a distorting system to a device or process that cannot perform restoration or reconstruction. In this case, the desired output image $y(m, n)$ is known, as is the nature of the operator D. The available input $u(m, n)$ is identical to the desired output $y(m, n)$. Simply applying $u(m, n)$ to D will not give $y(m, n)$, since D is a distorting operator. No attempt is made to invert or change the effects of D by filtering $y(m, n)$ to give some estimate of $u(m, n)$. Instead, a study of the effects of D is conducted, and an attempt is made to design the input signal that will provide the desired output. This is illustrated in Figure 12.12.

An example of an application using image construction is the design of images to be fed to systems for producing masks to be used in microlithography.[9] We start with the image pattern designed to be reproduced as a microlithographic mask. It is then fed to the system, which may consist of a microlithographic camera or laser printer. The camera is represented by a linear system. It is diffraction-limited (hence band-limiting) and is operating near its resolution limit. The film is high-contrast and operates as a hard limiter.

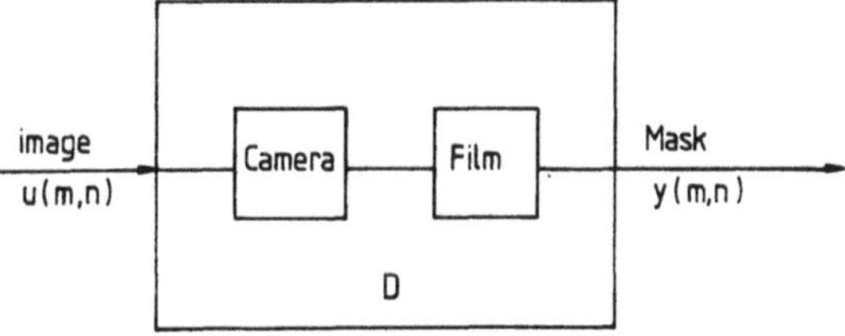

Figure 12.12. Microlithographic image-production system.

The simplest approach would be to use the desired pattern itself as the input, and hope that the final mask will produce an output that is exactly the same as the input. However, because of the distortion described previously, this will not be the case and $y(m, n)$ will be different from $u(m, n)$. The required image is a binary image with the mask at specified locations. Therefore, information is available on the zero crossings of the output $y(m, n)$, and it is known that the image is band-limited. Image $u(m, n)$ is assumed to possess at least these properties too. Image construction therefore consists of generating or somehow finding the gray scale image $u(m, n)$ with the specified zero crossings and bandwidth which, when applied to the system, produces the desired binary image $y(m, n)$ at the output.

Thus, if we regard $u(m, n)$ as the target image to be reconstructed from $y(m, n)$, then we have partial information in the space domain (zero crossings) and in the transform domain (band limiting).

A quick look at the history of image reconstruction shows that the basic approach to the problem has been two-pronged. In the first instance, researchers have tried to establish, theoretically, the uniqueness of the partial information present, i.e., is the partial information uniquely linked to the original image function?

The second aspect has been a consideration of different algorithms to perform the reconstruction. Many of the successful algorithms are iterative algorithms, and hence a further consideration is algorithm acceleration and convergence study.

Clearly, construction and reconstruction from partial information are identical problems with different emphasis.

12.5.2. Reconstruction Examples

This case study examines three examples of reconstruction from incomplete information. The first two are concerned with phase-only or magnitude-only reconstruction, and the third considers reconstruction of gray scale images from their polygonal approximations.

The phase and magnitude of the Fourier transforms of images play different roles, and it is well known[10] that many of the important features of a signal may be preserved if the phase only is retained. There are a number of important application areas where only one or other of the components of the Fourier transform of a signal can be measured directly. In such situations, the need arises for the reconstruction of the original signal from the available components. Reconstruction from magnitude is a fairly well-established problem that has also been called the "phase retrieval problem" and the "Fourier phase problem." Applications include X-ray crystallography, radioastronomy, and image processing.

Generally, an image cannot be uniquely defined in terms of only its Fourier phase or magnitude. There are, however, certain classes of sequences where this unique specification may be possible. For example, there is a Hilbert transform relationship between the phase and log magnitude of a minimum-phase sequence. This condition is restrictive and quite difficult to apply to images. Several uniqueness results have been obtained by research workers in this area, where the emphasis has been on developing equivalent statements of these conditions to enable them to be applied to practical images.

12.5.2.1. Phase-Only Algorithm

The iterative algorithm used is basically similar to the Gerchberg[6] and Papoulis[11] algorithm. These algorithms move alternately between the space and frequency domains at each iteration, imposing known constraints in each domain.

The basic phase-only reconstruction is illustrated in Figure 12.13 and consists of the following steps operating on a $2M \times 2N$ image, $y(m, n)$, obtained by padding the $M \times N$ images with zeros outside the region of support.

1. Make an initial guess of the unknown DFT magnitude $|Y_0(\omega_1, \omega_2)|$. Form the next estimate of the DFT, $Y_1(\omega_1, \omega_2)$, by combining this guess with the known phase

$$Y_1(\omega_1, \omega_2) = |Y_0(\omega_1, \omega_2)| \exp[j\phi_y(\omega_1, \omega_2)]$$

Compute the IDFT.

2. Apply the finite support constraint

$$y_p(m, n) = \begin{cases} y_p(m, n), & 0 \le m < M, \quad 0 \le n < N \\ 0, & \text{elsewhere} \end{cases}$$

and positivity constraint

$$y_p(m, n) = \begin{cases} y_p(m, n), & y_p(m, n) > 0 \\ 0, & y_p(m, n) \le 0 \end{cases}$$

It satisfied with the image, stop. Otherwise, continue. Compute the DFT.

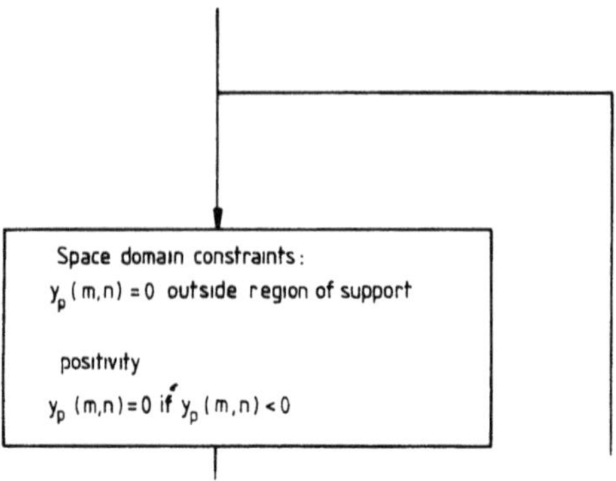

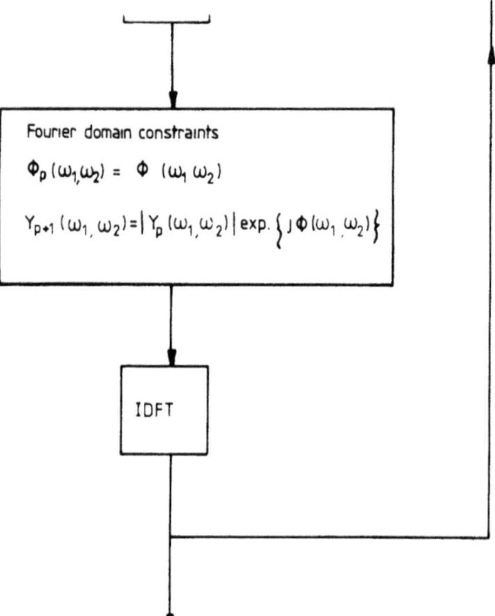

Figure 12.13. Basic phase-only reconstruction.

3. The magnitude $|Y_p(\omega_1, \omega_2)|$ of the pth iteration is used as the new estimate of $|Y(\omega_1, \omega_2)|$. The new DFT estimate is formed as

$$Y_{p+1}(\omega_1, \omega_2) = |Y_p(\omega_1, \omega_2)| \exp[j\phi_y(\omega_1, \omega_2)]$$

Compute the IDFT.
Go to step 2.

12.5.2.1a. Results. This algorithm was implemented and used to reconstruct the 128 × 128 pixel image shown in Figure 12.14, in which Figure 12.14a shows the original image, and Figure 12.14b shows the phase-only image formed by assuming the initial magnitude of the DFT to be given by a simple low-pass function. The reconstructed image after thirty iterations is shown in Figure 12.14c.

Figure 12.14a. Original image.

Figure 12.14b. Phase-only image—initial estimate.

Figure 12.14c. Reconstructed image after 30 iterations.

12.5.2.2. Magnitude-Only Reconstruction

As before, the magnitude-only reconstruction algorithm is based on the Papoulis algorithm, first proposed to solve the two intensity phase retrieval problem.[11] The algorithm used proceeds as follows:

1. Make an initial guess $\phi_0(\omega_1, \omega_2)$ of the unknown phase function $\phi(\omega_1, \omega_2)$. Form the next estimate of the DFT by combining the phase guess with the known magnitude function

$$Y_1(\omega_1, \omega_2) = |Y(\omega_1, \omega_2)| \exp[j\phi_0(\omega_1, \omega_2)]$$

   Compute the IDFT.
2. Apply known space-domain constraints to form the current space-domain estimate $y_p(m, n)$. The finite region of support gives

$$y_p(m, n) = \begin{cases} y_p(m, n), & 0 \le m < M, \quad 0 \le n < N \\ 0, & \text{elsewhere} \end{cases}$$

   and the positivity constraint

$$y_p(m, n) = \begin{cases} y_p(m, n), & y_p(m, n) > 0 \\ 0, & y_p(m, n) \le 0 \end{cases}$$

   Compute the DFT.

3. Form the new DFT estimate using the phase of this DFT

$$Y_{p+1}(\omega_1, \omega_2) = |Y(\omega_1, \omega_2)| \exp[j\phi_p(\omega_1, \omega_2)]$$

Compute the IDFT.
Go to step 2.

12.5.2.2a. Results. In general, this algorithm does not lead to an intelligible image when no phase information is available. Even after 1000 iterations the reconstructed image was an extremely poor representation of the original object when the initial phase specified is purely random.

12.5.2.3. Reconstruction from Contours

In the previous examples, we have been concerned with reconstructing images given partial specifications in the frequency domain. In many ways, the problem of reconstructing images from partial specification in the space domain is similar. For example, phase-only images are visually related to edge-detected or contour images. It is thus possible to envisage reconstruction schemes from contours, analogous to phase-only reconstruction.

This final example considers the problem of such a reconstruction scheme. As discussed in the background to this study, there are applications, such as preparing images for microlithography, where such image construction is desirable. Another even more important application is the possibility of achieving high-bandwidth reduction through transmitting only contour information and reconstructing gray scale images of sufficient intelligibility from this partial information.

Furthermore, if reconstruction from contours is achievable, there is interest in reconstruction from polygonal approximations of contours with increased data compression.

In the example considered here, the partial information is represented by the contour points and the values of the corresponding pixel intensities.

The missing texture is estimated using an iterative reconstruction algorithm. The partial information available is the contour and parameterized texture. *A priori* information in the space domain is the known contour and noncontour areas.

The iterative procedure described earlier is used to move between the space domain and another domain (the parameterized texture domain) where some values are known.

In this case, the knowledge is that we have smooth regions between the contours, and the goal in generating new possible values in these regions is to maximize their smoothness and prevent the introduction of any contours. This provides constraints that can be used to guide an iterative procedure that attempts reconstruction. One way of achieving this smoothness would be to low-pass filter (or to band-limit) the image at each iteration. Using this approach would provide a similar structure to those studied previously, with alternation between space and frequency domain and imposition of constraints.

The same effect, however, may be achieved by a form of spatial filtering. We consider a measure of image variation given by

$$C = \sum_{m=0}^{M} \sum_{n=0}^{N} \{[y(m,n) - y(m,n-1)]^2 + [y(m,n) - y(m-1,n)]^2\} \tag{12.2}$$

This is minimized if

$$\frac{\partial C}{\partial y(m,n)} = 0 \tag{12.3}$$

Defining the Laplacian $\Delta y(m, n)$ as

$$\Delta y(m,n) = y(m-1,n) + y(m,n-1) + y(m+1,n) + y(m,n+1) - 4y(m,n) \tag{12.4}$$

Equation (12.3) is satisfied if $\Delta y(m, n)$ is zero at all points except the contour points. The above finite-difference equation may be solved iteratively with the next iterate being formed as

$$y_{i+1}(m,n) = y_i(m,n) + \lambda \Delta y(m,n) \tag{12.5}$$

where λ is a relaxation parameter.

The reconstruction algorithm thus becomes:

1. Form $y_0(m, n)$, the initial estimate, by using the given values of the texture and an estimate for the rest of the image. Possible choices include 0, random values, or values generated using an autoregressive moving-average model.
2. Form $y_{i+1}(m, n)$, the next iterate, using equation (12.5).
3. Impose the known values of $y(m, n)$ at the contour points. If satisfied with the image, stop—otherwise, go to step 2.

The iteration tries to prevent the introduction of edges (high frequencies) in nonedge regions. It can thus be modeled as a repeated low-pass filtering operation. This algorithm is illustrated in Figure 12.15.

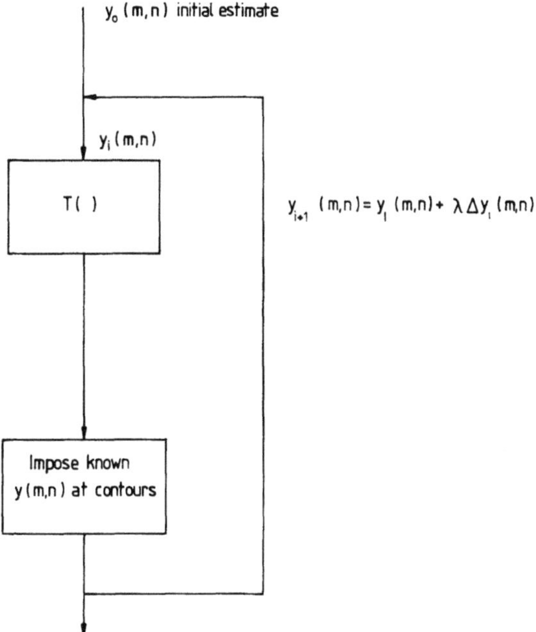

Figure 12.15. Iterative reconstruction algorithm.

12.5.3. Results

Figure 12.16a shows the contour image and Figure 12.16b the polygonal approximations obtained using the algorithms described. Figure 12.17a shows the reconstructed images using spline interpolation and Figure 12.17b that using the iterative approach outlined above.

The results indicate that subjectively acceptable reconstructions from contour and texture parameterized information are possible. The main factors on which success depends are the correct choice of edge image, and extraction of the contours.

12.5.4. Summary

This case study has considered the problem of reconstruction of images from partial information. Reconstruction has been treated as an inverse problem, and the use of an iterative scheme to implement the required filtering operation has been illustrated. These iterative techniques are applicable to a variety of partial information problems. The same general type of reconstruction algorithm has been applied to space- and frequency-domain partial information.

Figure 12.16a. Contour image.

Figure 12.16b. Polygonal approximations.

Figure 12.17a. Image formed by spline interpolation from Figure 12.16.

Figure 12.17b. Image formed by iterative reconstruction from Figure 12.16.

REFERENCES

1. R. W. Schafer, R. M. Mersereau, and M. A. Richards, Constrained iterative restoration algorithms, *Proc. IEEE* **69,** 432–450 (1981).
2. D. C. Youla and H. Webb, Image reconstruction by the method of convex projections; Part 1: Theory, *IEEE Trans. Med. Imaging* **MI-1,** 81–94 (1982).
3. M. I. Sezan and H. Stark, Image restoration by the method of convex projections; Part 2: Applications and numerical results, *IEEE Trans. Med. Imaging* **MI-1,** 95–101 (1982).
4. M. R. Civanlar and H. J. Trussell, Digital signal restoration using fuzzy sets, *IEEE Trans. Acoust., Speech, Signal Process.* **ASSP-34,** 919–936 (1986).
5. R. King, K. Singarajah, and A. S. Kwabwe, A Hybrid Technique to Restore the Phase and Magnitude of Noisy Linearly Degraded Images, IEEE International Conference on Acoustics, Speech and Signal Processing, Boston (1983).
6. R. W. Gerchberg, Super-resolution through error energy reduction, *Opt. Acta* **21,** 709–720 (1974).
7. A. S. Kwabwe, *Image Reconstruction from Incomplete Information,* PhD Thesis, Imperial College, University of London (1984).
8. D. E. Dudgeon and R. M. Mersereau, *Multidimensional Digital Signal Processing,* Prentice-Hall, Englewood Cliffs, NJ (1984).
9. S. I. Sayegh, Y-L. Kok, and J-H. Hong, An algorithm to find two-dimensional signals with specified zero crossings, *IEEE Trans. Acoust., Speech, Signal Process.* **ASSP-35,** 107–111 (1987).
10. A. V. Oppenheim and J. S. Lim, The importance of phase in signals, *Proc. IEEE* **69,** 529–541 (1981).
11. A. Papoulis, A new algorithm in spectral analysis and band-limited extrapolation, *IEEE Trans. Circuits Syst.* **CAS-22,** 735–742 (1975).

Index

www.ingramcontent.com/pod-product-compliance
Ingram Content Group UK Ltd.
Pitfield, Milton Keynes, MK11 3LW, UK
UKHW022321190726
13856UKWH00001B/144

* 9 7 8 1 4 8 9 9 0 9 1 9 0 *